T0269152

CAMBRIDGE LIBRARY COLLECTION

Books of enduring scholarly value

Technology

The focus of this series is engineering, broadly construed. It covers technological innovation from a range of periods and cultures, but centres on the technological achievements of the industrial era in the West, particularly in the nineteenth century, as understood by their contemporaries. Infrastructure is one major focus, covering the building of railways and canals, bridges and tunnels, land drainage, the laying of submarine cables, and the construction of docks and lighthouses. Other key topics include developments in industrial and manufacturing fields such as mining technology, the production of iron and steel, the use of steam power, and chemical processes such as photography and textile dyes.

Journal of a Voyage to Australia and Round the World for Magnetical Research

This work by William Scoresby (1789–1857) was edited by Archibald Smith (1813–1872) and published posthumously in 1859. It is the account of Scoresby's final voyage and last scientific study, which took place between February and August 1856. Scoresby made his Australian voyage on board the *Royal Charter*, owned by the Liverpool and Australia Steam Navigation Company. He wished to observe the changes that take place in the magnetic state of iron ships travelling on a north-to-south magnetic latitude, and to assess how magnetic changes affect the working of a compass so that he could discover the most reliable location for it on board ship. The first part of the work is an exposition of magnetic principles, followed by the results and conclusions of Scoresby's experiments. The second part contains a travel account of the actual voyage. It is a key work of nineteenth-century navigation science.

Cambridge University Press has long been a pioneer in the reissuing of out-of-print titles from its own backlist, producing digital reprints of books that are still sought after by scholars and students but could not be reprinted economically using traditional technology. The Cambridge Library Collection extends this activity to a wider range of books which are still of importance to researchers and professionals, either for the source material they contain, or as landmarks in the history of their academic discipline.

Drawing from the world-renowned collections in the Cambridge University Library, and guided by the advice of experts in each subject area, Cambridge University Press is using state-of-the-art scanning machines in its own Printing House to capture the content of each book selected for inclusion. The files are processed to give a consistently clear, crisp image, and the books finished to the high quality standard for which the Press is recognised around the world. The latest print-on-demand technology ensures that the books will remain available indefinitely, and that orders for single or multiple copies can quickly be supplied.

The Cambridge Library Collection will bring back to life books of enduring scholarly value (including out-of-copyright works originally issued by other publishers) across a wide range of disciplines in the humanities and social sciences and in science and technology.

Journal of a Voyage to Australia and Round the World for Magnetical Research

WILLIAM SCORESBY
EDITED BY ARCHIBALD SMITH

CAMBRIDGE
UNIVERSITY PRESS

CAMBRIDGE UNIVERSITY PRESS

Cambridge, New York, Melbourne, Madrid, Cape Town,
Singapore, São Paolo, Delhi, Tokyo, Mexico City

Published in the United States of America by Cambridge University Press, New York

www.cambridge.org
Information on this title: www.cambridge.org/9781108072489

This edition first published 1859
This digitally printed version 2011

ISBN 978-1-108-07248-9 Paperback

JOURNAL

OF A

Voyage to Australia and Round the World,

FOR

MAGNETICAL RESEARCH.

H. Adlard, sc.

London : Longman & Cº.

JOURNAL

OF A

VOYAGE TO AUSTRALIA

AND

ROUND THE WORLD,

FOR

MAGNETICAL RESEARCH.

BY

THE REV. W. SCORESBY, D.D., F.R.S., ETC.

CORRESPONDING MEMBER OF THE INSTITUTE OF FRANCE.

EDITED BY

ARCHIBALD SMITH, ESQ., M.A., F.R.S.

LATE FELLOW OF TRINITY COLLEGE, CAMBRIDGE.

LONDON:

LONGMAN, GREEN, LONGMAN, & ROBERTS,

PATERNOSTER ROW.

1859.

CONTENTS.

ERRATA.

Introduction, p. xxi, line 9, *for* $\tan^{-1} \frac{F}{C}$, *read* $\tan^{-1} \frac{f}{c}$

———————— p. xxii, equation 5, *for* C sin. ζ', *read* C cos. ζ'

INTRODUCTION.

THE voyage of which this volume contains the journal was undertaken by Dr. Scoresby with the object of observing the changes which take place in the magnetic state of an iron ship proceeding from a northern to a southern magnetic latitude, and of deciding certain questions as to the best mode of correcting the deviations of the compass in such a ship. The object was one of great importance, not only as regards science but as regards humanity. To the love of science and the devotion to the cause of humanity which led Dr. Scoresby to undertake this voyage at an advanced period of his life, he may be said to have fallen a martyr. He did not live to prepare his journals for the press; and at the request of Major-General Sabine, in whose hands his papers were placed by Mrs. Scoresby and his executors, I have undertaken the office of Editor.

The materials which Dr. Scoresby left consisted, besides the record of his magnetical observations, of his journal, written from day to day, and of an introductory dissertation on magnetism, of which the MS. was nearly complete. With regard to the publication of the journal some hesitation has been felt. A detailed narrative of a voyage made on a well-known and frequented track, and diversified by no unusual incidents, may seem hardly of sufficient interest for publication; but there are considerations which lead to the conclusion that the journal may be published, and with advantage, nearly in the shape in which it was written. Dr. Scoresby had some peculiar advantages as the narrator of such a voyage. He had been a sailor by profession, and had commanded several ships in voyages to the Arctic Seas. He was afterwards a landsman for thirty years. In this interval great changes had taken place in the material, form, size, and management of vessels. A new class of vessels—the iron clipper, with an auxiliary screw—had come into existence. In making a voyage in such a vessel to the southern hemisphere, Dr. Scoresby united the fresh interest in details which a seaman can seldom have, and the nautical knowledge which enabled him not only to observe and describe accurately what was done, but to understand and point out why it was done. The voyage, too, is one of great interest to many of our countrymen. And if it cannot be expected that the narrative will have much general interest, it cannot be doubted that there are many to whom even the most minute and trivial details will have a special interest.

The particular questions to which Dr. Scoresby's attention was directed during the voyage had been the subject of controversy between him and a mathematician and philosopher of the very first rank—the Astronomer-Royal, and I cannot place my readers in a position to judge of the motives and results of Dr. Scoresby's voyage without entering in some degree into the points in difference between them. These were of great importance, both theoretical and practical. They concerned the nature of the magnetism of the rolled and hammered iron of which an iron vessel is chiefly composed, the changes which that magnetism undergoes, and the mode of correcting the deviation of the compass caused thereby, and particularly the well-known mode of correcting the deviation by the application of magnets and soft iron proposed by Mr. Airy in the year 1839, and extensively used in the mercantile navy. This mode of correction Dr. Scoresby, in common with many or most of those who have examined the question, among whom I may rank myself, considered to be not only erroneous in principle but dangerous in practice. He protested strongly against it on all occasions, and his sense of its danger was one of the motives which induced him to undertake the voyage. Under these circumstances it will not, I trust, be considered that I am overstepping the bounds of my duty as Editor if I endeavour to ascertain what was correct or mistaken on each side of the controversy. As will be seen in the sequel, I have been led to think that Dr.

Scoresby has, with others, misunderstood Mr. Airy's meaning in a most important particular, and that his own theoretical views were thereby to some extent diverted from considerations to which he might otherwise have attributed greater importance. If I succeed in any degree in my own object, I am confident that I shall at least be advancing what each of the parties in the controversy valued far more than the acceptance of any particular theoretical views. The points in controversy will be more easily understood if I give a short sketch of the history of our knowledge of the " Deviation of the Compass."

The history of the " deviation of the compass " dates from the commencement of the present century. It had, indeed, been noticed by navigators of the last century, and its cause suggested, but no systematic attempt had been made to investigate its laws or to discover the means of correcting it until the voyage of Captain Flinders to Australia in the years 1801–1803. Captain Flinders observed (Voyage, vol. II., App. II.), that when his ship's head was directed to the N. or S. (magnetic) there was no deviation; when to the E. or W. (magnetic), a maximum deviation;—that the amount of deviation had a close connection with the dip of the needle— that in northern magnetic latitudes the *north* end of the needle of a compass in the usual place in the ship was drawn towards the bow—that at the equator there was no deviation—that in Bass's Strait, where the southern dip is nearly as great as the northern dip in the English Channel, the *south* end of the needle was drawn towards the bow. He attributed the deviation to its true cause—the attraction of the iron in the ship magnetised by induction from the earth. He gives a correct explanation of the mode of action, and from comparing observations made in different magnetic latitudes he deduced the conclusion that the deviation is proportional to the *dip* multiplied by the sine of the azimuth of the ship's head from the magnetic meridian. In this conclusion Captain Flinders was mistaken. The deviation in such vessels as Captain Flinders's, viz., wooden sailing ships, is in truth more nearly proportional, being directly proportional to the vertical force of the earth (its cause), and inversely proportional to the horizontal force (the directive force on the needle) to the *tangent* of the dip. The difference, however, is not material, except in high magnetic latitudes. Captain Flinders further leaves out of account that part of the deviation which will hereafter be referred to under the term "quadrantal deviation." But this is of very small amount in vessels like Captain Flinders's; so that to wooden sailing ships in low or middle magnetic latitudes, to which alone Captain Flinders refers, his formulæ and the mode of correction which he proposes are still applicable. He observes that the trouble of correction may be saved if a place can be found near the taffrail where the attraction of the iron at the stern will counteract by its greater vicinity the more powerful attraction of the centre and fore parts of the ship, and that *should the attraction be too weak, it may be increased by fixing one or more upright stancheons or bars of iron in the stern.* He says, " If a neutral station can be found or made exactly amidships, and of a convenient height for taking azimuths and bearings, let a stand be there set up for the compass, and if the stand must of necessity be moveable, make permanent marks that the exact place and elevation may always be known. Observations taken here should never undergo any change from altering the direction of the ship's head at any dip of the needle, but it will be proper to verify occasionally and to compare the azimuths and bearings with others taken on the binnacle. The course should also be marked from this compass though the ship be steered by one before the wheel, a quarter or half point being allowed to the right or left according as the two may be found to differ."

I have quoted these observations at length because, as regards the deviation in wooden sailing ships, little can be added to them in the present day. The mode of correction suggested, viz., by vertical iron bars, is certainly preferable to that afterwards proposed by Mr. Barlow.

The large deviations of the compass experienced in ships in the Arctic regions attracted the attention of Dr. Scoresby in the several voyages made by him in command of whaling ships. The results of his observations and of his consideration of the subject will be found in the Philosophical Transactions for 1819, p. 96; in the " Account of the Arctic Regions," vol. II., p. 537; and in the " Journal of a Voyage to the Northern Whale Fishery," etc., p. 89.

In the second of these, p. 547, Dr. Scoresby very nearly obtained the true law connecting the deviation arising from induced magnetism with the dip; for he observed that the increased deviation is owing not only to the increase of local attraction but to the diminution of the directive force of the earth on the needle.

The subject attracted the attention of the officers who accompanied Sir John Ross and Sir Edward Parry in the several voyages for the discovery of a north-west passage, between the years 1818 and 1824, and among others, of Major-General (then Captain) Sabine, R.A., who accompanied these expeditions as astronomer. The results of Captain Sabine's observations are given in a paper in the Philosophical Transactions for 1819, p. 120. In this paper Captain Sabine points out that the deviation depends as well on the diminution of the directive force on the needle as on the increase of the force causing the deviation—but without expressly deducing the mathematical expression for the law of dependence of the principal coefficient of the deviation on the dip.

This, so far as I am aware, was first done by Dr. Thomas Young, in a paper in Brande's Quarterly Journal for 1820, vol. IX., p. 372. In this paper Dr. Young showed that the deviation produced by the permanent magnetism of the ship varies inversely as the earth's horizontal force, and that the deviation produced by the vertical force acting on a mass of soft iron varies as the tangent of the dip. Dr. Young went on, and he was, I believe, the first who did so, to consider the effect of the horizontal force in producing the part of the deviation which has been termed "quadrantal;" but in doing so, if I have not misunderstood his process, his results are vitiated by the accidental use of a sine for a cosine.

In the year 1820 Mr. Barlow published his " Essay on

Magnetic Attractions." In this work he proposed to correct
the deviations of a ship's compass by the application to it of
a ball or plate of soft iron, found by experiments made with
it on a compass on shore to produce deviations of the same
amount and direction as the deviations produced on the compass on board by the iron of the ship. He appears to have
at first proposed to destroy the deviations of the compass in
the ship by applying the plate permanently to the compass
on board, in an opposite direction to that in which it produced the same deviations as the iron of the ship; but finding
that the deviation could not thus be entirely destroyed, he
then proposed to apply the plate temporarily in the position
in which it produced the same deviation as the iron of the
ship, so as to double the deviation caused by the ship. The
amount of the deviation so obtained was then to be applied in
the opposite direction as a numerical correction. The cause
of the failure of Mr. Barlow's plate in the first position has
been pointed out by M. Poisson and Mr. Airy. The deviation consists mainly of two parts;—first, that part described
by Captain Flinders, which varies as the sine and cosine of
the azimuth of the ship's head from the magnetic north,
and which may be called the "semicircular" deviation;—and
secondly, that part which varies nearly as the sine and cosine
of twice the same angle, and which part has been called the
"quadrantal" deviation. A plate which is found by experiment
to produce the same semicircular and quadrantal deviation as
the iron of a ship, will, when applied in the opposite direction,
destroy the semicircular deviation, but will double the quadrantal. From this objection the vertical bar of iron proposed by Captain Flinders is free: it corrects the semicircular
deviation without affecting the quadrantal deviation. The
substitution of the plate for the bar by Mr. Barlow seems,
therefore, in this respect, to have been a retrograde step, nor
am I aware that there are any other advantages in the plate
over the bar which justify the high encomiums which have
been generally bestowed upon it.

In the year 1824 Poisson's two Memoirs on the Theory
of Magnetism were communicated to the Institute, and pub-

lished in the fifth volume of the Memoirs of the Institute. In these memoirs M. Poisson founded a mathematical theory of transient induced magnetism on the physical theory of Coulomb, that by induction each particle of soft iron becomes a magnet, having an intensity proportional to that of all the forces which act on it, including the force of the magnetism developed by induction in all the other particles of the mass. In the second memoir M. Poisson gives, to express the action of the soft iron on the compass, formulæ involving coefficients to be determined by observation. In another memoir, communicated to the Institute in 1839, and printed in the 16th volume of the Memoirs, "Sur les Deviations de la boussole produites par le fer des vaisseaux," M. Poisson has adapted the formulæ to observations made on shipboard sufficient in number to determine the coefficients in the particular case of the soft iron being symmetrically placed on each side of the principal section of the vessel. The writer of this notice afterwards modified Poisson's formulæ so as to adapt them to the form in which the data generally present themselves, viz., a vessel having hard as well as soft iron—both unsymmetrically distributed—and observations of deviation being made in a certain number of equidistant points as shown by the compass affected by deviation. In this case we have generally more equations than are necessary to determine the coefficients; but the application of the method of least squares, owing to certain well-known properties of trigonometrical series, gives the values of the coefficients explicitly in a very simple form. These forms will be found in a pamphlet, entitled "Instructions for the Computation of a Table of the Deviations of a Ship's Compass," etc., published by order of the Lords Commissioners of the Admiralty, as a supplement to the "Practical Rules for ascertaining the Deviations of the Compass;" and also in the Philosophical Transactions for 1848, p. 347.

The subsequent part of this sketch will be better understood if some of the conclusions derivable from Poisson's (Coulomb's) theory are here indicated.

The problem of finding the attraction, on a magnetic particle at any given place, of a mass of soft iron of given shape magnetised by induction, or of a mass of hard iron of given shape permanently magnetised according to any given law, is one in general of insuperable difficulty.

But the problem we have to deal with is much simpler. The compass-needle may be considered as infinitesimally small compared to the distance from it of the nearest iron; the iron retains the same place relatively to the compass and to the ship; and lastly, the data of the problem are, not the shape and position of the iron, but its effect on the needle in a certain number of positions of the ship as shown by the disturbed compass. Taking advantage of these circumstances, the problem admits of an easy and complete solution.

In the mathematical treatment of this problem, it is necessary (at least in the first instance) to make the assumption that every portion of the iron in the ship is, as regards magnetism, of one or other of the two extreme qualities called "soft" and "hard." "Soft" iron is iron which becomes instantly magnetised to its full capacity when exposed to the influence of any magnetised body, and which loses its magnetism instantly when the influencing body is removed. "Hard" iron is iron which does not become magnetised by ordinary induction, but which when magnetised retains its magnetism unaffected by the influence of other magnetic bodies. The assumption is not strictly true of any iron. All iron appears to be in an intermediate state, requiring time and the molecular disturbance produced by blows, bends, or twists, to receive or lose magnetism. This intermediate state the mathematician must deal with by supposing the coefficients in the expressions derived from the supposition we have mentioned to undergo a change, or by some expedient of the same kind. The magnetism of such iron is called by Dr. Scoresby "retentive," by the Astronomer-Royal "sub-permanent."

The effect of hard iron on the compass is very simple. It gives rise to a single disturbing force of constant amount acting in a constant direction in the vessel, or, as we may

consider it, to three disturbing forces of constant amount,—
one acting *fore and aft*, one acting *athwartship*, and the
third acting *vertically*.

The first will produce a deviation proportional (the deviations being small) to the sine of the azimuth of the ship's
head; the second will produce a deviation proportional to
the cosine of the same azimuth. The third will produce no
deviation when, as we here suppose, the ship is on even keel.
The whole effect is, therefore, what I have called a "semicircular" deviation. The force which causes this deviation
being constant, the deviation produced, for a given azimuth
of the ship's head, will be inversely proportional to the horizontal force of the earth at the place..

The magnetism induced in the soft iron of the ship by the
vertical force of the earth will produce a disturbing force of
the same nature, as regards the constancy of its direction in
the ship, but of which the magnitude for a given azimuth is
proportional to the earth's vertical force. It therefore produces a semicircular deviation, the amount of which is
directly proportional to the earth's vertical force, and inversely proportional to the earth's horizontal force, *i. e.*,
which is directly proportional to the tangent of the dip.

The deviation produced by the horizontal force of the
earth is less simple in its law. For the single force we may
substitute its two resolved parts: one directed towards the
ship's head, and equal to the horizontal force multiplied by
the cosine of the azimuth of the ship's head; the other
directed to the starboard side, and equal to the horizontal
force multiplied by the sine of the azimuth of the ship's
head. Each will produce a disturbing force proportional
to the resolved part of the horizontal force, but not generally in its direction. The combination of the two resolved
parts gives a disturbing force acting fore and aft proportional
to the horizontal force, and varying partly as the cosine and
partly as the sine of the azimuth, and to a similar disturbing
force acting athwartship.

The fore and aft force proportional to the cosine of the
azimuth, and the athwartship force proportional to the sine

of the azimuth, produce no deviation when the ship's head is on any of the four cardinal points, because on each the force is either zero or acts in the direction of the needle, and they produce a maximum deviation when the ship's head is near any of the intermediate points, N.E., S.E., S.W., N.W. They therefore produce a "quadrantal" deviation proportional nearly to the sine of twice the azimuth.

The fore and aft force proportional to the sine of the azimuth produces no deviation when the ship's head is N. or S., and a maximum deviation when E. or W.; but this is not, as at first sight it might seem, a "semicircular" deviation, because the two maxima act in the same direction. The mean is, therefore, a constant deviation in one direction; and the variable part is a quadrantal force, of which the maxima occur when the ship's head is on the four cardinal points, proportional nearly to the cosine of twice the azimuth. The same observations apply to the deviation caused by the athwartship force proportional to the cosine of the azimuth.

The last mentioned disturbing forces and the directive force on the needle being each proportional to the horizontal force, the deviations produced by them are independent of the earth's force, and will be the same in any part of the globe.

The following then are the results:—

The permanent magnetism of the hard iron causes a semicircular deviation inversely proportional to the horizontal force.

The transient magnetism induced in the soft iron by the vertical force of the earth causes a semicircular deviation proportional to the tangent of the dip.

The transient magnetism induced in the soft iron by the horizontal force of the earth causes a constant deviation and also a quadrantal deviation, both independent of the latitude.

These results are thus expressed mathematically:—If δ be the deviation, reckoned positive when the N. point of the compass deviates to the E.; θ the dip of the needle; H the earth's horizontal force; ζ' the azimuth (by compass) of the ship's head, reckoned from the magnetic north to east:—

$$\delta = A + \left(B_1 \tan. \theta + \frac{B_2}{H}\right) \sin. \zeta' + \left(C_1 \tan. \theta + \frac{C_2}{H}\right) \cos. \zeta'$$
$$+ D \sin. 2\zeta' + E \cos. 2\zeta'$$

A B_1 B_2 C_1 C_2 D E being constants, depending only on the quantity, quality, and position of the iron in the ship.

It is useful to familiarise ourselves with the consideration of masses of soft iron which may produce the several terms in this expression.

For this purpose the most convenient shape to consider is a thin rod or bar.

Such a bar held in the direction of the dip receives by induction a magnetism proportional to the total force of the earth's magnetism at the place; if held at any inclination to the dip, a magnetism proportional to the total force multiplied by the cosine of that inclination. A vertical bar, therefore, receives a magnetism proportional to the total force multiplied by the sine of the dip, *i. e.*, to the vertical force. A horizontal bar in the magnetic meridian receives a magnetism proportional to the total force multiplied by the cosine of the dip, *i. e.*, to the horizontal force. A horizontal bar in any other azimuth receives a magnetism proportional to the horizontal force multiplied by the cosine of the azimuth.

It will easily be understood from what precedes that a vertical bar of soft iron below the level of the compass and before it, or above and abaft, gives a positive B_1; above and before, or below and abaft, a negative B_1; below and to starboard, or above and to port, a negative C_1; above and to starboard, or below and to port, a positive C_1. A horizontal bar of soft iron directed towards the compass, and before or abaft, gives a positive D; on the starboard or port side, a negative D; to N. E. or S. W., a negative E; to S. E. or N. W., a positive E; and two bars placed in the direction in the fig. (1),

Fig. (1)

a constant easterly deviation;

o

Fig. (2)

placed in the direction in fig. (2),
a constant westerly deviation.

o

The results at which we have arrived are exact, provided
the several coefficients are so small that their squares and
products may be neglected.

The exact expression, whatever be the value of the coefficients, may be obtained by mathematical operations of so
great simplicity, that I may, perhaps, be excused for repro-
ducing them here.

Take the centre of the compass for the origin of co-ordinates. Let X Y Z represent the total force of the earth's
magnetism resolved in the direction of (1) the head of the
ship, (2) the starboard side, (3) vertically downwards.
Each of these resolved forces will induce a force proportional
to, but not generally in the direction of, the force which
causes it, and therefore the resolved parts in the three
directions of co-ordinates may be represented by $a\,X + b\,Y
+ c\,Z$ along the axis of X; $d\,X + e\,Y + f\,Z$ along the
axis of Y; and $g\,X + h\,Y + k\,Z$ along the axis of Z. Let
P Q R represent the resolved force of the permanent magnetism of the hard iron in the same directions. Then
calling the whole forces which act on the needle in these
direction X', Y', and Z', we have evidently

$$X' = X + a\,X + b\,Y + c\,Z + P$$
$$Y' = Y + d\,X + e\,Y + f\,Z + Q$$
$$Z' = Z + g\,X + h\,Y + k\,Z + R$$

The 12 coefficients, $a\,b\,c\,d\,e\,f\,g\,h\,k$ P Q R, may be determined by observations made with a sufficient number of
corresponding values of X' Y' Z', X Y Z.

Observations with different values of X and Y are of course
easily obtained by the ordinary process of swinging the ship.

Observations with different values of Z can only be obtained by heeling the ship, or by observations made in different magnetic latitudes.

The expression for the deviation is most conveniently obtained by the use of polar co-ordinates.

Let H represent the horizontal intensity of the earth's magnetism, so that $H = \sqrt{X^2 + Y^2}$

Let θ represent the dip.

Let ζ represent the azimuth of the ship's head, reckoned to the east of the magnetic north, or the "correct magnetic course."

Let $H' = \sqrt{X'^2 + Y'^2}$ represent the horizontal intensity of the force, composed of the earth's force and of the force of the iron in the ship, which acts on the needle.

θ' the dip of the needle acted on by the iron of the ship.

ζ' the azimuth of the ship's head, reckoned east from the N. point of the disturbed needle, or the "compass course."

Let $\delta = \zeta - \zeta'$ represent the easterly deviation of the needle.

And, for convenience, let

$$a = M + D \qquad\qquad b = E - A$$
$$e = M - D \qquad\qquad d = E + A$$
$$c \tan. \theta + \frac{P}{H} = B \qquad f \tan. \theta + \frac{Q}{H} = C.$$

Then we have the following expressions for the deviation and horizontal force in terms of the *compass* course :—

$$(1 + M) \sin. \delta = A \cos. \delta + B \sin. \zeta' + C \cos. \zeta'$$
$$+ D \sin. (\zeta + \zeta') + E \cos. (\zeta + \zeta') (1).$$

$$\frac{H'}{H} = (1 + M) \cos. \delta + A \sin. \delta + B \cos. \zeta' - C \sin. \zeta'$$
$$+ D \cos. (\zeta + \zeta') - E \sin. (\zeta + \zeta') \quad . \quad . \quad (2).$$

And the following expressions for the same, in terms of the *correct magnetic* course :—

$$\frac{H'}{H} \sin. \delta = A + B \sin. \zeta + C \cos. \zeta + D \sin. 2\zeta + E \cos. 2\zeta$$
$$(3).$$

$$\frac{H'}{H} \cos. \delta = (1 + M) + B \cos. \zeta - C \sin. \zeta + D \cos. 2\zeta - E \sin. 2\zeta$$
$$(4).$$

b 2

These expressions are quite accurate on our assumption as to the quality of the iron.

The following are the conclusions derived from them as to the forces which act on the horizontal needle. They are :—

1. The horizontal force of the earth $=$ H, depending on the geographical position of the ship, and acting towards the magnetic north.

The following disturbing forces, which are caused by the induction of the Horizontal Force on the soft iron in the ship, viz :—

2. A force M H, acting towards the magnetic north, and the effect of which is simply to increase or diminish the directive force on the needle, according as M is $+$ or $-$, and thus indirectly to diminish or increase all the deviations rateably.

3. A force A H, acting towards the magnetic east, and causing the constant part of the deviation.

Forces 2 and 3 may be combined into a force $= \sqrt{M^2 + A^2}$ H, acting towards a point whose easterly azimuth is $\tan^{-1} \frac{A}{M}$

4. A force D H, acting towards the ship's head when the head is north, but of which the azimuth increases by twice as much as the azimuth of the ship's head.

5. A force E H, acting towards the starboard side when the ship's head is north, but of which the azimuth increases by twice as much as the azimuth of the ship's head.

Forces 4 and 5 may be combined into a force $= \sqrt{D^2 + E^2}$ H, acting towards a point whose azimuth is $\tan^{-1} \frac{E}{D}$ when the ship's head is north, but of which the azimuth increases by twice as much as the azimuth of the ship's head, giving rise to the quadrantal deviation.

The deviations caused by forces 2, 3, 4, 5, are independent of the earth's force and of the dip, and ought therefore to be the same for a compass in the same position in the same ship in all latitudes.

The following disturbing forces, which are caused by the induction of the Vertical Force on the soft iron in the ship, viz. :—

6. A force c Z, acting towards the ship's head.

7. A force f Z, acting towards the starboard side.

Forces 6 and 7 may be combined into a force $= \sqrt{c^2 + f^2}$ Z, acting in a direction $\tan^{-1} \frac{F}{C}$ to the right of the ship's head. The deviation caused by these forces is proportional to the tangent of the dip.

The following disturbing forces, which are caused by the independent magnetism of the hard iron of the ship, viz. :—

8. A force P, acting towards the ship's head.

9. A force Q, acting towards the starboard side.

Forces 8 and 9 may be combined into a force $= \sqrt{P^2 + Q^2}$, acting in a direction $\tan^{-1} \frac{Q}{P}$ to the right of the ship's head. The deviation caused by these forces is inversely proportional to the horizontal force.

Forces 6 and 8 may be combined into a force B H $= c$ Z $+$ P acting towards the ship's head, and forces 7 and 9 into a force C H $= f$ Z $+$ Q, acting towards the starboard side, and these into a force $= \sqrt{B^2 + C^2}$ H, acting in a direction $\tan^{-1} \frac{C}{B}$ to the right of the ship's head.

B is therefore the proportion which the force directed towards the ship's head, arising from the magnetism of the hard iron and from the magnetism induced in the soft iron by the vertical part of the earth's force, bears to the horizontal force of the earth at the place.

C is the proportion which the force directed towards the starboard side, arising from the same causes, bears to the horizontal force of the earth at the place.

The forces 6, 7, 8, 9, give rise to the semicircular deviation, which therefore consists of two parts, one depending on the soft iron and proportional to the tangent of the dip, and the other depending on the hard iron and inversely proportional to the horizontal force.

The quantities M A D E depend on the soft iron, and their permanence may be relied on with some confidence.

P and Q, and therefore B and C, depend on what I have called the permanent magnetism of the hard iron, but which magnetism is, in fact, to a great extent only sub-permanent or retentive. They are therefore liable to undergo a change not depending only or in any determinable way on the geographical position, and making any general mode of correction of the compass, either mechanical or tabular applicable to all latitudes, impossible.

Of the equations (1) (2) (3) (4), the first is in general the most convenient, as it does not require any observations of the horizontal force.

If the deviation be so small that the squares and products of the coefficients may be neglected, we have

$$\delta = A + B \sin. \zeta' + C \sin. \zeta' + D \sin. 2\zeta' + E \cos. 2\zeta' \quad (5)$$

In this expression the coefficients are of course expressed in arc.

It must be observed, that although the coefficient A may have a real value caused by any elongated horizontal mass of iron, as a long iron gun, the axis of which does not pass through the compass, yet it may also have an apparent value arising from an index error in the compass on board; or an index error, or an error caused by local attraction, affecting the compass on shore with which the compass on board is compared. In general, A is small and not greater than may be supposed to be caused by such errors; but I am assured by Mr. Evans, the Superintendent of the Compass Department of the Admiralty, that in some vessels, particularly in iron gun-boats, the value of A is too large to be attributed to such errors. The value of E is also, in every case which I have computed, so small, that it might be neglected, giving us, therefore, the following formula, which will in general express with great accuracy the apparently most irregular deviations:—

$$\delta = B \sin. \zeta' + C \cos. \zeta' + D \sin. 2\zeta' \quad . \quad . \quad (6)$$

This expression, or the expression

$$\delta = A + B \sin. \zeta' + C \cos. \zeta' + D \sin. 2\zeta' + E \cos. 2\zeta',$$

which is sufficiently accurate for all ordinary purposes, has the great advantage that from observations made on any number of equidistant parts the values of the coefficients are obtained with the greatest possible ease and expedition. Forms for computing these from observations made on 8, 16, or 32 points, will be found in the supplement to the Practical Rules already referred to; and using equation (6), the coefficients B C D may be obtained with the greatest readiness from observation of the bearing of a distant object made with the ship's head on the N.E., S.E., S.W., and N.W. points, and from them a table of deviations from all other points computed or traced graphically. I now continue the sketch.

In the year 1839 a paper was presented to the Royal Society by Mr. Airy, and published in the Philosophical Transactions, entitled "Account of Experiments on Iron-built Ships, instituted for the purpose of discovering a Correction for the Deviation of the Compass produced by the Iron of the Ships." This paper exercised so great an influence on the mode of correcting the compasses of ships practised in this country, and is of so much importance with reference to the controversy between Mr. Airy and Dr. Scoresby, to which I have referred, and to which it will be necessary to revert, that I trust I am not going out of my way in examining in some detail Mr. Airy's method and results. In this paper Mr. Airy, for what reason I do not quite understand, does not at first avail himself of those circumstances in the problem which give the solution on Poisson's theory with so much readiness, but applies himself in the first place to the solution of the problem of finding the attraction on the needle of a mass of soft iron of given shape magnetised by induction. In doing so Mr. Airy observes that it would have been desirable to make the calculations on Poisson's theory, and that that theory possesses greater claims on our attention, as a theory representing accurately the facts of some very peculiar cases, than any other, but that the difficulties in the application of that theory to complicated cases are great, perhaps insuperable. To avoid these difficulties Mr. Airy introduces the supposition that by the action of terrestrial magnetism every particle of soft iron is converted into a magnet, whose direction is parallel to that of the dipping-needle, and whose intensity is proportional to the intensity of terrestrial magnetism; in other words, that the particles of soft iron magnetised by induction do not induce magnetism in each other. This supposition furnishes readily expressions for the coefficients $a\,b\,c\,d\,e\,f\,g\,h\,k$, in the form of definite integrals; and Mr. Airy observes that in ordinary cases this simple theory will give the same comparative though not the same absolute results as Poisson's.

It is to be observed, however, that the actual arrangement of the iron of an iron ship is almost entirely of that peculiar and complicated kind to which Mr. Airy here alludes, in which Poisson's theory represents accurately the facts, but in which the simpler theory used by Mr. Airy gives results differing very widely from the facts. For instance, almost the whole of the very powerful magnetism induced in bars of soft iron held in or near the direction of the dip is caused by that mutual induction of the particles of soft iron which Mr. Airy's theory leaves out of account. According to that theory a bar of soft iron held in the direction of the dip would exert no greater force on the needle than the same bar condensed into a disc, a result which the simplest experiment with a bar held in the line of dip will show to differ very much from the fact.

Mr. Airy does not, however, make use in actual computation of the definite integrals so obtained. He substitutes for them literal coefficients corresponding to the $a\,b\,c\,d\,e\,f\,g\,h\,k$ of Poisson's method, which coefficients, like Poisson's, are to be determined by observations of deviation, force, etc. Mr. Airy's results would, therefore, be free from the objections to which his theory is open, were it not that his theory gives three relations $b = d$, $c = g$,

$f = h$,* which do not exist in nature, and the nine independent coefficients of the more perfect theory are thus reduced to six, and the constant term A, which, as we have seen, has sometimes a real value, becomes equal to zero.

The last objection, for the reasons before mentioned, is rather a theoretical than a practical objection, but there is a more serious objection to some of Mr. Airy's conclusions.

It is this.   It follows, from the less perfect as well as from the more perfect theory, and Mr. Airy points out that we have no means of determining separately the parts of which the semicircular deviation consists, viz., that which arises from induced magnetism and that which arises from permanent magnetism, by observations made at one place, with the ship on even keel; but Mr. Airy considering that as in the first instance the correction must be effected in a British port, it was desirable to form an *à priori* conjecture as to these magnitudes, accordingly made the *conjecture* that the part of the semicircular deviation arising from induced magnetism is so small that it may be neglected.

Mr. Airy's reasons, as I collect from the paper in the Philosophical Transactions for 1839, to which I have referred, and also from a paper by the same author, entitled " Results of Experiments on the Disturbance of the Compass in Iron-built Ships," Weale, 1840, are of this nature :—

" Although the amount of that part of the semicircular deviation which is caused by induced magnetism cannot be determined by means of observations at any one place, we know that it is a quantity of the same order as the quadrantal deviation.   It may be somewhat greater or somewhat smaller, but not remarkably greater ; and with some possible distributions of the soft iron, it may be zero.   Of the two iron vessels examined, one, the *Rainbow*, with a semicircular deviation of 50°, had a quadrantal deviation of only

---

* The mathematician will easily see that in this case these coefficients are the second partial differential coefficients of the potential, and that the equalities referred to are the same as

$$\frac{d^2 V}{dx\,dy} = \frac{d^2 V}{dy\,dx} \quad \frac{d^2 V}{dx\,dz} = \frac{d^2 V}{dz\,dx} \quad \frac{d^2 V}{dy\,dz} = \frac{d^2 V}{dz\,dy}$$

1°; the other, the *Ironsides*, with a semicircular deviation
of 30°, had a quadrantal deviation of 1° 6′. There is, therefore, good reason to conclude that the part of the semicircular deviation caused by induced magnetism is in all cases
(except in very high magnetic latitudes) extremely small,
and that, so far as observations with the best compasses can
go, the correction of the semicircular deviation made by
fixed magnets in one latitude will be perfectly correct in
every other latitude."

Of these reasons it may be observed, that in many cases, of
very frequent occurrence in iron ships, there is no relation
between the two things compared. A single iron stancheon
may give a large amount of semicircular deviation and no
perceptible amount of quadrantal deviation. A deck-beam
may give a large amount of quadrantal deviation and no
perceptible semicircular deviation.

But supposing the analogy to hold, the inference to be
deduced from iron ships in general differs very widely from
that deduced from the *Rainbow* and *Ironsides*. Let us
suppose a horizontal and a vertical bar of soft iron placed
respectively at the same distance from the compass. The vertical bar, in the position in which it produces the maximum
of semicircular deviation, is acted on by the whole vertical
force, and acts at right angles to the needle; the horizontal
bar, in the position in which it produces the maximum of
quadrantal deviation, is acted on by the horizontal force
multiplied by the sine of 45°, and acts at an angle of 45°.
The semicircular deviation produced by the vertical bar
would, therefore, *ceteris paribus*, exceed the quadrantal
deviation produced by the horizontal bar in the proportion
of the vertical force to $\frac{1}{2}$ the horizontal force, or of 2 tan.
dip : 1, or (in England, where the tangent of the dip $= 2\cdot75$)
of $5\frac{1}{2}$: 1. Now, in general, the semicircular deviation in England does not bear a greater ratio than this to the quadrantal
deviation, and, therefore, the inference to be drawn (if any)
from this analogy would rather be that the *whole* of the
semicircular deviation in iron ships arises from induced magnetism. Thus, in the case of several iron steamers swung in

England, whose semicircular and quadrantal deviations are given by Mr. Airy in a paper in the Philosophical Transactions for 1853, p. 53, we have the following results:—

	Quadrantal Deviation.	Semicircular Deviation.
Bloodhound	3° 48'	15° 20'
Jackall	4° 14'	17° 50'
Trident	3° 35'	20° 40'
Vulcan	3° 36'	9° 5'
Simoom	4° 31'	22° 0'

In all of these the semicircular deviation is less than 5½ times the quadrantal; and unless I have misapprehended Mr. Airy's reasons, the conclusion to be drawn from these vessels would be that nearly the whole of the semicircular deviation was caused by induced magnetism. That this, however, was not the case is shown by the observations made in other latitudes in the same iron vessels, showing that a large amount (in the *Trident* nearly the whole) of the deviation arose from permanent magnetism.

The conclusion to which we are led, then, is the negative one, that no *à priori* conjecture, having the least probability of correctness, as to the relative proportions of the induced and permanent magnetism which give rise to the semicircular deviations, can be formed.

There are other considerations which lead to the same result. We may suppose the compass to be placed at a position in the ship in which the induced magnetism compensates itself so as to produce no semicircular deviation. In that case all the semicircular deviation will be caused by permanent or retentive magnetism; but we may in the same manner suppose a compass to be placed where all the permanent or retentive magnetism will compensate itself, in which case the whole semicircular deviation will arise from transient induced magnetism; and there was, in fact, no phenomenon observed by Mr. Airy in the *Rainbow* and *Ironsides* which might not have been caused by the transient induced magnetism of the soft iron in these ships.

This conjecture was one on the correctness of which the mode of correction originally proposed by Mr. Airy, viz., by *fixed* magnets, depended. Mr. Airy subsequently proposed to correct the semicircular part of the deviation by *adjustible* magnets, the position of which is to be altered by the Captain according as the changes in dip, in the sub-permanent magnetism of the ship, or in the magnetism of the correcting magnets, introduces any serious amount of semicircular deviation. This mode of correcting the semicircular deviation would no doubt succeed in skilful hands, except in the case of sudden changes in the sub-permanent magnetism of newly-built iron ships, but it may be doubted how far it is prudent to trust such delicate manipulation to unskilful hands. I am not aware whether this mode of correction has been practised, or whether it has succeeded in practice.

I have stated that the conclusion of Mr. Airy as to the relative amounts of permanent and transient induced magnetism is put forward by him merely as a conjecture; but it is necessary to observe, in order to explain the mistake into which I have already alluded as one into which I cannot help thinking Dr. Scoresby fell, that some expressions used by Mr. Airy in this and subsequent papers appear to be liable to be misunderstood, and to have been misunderstood not only by Dr. Scoresby but by other persons.

Their mistake was this. Mr. Airy having made the conjecture I have referred to, afterwards assumes the correctness of the conjecture, and uses the term "permanent magnetism" to express the force which causes semicircular deviation— in other words, he uses the term "permanent magnetism" as equivalent to "permanent magnetism" $+$ "transient magnetism induced by the vertical part of the earth's force." The same form of expression is used by Mr. Airy in the paper published by Mr. Weale, to which I have referred, and in a letter to the editor of the Nautical Magazine, published in the Nautical Magazine for December 2, 1843. When, therefore, Mr. Airy in those papers speaks of his investigations and observations as showing that the deviation (other than the quadrantal) was caused by "permanent magnetism," and proposed to correct such deviation by fixed magnets, persons who looked into the paper only for the results of the mathematical investigation without actually repeating or following the mathematical operations, easily fell into the mistake of supposing that Mr. Airy had demonstrated, or professed to have demonstated, that the semicircular deviation did arise entirely or almost entirely from permanent magnetism, and that no part, or only a small and unimportant part, arose from transient induced magnetism.

Mr. Airy himself must of course be acquitted from any theoretical mistake on this point. The less perfect theory, of which he made use advisedly and with the object I have mentioned, agreed with the more perfect theory of M. Poisson, in showing that the semicircular deviation consists of two parts which cannot be separated by observations of the horizontal needle at any one place, and this Mr. Airy points out, as well as that the effect of the impracticability of separating these parts is to render the compass incorrect in one magnetic latitude when it has been made correct in a different magnetic latitude. Mr. Airy's mistake, if I may be excused for so designating it, was in his " conjecture " that the part of the semicircular deviation which depends on the induced magnetism is so small that it may be neglected except in the case of a ship sailing very near the magnetic pole, and in basing on this conjecture, which is certainly in many cases incorrect, a general mode of correcting the deviation by fixed magnets.

The following passage in Dr. Scoresby's Magnetical Investigations, vol. ii., p. 239, seems to show that he had misunderstood Mr. Airy's results:—" In his experiments on board the iron-built steam ship the *Rainbow*, Mr. Airy found that some other quality of magnetism besides that simply due to ordinary terrestrial induction was not only present but particularly influential. The proportions, too, in which the transient and partially enduring magnetism, as affecting the compass in the vessel respectively existed were inquired into, and found to place the latter in a highly dominant position." This, as we have seen, is a mistake. Mr. Airy's observations did not *prove* that there was any other cause operating than terrestrial induction, and *à fortiori* did not establish any relation between the amounts of the transient and permanent or retentive magnetism.

I have insisted on this point, because it appears to me not improbable that Dr. Scoresby's mistake influenced in some degree his subsequent speculations and investigations; and that believing Mr. Airy to have demonstrated that the semi-circular deviation in iron ships was not to any appreciable degree due to transient induced magnetism, and yet finding from observations made in different latitudes that the magnetism of the ship was not permanent, he was led to attribute the semicircular deviation in iron ships almost entirely to magnetism of the kind which he denominated "retentive," *i.e.*, magnetism which may be considered permanent in the ordinary process of swinging a ship, but which changes under a change in the inducing force, and with great readiness when aided by mechanical violence.

Dr. Scoresby was thus led to investigate practically and theoretically the effect on the compass of "retentive" magnetism, and to call attention to a cause which mathematicians had to a great degree overlooked; and he showed, from the changes which take place in the magnetism of iron ships without a change of latitude, that a great part of the deviation arises neither from transient nor permanent. but from "retentive" or "sub-permanent" magnetism.

Dr. Scoresby approached the question of the deviation of the compass from an entirely different side from M. Poisson and Mr. Airy. He investigated the subject experimentally, not mathematically. He was free from the temptation to which the mathematician is exposed, of adapting his theory to the powers of his analysis rather than to the phenomena observed. His experiments were made with iron of a like quality with that of which iron ships are composed, viz., plates prepared for ship building, and the result of his experiments was to show that the state of the iron was very different from either of the states which had been supposed by mathematicians. The iron was not "soft," because it had acquired by hammering a great degree of magnetism, which it retained while the ship was swung in different positions;

but neither was it "hard," because in a new position and under the influence of a new inducing force and the mole-cular disturbance caused by blows and strains, and even time alone, it changed. On this feature in the iron Dr. Scoresby strongly insisted, and to him is principally due what we know of iron in this state.

Retentive magnetism is the last kind to which the mathe-matician has recourse in order to explain phenomena. He cannot deal with it directly in his analysis. He can only treat it is as permanent magnetism which undergoes a change, or as transient magnetism which requires time and molecular disturbance to enable it to take up its new estate. Poisson does not notice it. Mr. Airy, until the subject was brought forward by Dr. Scoresby, did not notice it. Its effect on the deviation of the compass and ship was first, I believe, suggested by General Sabine in the Philosophical Trans-actions for 1843, p. 153, in the reduction of the magnetic observations made in the ships of Sir James Ross's Antarctic expedition. He inferred from the observations, "that when a ship changes her magnetic latitude, the corresponding change in the magnetism of the ship, or more strictly in that portion of it which is derived from induction, follows, but does not always, or altogether, take place instantaneously;" * * "that some portion of the iron might be of a quality inter-mediate between that of perfectly soft iron which undergoes instantaneous change, and that of iron which acquires per-manent magnetism, and that such portions should be liable, in regard to their magnetic condition, to be more or less in arrear of the ship's magnetic position."

Dr. Scoresby's attention had been early directed to the effect of percussion, and twists and strains generally, in ren-dering iron susceptible of induction. In the Transactions of the Royal Society of Edinburgh for 1821, there is a paper by him entitled, "Description of a Magnetometer, being a new Instrument for measuring Magnetic Attractions and finding the Dip of the Needle, with an Account of Experiments made by it." The instrument was an iron bar moved in the plane of the magnetic meridian till it ceased to affect the needle, when its inclination measured the dip. In this paper Dr. Scoresby describes a number of experiments and gives a number of conclusions as to the effect of percussion, filing, bending, and twisting, in rendering bars susceptible of induction.

In the Philosophical Transactions for 1822, p. 241, there is a paper by Dr. Scoresby, " Experiments and Observations on the Development of Magnetic Properties in Iron and Steel by Percussion." This paper was an application of the principles of the last paper to the construction of artificial magnets.

A continuation of the same paper, dated " On board the ship *Baffin,* off Iceland, August 27, 1123," is contained in the Philosophical Transactions for 1824, p. 197.

In a paper " On the Defects and Dangers arising from the use of Corrective Magnets for Local Attraction in the Compasses of Iron-built Vessels," read before the British Association in 1847, but of which I have not been able to find any account, Dr. Scoresby, I believe, called the attention of the Association to this quality of the iron composing iron ships.

In the fourth part of the "Magnetical Investigations" Dr. Scoresby considered the magnetism of iron-built ships with relation to the action of the compass and the changes to which their magnetic condition is liable. This paper sets out with referring to Mr. Airy's paper of 1839, and the mode of correction proposed by it as sufficient if the iron were all of the two qualities supposed by Mr. Airy (which we have seen is a mistake), and he then refers to the third quality, which he denominates " retentive " magnetism, as playing an important part in the deviation of the compasses of iron ships, and as not met by Mr. Airy's proposed mode of correction.

In the second chapter Dr. Scoresby gives the results of experiments on the rolled iron plates used in iron ship-building. His experiments led him to the conclusion that the plates of which a ship is built do not, from the mere process of manufacture, independently of the hammering received in shipbuilding, receive any large amount of permanent magnetism. In one plate the transient induced magnetism, with the plate placed upright, exceeded the per-

manent magnetism in the proportion of about 9 to 1. The
same plate, when hammered in its vertical position, had an
amount of permanent magnetism developed amounting to
not more than two-sevenths of the transient induced magnetism, and this was easily inverted by a few blows when the
plate was reversed. From these experiments Dr. Scoresby
deduces the inference that a great part of the deviation in
iron ships is caused by the retentive magnetism, which at
first is very powerful from the great amount of hammering
to which the iron of such a vessel is subjected in building,
but which is altered as soon as the vessel is exposed to blows
or strains. The experiments seem further to show that with
such retentive magnetism is united a large amount of transient induced magnetism.

At the meeting of the British Association at Liverpool, in
1854, Dr. Scoresby read a paper "On the Loss of the
Tayleur, and the Changes in the Compasses in Iron Ships."

The *Tayleur* was a new iron ship, about 2000 tons burthen.
Her steering compass had originally a maximum deviation of
60°; this was corrected by a magnet. She met severe
weather in going down the Channel; and if, as Dr. Scoresby
supposes, the effect was to "shake out" the magnetism of
building and give a new magnetism, this would leave the
correcting magnets to produce a deviation which threw her
on the Irish Coast.

This Dr. Scoresby explained in this paper, and insisted
much on the condition of iron, viz., the retentive, which
causes an intensely high state of magnetism to be induced
which cannot be retained after the ship has been exposed to
different inducing forces and blows.

This paper called forth remarks from Mr. Airy, published
in the Athenæum, and in the Mercantile Marine Magazine
for December, 1854. In this paper Mr. Airy still treats as
inconsiderable the part of the semicircular deviation which
arises from induced magnetism, and speaks of the paper of
1839 as having established that the greater part of the deviation depends on the permanent or polar magnetism, and
urges that although it was evident there were causes in

action tending to produce effects like those produced by
Dr. Scoresby's experiments, yet that it was equally evident
that the action of those causes must be exceedingly slow.
That the plates of iron of a ship receive no shocks from the
waves, but that the direct effect of the most violent sea is
that the plates of iron are, in the course of two or three
seconds of time, plunged five or six feet deeper in water and
sustain a corresponding hydrostatic pressure. That the change
to be expected in a ship's sub-permanent (retentive) magnetism in sailing from England to the Cape of Good Hope
does not essentially depend on her passing into another magnetic hemisphere; but that, supposing her to have been built
with her head north, she is turned with her head south
and exposed to slight tremours for some months. That
as regards the loss of the *Tayleur*, it was not likely or even
possible that in two days the magnetism of the ship would
be so much changed that the compass would be disturbed
through an angle of two points or even one-tenth of that
amount. That in Her Majesty's ship *Trident* the deviations
at the Thames and at Malta were nearly in the inverse proportion of horizontal intensity, and show that the disturbing
force had remained unaltered. That the correction by permanent magnets should be generally adopted, but that for
voyages to the southern hemisphere the Captain should be
furnished with the means of adjusting the magnets.

This paper was answered by Dr. Scoresby in a paper
published in the Athenæum, and in the Mercantile Marine
Magazine for January, 1855. In this paper Dr. Scoresby
contended strongly that Mr. Airy's mode of compass correction could not be correct, and must in some cases dangerously mislead. Dr. Scoresby, in opposition to Mr. Airy,
contends that a ship is "*most* liable to undergo a change of
magnetism at any time." That the shocks of the sea on the
iron plates of a ship have sufficiently the quality of impact
to render his (Dr. Scoresby's) experiments by percussion
applicable for indicating the probable result of magnetic
changes. He points out that Mr. Airy's conclusion as to the
Tayleur is inconsistent with the observed facts, viz., that two
corrected compasses which agreed at starting differed by two

points two days afterwards. That this was to be expected from known facts of observation. That in ships going to southern latitudes adjustment by magnets will generally aggravate the error, and that the errors might be corrected by a compass at the masthead.

A reply by Mr. Airy to this paper will be found in the Mercantile Marine Magazine for March, 1855. In this Mr. Airy, referring to his conjecture that the induced part of the semicircular deviation was so small that it might be neglected, states that he is inclined to think that his estimate of the possible magnitude of that term had been too low. He insists that the blow from water, however great, is still not of the nature of impact but pressure. That there is no sufficient evidence of sudden changes of great magnitude.

In this paper there are the following important passages:—

" One general law seems to apply to ships going into the southern hemisphere,—that on returning to England their compasses are, with very trifling errors, as correct as when they left England. But the state of their compasses in the southern hemisphere varies greatly; some are perfectly correct, others are very erroneous. I do not imagine that in any of these cases the sub-permanent (retentive) magnetism has undergone any particular change. I think it far more probable that the error arises from transient induced magnetism. Though the original theory was correct the application of it has been incorrect, from throwing the correction exclusively on magnets, and not introducing also the action of a mass of soft iron below the level of the compass."

To this paper there was a rejoinder by Dr. Scoresby in the Mercantile Marine Magazine for April, 1855.

In this Dr. Scoresby points out how much Mr. Airy had under estimated the impact of waves striking a ship. He insists on the observations which show sudden changes in the magnetism of vessels without change of geographical position, which can only arise from changes in the retentive magnetism.

Dr. Scoresby alludes but does not enter minutely into Mr. Airy's expression of opinion that the changes in vessels going to the south arise from induced not retentive magnetism.

In doing so I cannot help thinking that Dr. Scoresby was

still under the influence of the mistake which I suppose him to have made, as to what Mr. Airy had accomplished in his paper of 1839, and that believing Mr. Airy to have *demonstrated* that the part of the semicircular deviation produced by induced magnetism was insensible, he sought for the cause of the whole change actually produced in the retentive magnetism.

These differences of opinion could not be removed without actual observations in the same vessel in different latitudes; but it must be observed, that such observations, although deciding some would not decide all the points in question. If, on the return of a ship from the southern hemisphere, her magnetic state is found to be very different from her state at her departure, the inference is inevitable that a great change has taken place in her retentive magnetism. But if a compass corrected by magnets has a large error in the southern hemisphere, there are no obvious means of determining how much of the error arises from retentive or how much from transient induced magnetism.

In order to investigate these interesting and important questions Dr. Scoresby undertook the voyage of which the narrative follows. It remains for me to show what the result of the voyage was in furnishing an answer to the questions.

In some respects a smaller and more manageable vessel would have had advantages over the *Royal Charter*.

Dr. Scoresby's researches required that the ship should be carefully swung whenever an opportunity offered; but the difficulty of swinging so large a ship, and the importance of every hour that could be saved in the voyage out and home, prevented the magnetic state of the *Royal Charter* being observed with the accuracy which was desirable, and it was only with the greatest and most resolute exertions on the part of Dr. Scoresby that observations of any value were made.

The *Royal Charter* was swung at Liverpool on the 4th of January 1856, and the results of the observations will be found at page 13 of the narrative of the voyage.

She was again swung at Hobson's Bay, Port Phillip, on the 2nd of May, as described in page 182 of the narrative. The following are the results of this swinging:—

Deviations of the Compasses of the Royal Charter, swung at Hobson's Bay, Port Phillip, Victoria, 2nd May, 1856.

Correct Magnetic Course.	Course by Admiralty Compass.	Deviations { E denoting an Easterly Deviation. W denoting a Westerly Deviation.			
		Admiralty Compass.	Masthead Compass.	Steering Compass.	Companion Compass.
N. 3°49' E.	North	W. 11·	W. 2·30	E. 9·45	E. 6·15
2·15	N. by E.	W. 9·	W. ·30	E. 11·15	E. 7·15
15·	N. N. E.	W. 7·	E. 1· 0	E. 11·45	E. 7·45
29·	N. E. by N.	W. 4·45			
42·	N. E.	W. 3·	W. ·15	E. 10·15	E. 6·15
55·	N. E. by E.	W. 1·15			
66·30	E. N. E.	W. 1·30	W. 1· 0	E. 5·	E. 2·30
77·	E. by N.	W. 1·45			·
89·	East	W. 2·30	W. 2·15	W. 5·45	W. 3.30
98·30	E. by S.	W. 2·45			
109·	E. S. E.	W. 3·30	W. 3·30	W. 12·	W. 7· 0
120·15	S. E. by E.	W. 3·30	W. ·45		
132·15	S. E.	W. 2·45	W. 1·30	W. 15·	W. 9· 0
144·45	S. E. by S.	W. 1·30	· ·	W. 19·15	W. 15·30
158·45	S. S. E.	E. ·15	W. 1·15	W. 18·15	W. 13·30
174·45	S. by E.	E. 4·30	E. ·30	W. 14·15	W. 10·15
187·30	South	E. 7·30	E. ·30	W. 11·45	W. 8·30
201·45	S. by W.	E. 10·15	W. ·45	W. 9·30	W. 6.15
215·15	S. S. W.	E. 12·45	W. 1·30	W. 10·15	W. 5·15
226·	S. W. by S.	E. 12·15	W. 1·30	W. 5·15	W. 3·45
236·15	S. W.	E. 11·15	· ·	W. 3·30	W. 3·
245·30	S.W. by W.	E. 9·15	W. ·30	W. 2·45	W. 2·15
254·30	W. S. W.	E. 7·	W. ·30	W. 1·45	W. 1·15
262·30	W. by S.	E. 3·45	W. ·45	W. 1·	W. ·45
270·15	West	W. ·15	E. ·15	E. ·30	E. ·30
277·30	W. by N.	W. 3·45	W. 1· 0	E. 3· 0	E. 1·45
285·45	W.N. by W.	W. 6·45	E. 1·45	E. 2·15	E. 2·15
293·45	N.W. by W.	W. 10·	W 1·15	E. 3.30	E. 2·45
301·45	N. W.	W. 13·15	W. 2· 0	E· 2·15	E. 1·30
313·	N.W. by N.	W. 13·15		E. 6.30	E. 5·
325·30	N. N. W.	W· 12·		E. 7·45	E. 6·
337·	N. by W.	W. 11·45		E. 9·	E. 7·

On the return of the *Royal Charter* to Liverpool, on the 13th or 14th of August, the deviations of her compasses were observed by Dr. Scoresby, with the aid of Mr. Rundell, the Secretary of the Liverpool Compass Committee, in the manner described in the Narrative of the voyage, and the following were the results :—

Deviations of the Compasses of the Royal Charter, swung at Liverpool August 13 *and* 14, 1856.

Correct Magnetic Course.	Deviations { E denoting an Easterly Deviation. W denotes a Westerly Deviation.		
	Admiralty Compass.	Steering Compass.	Companion Compass.
N. 8·30 E.	E. 0·45	E. 20·30	E. 17·45
20·30	E. 0·	E. 19· 0	E. 16·
40·15	E. 1·30	E. 21·30	E. 18·
53·45	E. 2·	E. 19·	E. 17·
67·15	E. 1·	E. 13·30	E. 15·
178·45	. .	W. 22·15	
180·	E. 1·	W. 22· 7	W. 17·
183·	E. 1·30	W. 22·30	W· 17·
199·	E. 6·	W. 18·45	W· 14·15
203·	E. 7·45	W. 18·15	W. 14·30
204·45	E. 8·	W. 17·30	W. 14·45
208·	. .	W. 16·	W 13·
217·15	E. 9·	W. 14·	W. 12·
230·45	E. 8·45	W. 10·	W. 9·15
247·15	E. 6·45	W. 8·30	W. 3·30
259·	E. 5·30	W. 2·30	W. 1·
266·15	E. 3·	W. 1·45	W. 2·30
276·30	W. 1·30	E. ·15	E. ·45
288·45	W. 6·	E. 3·15	E. 3·15
305·30	W. 8·45	E. 7·45	E. 7·45
321·15	W. 9·	E. 12·30	E. 12·45
341·30	W. 4·30	E. 17·30	E. 17·30
348·30	W. 0·	E. 22·30	E. 22·30

For the purpose of comparing the deviations at the different places, and freeing them as much as possible from accidental error, curves were drawn from these observations by Mr. Rundell, who obtained from them the following adjusted tables of deviations :—

Deviations of the Compasses of the Royal Charter, obtained by projecting the observed Deviations in Curves.

Course by each Compass.	ADMIRALTY COMPASS.			STEERING COMPASS, Corrected by Magnets, etc.			COMPANION COMPASS, Corrected by Magnets, etc.		
	Liverpool, Jan. 1856.	Melbourne, May, 1856.	Liverpool, Aug. 1856.	Liverpool, Jan. 1856.	Melbourne, May, 1856.	Liverpool, Aug. 1856.	Liverpool, Jan. 1856.	Melbourne, May, 1856.	Liverpool, Aug. 1856.
North.	− 23 45	− 11	− 0 45	+ 1 40	+ 12 0	+ 21 40	− 3 40	+ 7 10	+ 16 40
N. by E.	− 19 45	− 9 15	− 0 30	+ 3 20	+ 11 40	+ 21 10	− 1	+ 7 20	+ 17 20
N. N. E.	− 14	− 7 15	− 0 0	+ 4 15	+ 11	+ 20 30	0	+ 6 50	+ 17 30
N. E. b. N.	− 10 20	− 5	+ 1	+ 4 50	+ 10	+ 19	+ 0 50	+ 6 10	+ 17 10
N. E.	− 7 40	− 3	− 1 45	+ 4 50	+ 8 20	+ 16 30	1	+ 5	+ 16 40
N. E. by E.	− 5 40	− 1 15	− 2	+ 4 30	+ 6 15	+ 13	0 40	+ 3 40	+ 14 30
E. N. E.	− 4 45	− 1 15	− 1 20	+ 3 40	+ 3	+ 9 10	0	+ 1 40	+ 11
E. by N.	− 4 40	− 1 15	− 0 15	+ 3		+ 4 20	− 0 40	0	+ 7
East.	− 4 30	− 1 30	− 2	+ 2	− 4 30	− 0 40	− 1 20	− 2 30	− 2
E. by S.	− 3	− 2 30	− 3 40	+ 1	− 7 30	− 7	− 1 20	− 4 50	− 2
E. S. E.	− 0 30	− 3 30	− 4 40	+ 0 30	− 10	− 11 40	− 1 5	− 6 50	− 6
S. E. by E.	+ 1 30	− 3 30	− 4 50	0	− 12 50	− 16	− 1	− 8 50	− 11
S. E.	3 30	+ 2 45	− 4 20	− 1 40	− 15	− 19 30	− 0 50	− 10 40	− 13 40
S. E. by S.	6 45	1 30	− 3 10	− 1 40	− 16 40	− 21	− 0 40	− 12	− 15 15
S. S. E.	11	1 15	− 1 20	− 2 40	− 17 30	− 22	− 0 30	− 12 30	− 15 40
S. by E.	15	5	+ 1 20	− 3	− 18 10	− 22 10	− 0 20	− 12 20	− 15 50

Deviations of the Compasses of the Royal Charter, etc.—continued.

Course by each Compass.	ADMIRALTY COMPASS.			STEERING COMPASS, Corrected by Magnets, etc.			COMPANION COMPASS, Corrected by Magnets, etc.		
	Liverpool, Jan. 1856.	Melbourne, May, 1856.	Liverpool, Aug. 1856	Liverpool, Jan. 1856.	Melbourne, May, 1856.	Liverpool, Aug. 1856.	Liverpool, Jan. 1856.	Melbourne, May, 1856.	Liverpool, Aug. 1856.
South.	19	8	+ 3 45	— 3	— 17	— 22	— 10	— 11 15	— 15 40
S. by W.	— 21 45	10 35	5 40	— 2	— 15	— 21 20	0	9 30	— 15
S.S.W.	— 23 40	12 45	7 30	30	— 12	— 20 10	+ 1	7 10	— 14 40
S.W. by S.	— 23 50	12 15	8 40	+ 20	— 9	— 18 30	2 30	5 30	— 13 50
S.W.	— 22 40	11 15	9	— 1 30	— 6 40	— 15 40	3 30	4	— 12
S.W. by W.	— 19 30	9 30	8 45	— 2	— 4 20	— 12	— 4 20	2 30	— 9 20
W.S.W.	— 15 10	7	7 10	— 2 40	— 2 20	— 8	— 4 30	1 30	— 6 10
W. by S.	— 9 50	4	5 15	— 2 30	50	— 4 20	— 4 10	45	— 3
West.	— 4 20	2 20	2 30	— 2	+ 1	30	2 50	0	40
W. by N.	— 1 0	— 3 45	— 1 50	— 1	— 2 30	+ 1 40	1 20	+ 1 50	+ 3
W.N.W.	— 6 40	— 7 15	— 5 45	30	— 4	— 6 30	40	+ 2 15	5 30
N.W. by W.	— 12 25	— 10 20	— 8 10	0	— 4 30	— 10 10	— 2	2 30	9
N.W.	— 17 30	— 12 40	— 8 30	40	— 6 15	— 14	— 4	4 40	11 30
N.W. by N.	— 22	— 13 45	— 8	— 1	— 8 10	— 17 10	— 5 30	5 20	13 30
N.N.W.	— 24 30	— 13 30	— 6 30	— 1	— 9 20	— 20 20	— 5 40	6	16
N. by W.	— 24 45	— 12 30	— 3 30	0	— 11 20	— 20 50	— 5 10	6 50	17

Applying to these tables the formula (5), to which I have already referred, I obtain from them the following values of the coefficients A B C D E:—

	A.	B.	C.	D.	E.
ADMIRALTY COMPASS					
Liverpool, Jan. 1856	− 0° 18	− 3° 40	− 19° 35	7° 13	− 1° 18
Melbourne, May, 1856	− 1 27	− 1 20	− 9 3	6 58	− 0 25
Liverpool, Aug. 1856	− 3	− 1 30	− 3 22	5 58	+ 28
STEERING COMPASS, corrected by Magnets and Soft Iron.					
Liverpool, Jan. 1856	− 55	− 35	− 1 43	2 5	− 1 17
Melbourne, May, 1856	− 1 54	− 2 25	13 35	2 34	− 30
Liverpool, Aug. 1856	− 50	15	22 25	1 35	+ 25
COMPANION COMPASS, corrected by Magnets and Soft Iron.					
Liverpool, Jan. 1856	− 17	− 51	− 1 48	2 22	− 1 23
Melbourne, May, 1856	− 1 25	− 1 45	8 45	1 55	− 28
Liverpool, Aug. 1856	− 47	− 1 15	17 35	1 34	− 11

The following were the values of the coefficients for the steering and companion compasses at Liverpool in January, 1856, before compensation, obtained by Mr. Rundell:—

	A.	B.	C.	D.	E.
Steering Compass (before compensation)		− 27° 7	− 30° 7	5°	
Companion Compass (before compensation)		− 8	− 25		

The values of the same coefficients obtained by Mr. Rundell, as given in the Report of the Liverpool Compass Committee, p. 51, are :—

	A.	B.	C.	D.	E.
Admiralty Compass.					
Liverpool, Jan. 1856	− 0°27′	− 3°48′	− 19°42′	+ 6°59′	− 0°52′
Melbourne, May, 1856	− 1°27′	− 1°11′	− 8°59′	+ 6°23′	− 0°16′
Liverpool, Aug. 1856	− 0°3′	− 1°6′	− 3°22′	+ 6°10′	+ 0°56′
Steering Compass, corrected by Magnets and Soft Iron.					
Liverpool, Jan. 1856	+ 0°20′	+ 0°32′	+ 1°50′	+ 2°0′	− 1°20′
Melbourne, May, 1856	− 0°59′	− 2°30′	+ 13°53′	+ 2°34′	− 0°45′
Liverpool, Aug. 1856	− 1°12′	− 0°15′	+ 22°29′	+ 1°37′	+ 0°3′
Companion Compass, corrected by Magnets and Soft Iron.					
Liverpool, Jan. 1856	− 0°2′	− 0°56′	− 1°49′	+ 2°15′	− 0°50′
Melbourne, May, 1856	− 1°16′	− 1°59′	+ 9°2′	+ 1°56′	− 0°27′
Liverpool, Aug. 1856	+ 0°43′	+ 1°15′	+ 17°31′	+ 1°38′	− 0°6′

Mr. Rundell also obtained the following values of the coefficients for the mast compass :—

	A.	B.	C.	D.	E.
Mast Compass.					
Liverpool, Jan. 1856	− 0°35′	+ 0°46′	− 2°56′	+ 2°28′	+ 0°9′
Melbourne, May, 1856	− 1°	− 0°15′	− 0°9′	+ 0°33′	+ 0°11′

These results are, however, subject to great uncertainty, as the compass was too sluggish and its graduation too rough for exact experiments.

In comparing these values it will be seen that A and E
are small and irregular. They probably arise almost entirely
from errors of observation. D, which gives the principal
part of the quadrantal deviation, in theory ought not to
change with a change of latitude, and its constancy of value
is very satisfactory. The differences in its value may arise
partly from errors of observation and partly from change of
temperature.

B, as we have seen, represents the proportion which the
disturbing force of the ship towards the head bears to the
horizontal force at the place. C represents the proportion
which the disturbing force towards the starboard side bears
to the horizontal force.

The *Royal Charter* was built with her head in the
direction N. 50° W. (magnetic). When newly launched
there was, therefore, a strong tendency of the north point of
the compass to the port side, giving a large negative value
for C in her standard compass, and also in the steering and
companion compasses before being corrected by fixed magnets. The blows and strains which the *Royal Charter*
experienced in her voyage diminished or, so to speak,
"shook out" this inequality to such an extent that the
standard compass, which had originally a deviation of nearly
20° to the port side when the ship's head was north, as shown
by the value of C, had this deviation reduced to 3° 22' on
the ship's return to Liverpool; while the steering compass,
which originally had been rather over corrected, having a
C = 1° 43', returned with C = 22° 25', an increase of
20° 42'; while the companion compass, originally rather
under corrected, having a C — 1° 48', returned with a
C = 17° 35', a change of 19° 23'. This change was one
which had evidently taken place in the retentive magnetism
of the ship, and seems to justify Dr. Scoresby's anticipations,
and to show that Mr. Airy had very much under estimated
the change in the retentive magnetism.

This change shows the complete failure in such a voyage of
the correction by *fixed* magnets,—while the standard compass uncorrected had its deviation reduced to a great extent

during the voyage, the deviation of the corrected compasses grew up to more than two points.

With regard to B it seems difficult to say what inference should be drawn. It appears to have been considerable in the steering compass before correction, but when corrected to have changed very slightly. It seems to show an amount of permanency in the magnetism which attracted the north point of the compass to the stern for which it appears to me difficult to account, and which may perhaps show that if iron ships were built with the keel in the magnetic meridian, and so as to have no considerable deviating force to either side, permanent magnets might be applied as correctors with less risk than when there is an original large deviation to one side.

The smallness of the value of B, and the nature of the changes in C, seem to show that in the three compasses of the *Royal Charter*, as in the case of the standard compass of the *Trident*, to which I have already referred, Mr. Airy's conjecture as to the smallness of that part of the semicircular deviation which arises from transient induced magnetism was not far from being correct.

The following letter will show the inferences deduced by Mr. Airy from the observations in the *Royal Charter* :—

" ROYAL OBSERVATORY, GREENWICH,
" *October* 2*nd*, 1856.

" MY DEAR SIR,—The pressure of Observatory business has prevented me for a time from examining the observations of compass deviation in the *Royal Charter*, but I have now taken them up, and find the following results :—

" 1. Quadrantal deviation.

Liverpool, 1856, January	.	+ 6° 40'
Melbourne ,, May	.	.	5° 50'
Liverpool ,, August	.	5° 51'

These agree well. Their magnitude shows that, in the operation of mechanical correction of the compass, the use of a mass of unmagnetized iron (to produce its correcting effect by its transient induced magnetism) must never be omitted. I fear that persons who profess to correct compasses *have* too often omitted it.

"2. Ship's magnetism towards the head, estimated in absolute measure (whole horizontal force of terrestrial magnetism at Liverpool = 3·54 ; at Melbourne = 5·94).

Liverpool, 1856, January . — 0·199

Melbourne   ,,   May .   . — 0·147

Liverpool   ,,   August . — 0·106

"3. Ship's magnetism towards the starboard side, estimated in the same scale.

Liverpool, 1856, January . — 1·156

Melbourne   ,,   May .   . — 0·895

Liverpool   ,,   August . — 0·206

"At Liverpool a deviation of 1° will be produced by ·062 in either of these magnetisms ; at Melbourne, a deviation of 1° will be produced by ·104.

"It appears, therefore, that the ship's magnetism has been steadily diminishing in both parts of the voyage. The headward magnetism (considered without reference to its sign) has diminished very uniformly : the starboard magnetism has diminished much more rapidly on the homeward voyage than on the outward voyage. I have the impression (but I know not whether correctly), that this is not the interpretation which you put on the changes : it is, however (small unimportant inaccuracies of calculation excepted), undoubtedly correct.   "I am, my dear Sir,

"Yours very truly,

"Rev. Dr. Scoresby.   "G. B. AIRY."

It may be mentioned that Dr. Scoresby's observations, made by applying a pocket compass to the iron plating at the "top sides" of the *Royal Charter*, showed that in the southern hemisphere their magnetism became inverted.

The observations made with the mast compass, though unfortunately their value was much diminished by the defects of that compass, seem sufficient to show that such a compass may be of the greatest possible value as a compass of reference and as a check on the standard compass.

The importance of these results may make us regret that Dr. Scoresby was not able to procure more accurate observations, and at a greater number of places, and may lead us to anticipate that still more valuable results might be derived from very careful observation of the deviation of the com-

passes and of the variations of the vertical and horizontal
force made on board an iron ship before leaving England in
the southern hemisphere, and after her return to England.

This sketch would be incomplete without some reference
to what has been done on this subject by public bodies.

About the year 1840 a Committee of scientific men and
naval officers was appointed by the Admiralty to consider
various questions connected with ship's compasses, and the
correction of the deviation. That Committee recommended
the improved form of compass now used as the standard
compass in the navy, and the correction of the deviation, not
by mechanical corrections, but by " swinging " the ship, and
obtaining a table of the deviations, to be afterwards applied
to correct the observed courses and bearings. This system
has been ever since followed in Her Majesty's ships, under
the able superintendence, first, of Captain E. J. Johnson, and
afterwards of Mr. F. J. Evans, the present Superintendent
of the Compass Department, and with such success that I
believe I am correct in saying that, with the single exception
of the *Birkenhead*, there is no reason to believe that any of
Her Majesty's ships have been wrecked in consequence of
the deviation of their compasses. A work on the deviation
of the compass, by Captain Johnson, contains a valuable
collection of the deviations observed in ships having compasses not corrected by magnets. Various sets of " Instructions " for the correction of the deviation have been published
by the Admiralty.*

---

* The following is a list of these Instructions:—

" Practical Rules for Ascertaining and Applying the Deviations of the Compass
caused by the Iron in a Ship." Potter, London, 1855.

First Supplement to the foregoing, being " Instructions for the Computation of
a Table of the Deviations of a Ship's Compass, from Observations made on 4, 8,
16, or 32 Points, and a Graphic Method of Correcting the Deviation of a Ship's
Compass." By Archibald Smith, Esq., M.A. Potter, London, 1855.

Second Supplement to the Practical Rules, being a " Graphic Method of
Correcting the Deviation." By Captain Alfred Ryder, R.N.

" Directions for the Use of a small Apparatus to be employed with a Ship's
Standard Compass, for the purpose of Ascertaining the Changing (Semicircular)
part of the Deviation." By Major General Edward Sabine, R.A. Potter, London,
1856.

In the mercantile marine the case is different. No uniform system has there obtained. The profession of a compass-adjuster has sprung up. Each compass-adjuster corrects in his own way. The merchant captain does not understand the process. He assumes that his compasses are correct till he finds out that they are in error, sometimes by shipwreck; and when he discovers the error at sea he is unable to correct it himself, and is obliged to navigate disregarding the compass.

On the occasion of the meeting of the British Association at Liverpool, in the year 1854, the attention of the shipowners and underwriters of Liverpool was strongly called to the subject by Dr. Scoresby and others. The result was the appointment of a local body under the name of the Liverpool Compass Committee, consisting of a number of influential shipowners, shipbuilders, underwriters, and men of science, who secured the services of a most competent secretary, Mr. W. W. Rundell, and who, in two Reports made to the Board of Trade in 1855 and 1856, and presented to both Houses of Parliament in the year 1857, collected, discussed, and published a most valuable series of observations on the deviation of the compass in vessels having their compasses corrected by fixed magnets. I may refer to this Report, pp. 47-55, for a full and able discussion by Mr. Rundell of the observations made in the *Royal Charter*. The Liverpool Compass Committee, having effected as much as could be effected by a private body, brought their labours to a close in the year 1857.

The Board of Trade has lately directed the attention of the mercantile marine to this subject in its publications,* but it has not yet taken up the subject in that systematic and continuous way to which the Report of the Liverpool Compass Committee points as desirable. That this might be done with great advantage, and with little more expense than that incurred by the Liverpool Compass Committee during its

---

* "A Circular on Deviation," by Rear-Admiral Fitzroy; and "Instructions for Correcting the Deviations of the Compass," Edited by Archibald Smith, Esq., M.A., F.R.S.

existence, does not seem to admit of doubt. Whatever difference of opinion may be entertained as to applying correctors to the steering compasses of iron ships, it can hardly admit of question that every iron ship should have at least one compass removed as much as possible from the influence of iron and not corrected by magnets, and should be swung at the beginning and end of every voyage of any length, and the deviation of the uncorrected and corrected compasses (if any) observed. No man is competent to command an iron ship who is not competent to make these observations; and if these observations were transmitted to the Board of Trade, and systematically reduced and discussed and published, most valuable results would certainly be obtained. It may be added that the observations so made would furnish a test of the care, skill, and good faith of the Captain who made the observations, as the process of reduction shows almost infallibly the genuineness and correctness of the observations.

A. S.

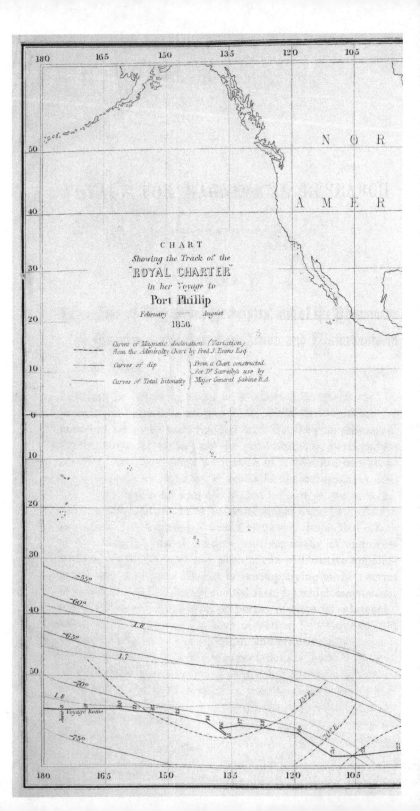

CHART
Showing the Track of the
"ROYAL CHARTER"
in her Voyage to
Port Phillip
February ——— August
1856.

——— Curves of Magnetic declination (Variation)
      from the Admiralty Chart by Fred. J. Evans Esq.
———— Curves of dip   } From a Chart constructed
                              } for Dr Scoresby's use by
······· Curves of Total Intensity } Major General Sabine R.A.

N O R

A M E R

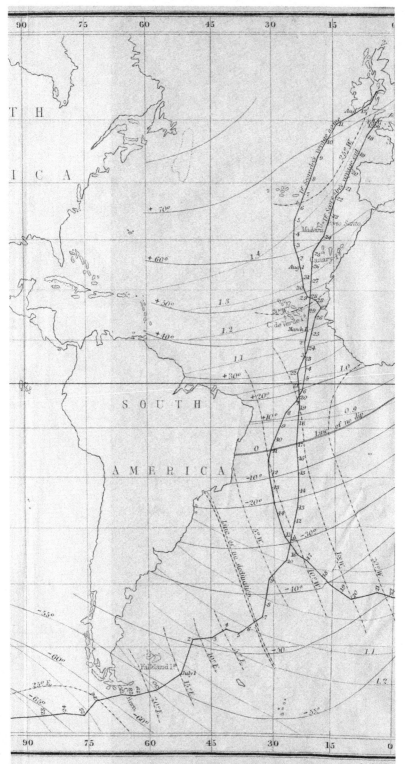

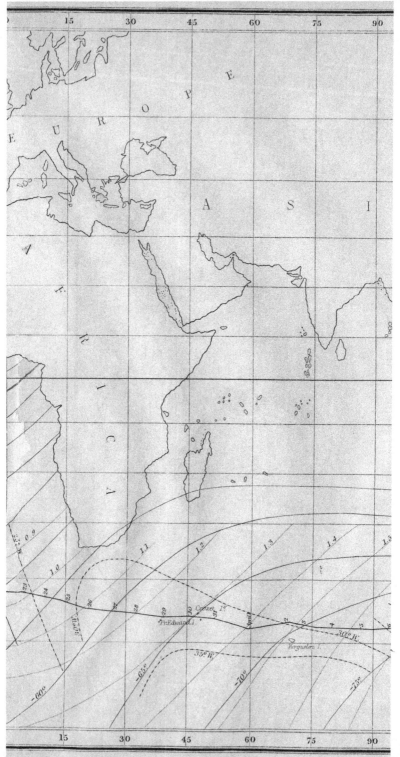

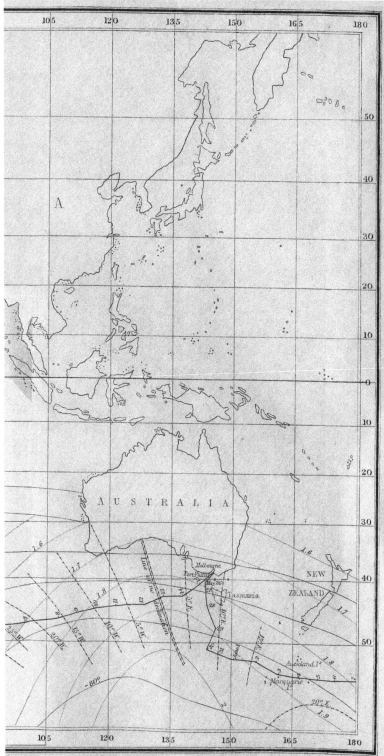

A

AUSTRALIA

Melbourne
Port Phillip
May 20.

Tasmania

NEW
ZEALAND

Auckland I.

Macquarie

VOYAGE FOR MAGNETICAL RESEARCH.

PART I.

Exposition of Magnetical Principles, and of the Phenomena of Magnetism and Compass-Action and Disturbance in Iron Ships.

It would be expecting too little, it is hoped, for the fate of the present publication, to assume that it would be looked into by none but scientific men familiar with the subjects discussed. On the contrary, it may not be presumptuous, perhaps, for me to expect, that many navigators of iron-ships, as well as young sea-officers, whether of scientific acquirements or not, will search herein for something that may be of use to them, of a practical nature, in their professional capacity. And it may be further assumed, I would venture to hope, that others of the general class of readers, and especially of enquirers into the wondrous laws and phenomena of Creative appointment, may have some interest in accompanying me in a series of descriptions of gradually elicited facts by which *magnetism*, one of the deeply mysterious properties of material substances, received, in respect to its laws of action, development, and changes in iron, very many striking and beautiful elucidations.

But, whilst indulging such expectations, it became obvious on very slight consideration, that, neither could the navigator profit duly from these researches, nor the general reader find that interest or instruction for which the researches might be reasonably adapted, without knowledge of the leading principles of magnetism, or without some measure of acquaintance

with the ordinary phenomena of magnetism in iron, especially
as developed and intensified in its *retentive* quality, by means
of mechanical action in iron-ships.

As, however, the requisite information—comprising, as it
were, the grammar of our subject, would probably be out of
convenient reach with many readers, or if, in reasonable measure and fulness attainable, might be found embodied in extensive treatises,—it appeared to me as being very desirable
that this publication should comprise within itself whatever
elementary information, or other peculiar information on the
principles investigated, might be needful for affording a clear
understanding of the special facts and phenomena brought
out in the course of the voyage of the *Royal Charter*.

Thus, whilst presenting to the reader a brief and compact
view of the general system of magnetic science, I am enabled
to set forth distinctively and without embarrassment by other
topics, the principles and phenomena most specially allied to
the subjects of research; to describe a variety of original
processes and elucidations in general magnetics, on practical
magnetisation and demagnetisation, not elsewhere to be met
with; to give the substance of elaborate researches for the
improvement of permanent magnets and compass-needles;
and to place on record an exposition of the theoretic views,
personally deduced from a long series of experimental researches on malleable and other species of iron when subjected
to mechanical violence,—the testing of which views constituted an important article in the inducements for my undertaking a voyage to Australia and round the world. A sufficient
description of these principles, phenomena, researches, and
theoretic deductions, it is presumed, will be found to be comprised in the Chapters immediately following.

CHAPTER I.

General Magnetical Principles and Phenomena.

1. **MAGNETISM**, from the mysterious character of its phenomena, has always been a subject of popular interest and experiment. As an *actuating* power, its relation to the wonderful performances of the electric-telegraph, in these modern times, has given to it an immense advance in the scale of practical sciences. As a *directing* agency, it has long been known and valued in navigation, mining, surveying, etc. as affording to the sailor, especially, under circumstances favourable to correct compass-action, safe and beneficial guidance in trackless seas, and under dark or obscure skies. It has also, on the other hand, assumed, especially in late years, a grave position of importance, as a *misguiding influence*, tending, by its disturbance of the compass and fitful changes of disturbing force, to the serious endangering, under circumstances not contemplated or understood, of life and property by a fallacious confidence.

2. **Universal prevalence of Magnetism in Physical Substances.** —Magnetism is an essential force or property in the constitution of all material substances, and is one of the mighty agencies by which nature is subordinated to the will, and performs the destiny and uses appointed of the Great Creator. It is inseparably associated with electricity (but is not, as has been popularly assumed, identical with it), as also with caloric or the matter of heat, and with other subtle influences productive of chemical affinities, solidity, etc. But, whilst thus existing, in common, in material substances, they are usually in a latent or neutral condition, so as, in the natural and quiescent state of things, not to be ordinarily perceptible.

3. **Apparent Relation of Magnetism, Electricity, Chemical Action, Light, Heat.**—None of these different principles— magnetism, electricity, chemical affinity, light, heat—appear to be of primary or independent nature, but rather attributes

or qualities of some unseen and unknown *master-agency*, the common parent of all. For that none is essentially *master*, we infer from the experimental fact, that each one, in turn, may be made to elicit or produce the others. Thus, by the agency of *magnetism* we can develope electricity, heat, chemical action, as well as light. So, by means of *electricity*, we can produce all the others. By *chemical action*, too, we obtain the like results. By heat, again, we produce chemical changes, magnetism, electricity, and light. And, by means of light, chemical action, heat, electricity, are producible; and, as the experiments of Morichini and Mrs. Somerville seem to prove, magnetism also.

4. **Iron, as a Material, the most highly susceptible of Magnetism.**—Magnetism has its most striking and powerful developments in iron, or other descriptions of ferruginous material. But certain other metals, such as nickel, cobalt, magnesia, and chromium, are, in inferior degree, susceptible of the magnetic condition.

5. **Two Qualities or Polarities always co-existing.**—Like electricity, magnetism has two qualities always co-existing in the same mass of material, and, in the case of magnetism, residing, apparently, in every molecule or particle of matter. The diverse qualities, which in electricity are named *positive* and *negative,* are, in magnetism, popularly distinguished as *northern* and *southern* polarities; more technically, they are designated *plus* and *minus;* otherwise, again, as *boreal* and *austral.* Another form of denomination of the two polarities, for reasons hereafter noticed, as occasion for deviation from the more popular names seem to require, is employed in the following pages.

6. **Poles of Magnets.**—In magnetic bars, or other forms of magnets in which the length of the integral plates or bars greatly exceeds the other dimensions, the polarities are the strongest at or near the extremities, and, weakest, indeed quite insensible, about the middle. Hence, because of the tendency to concentrated action there, the extremities are denominated the *poles;* whilst the neutral line or place in the middle, separating the two polarities, is considered as the *equator* of the magnetic body.

7. **Dissimilar Poles attract each other, and similar repel.**—
The two polarities, it should be always borne in mind, are
reciprocally *attractive* of one another, while each of them is
essentially *repellant* of its own kind. Thus the northern
attracts the southern, and the converse ; whilst the northern
repels the northern, and the southern repels the southern.
Any exception to this law can only be *apparent*. In the case
of a small or weak magnet being placed with one of its poles
in contact with the similar pole of a powerful magnet, or very
close to such pole, attraction may take place instead of repulsion; but this will be found to be due to the overwhelming
energy of the master-magnet in inverting the polarity of the
other, and so attracting it.

8. **Tests of Magnetism.**—The existence of the magnetic
condition, in its *sensible* form, is usually shewn by the tests
of attracting pieces of iron, or iron filings, and, more deli-
cately, of causing deflection or movement in the needle of a
compass. A sewing needle, when duly magnetised and suspended by some fibres of floss-silk, or a fine hair, becomes a
very delicate test of the existence of the magnetic condition.
The *denomination* of the polarity in any portion or place in a
magnetic body thus determined, is shewn by the *nature* of the
influence exerted on the known poles of the needle. Thus,
in the use of the poised magnetic needle, as a test, we may
learn with certainty these two things : 1st, that the bar or substance which influences the needle of the compass is actually
magnetic ; and 2ndly, that the part attracting the north of
the needle must have southern polarity, or, attracting the
south, northern.

9. **Magnetic Repulsion.**—Whilst the property of magnetic
attraction is so simply shewn by the familiar process of the
sustaining of magnetisable substances, *repulsion* is not indicated without the aid of a substance rigidly and positively
of contrary polarity. Thus any magnet will attract a piece of
soft iron with either pole alike; but when either pole is
placed in contact with the *same pole* of another equally hard
and powerful magnet, there will not only be no adhesion, but
a repellant action, sometimes of sensible force, and, in any

case, sufficient, if one of the magnets be so placed as to be freely moveable, to turn it aside, or quite round. Where one of the magnets is suspended by a thread, or on a pivot, as in the compass-needle, then the repulsion of similar poles will generally be shewn, however small or weak the moveable needle may be in comparison of the actuating magnet.

10. **The two Polarities not separable.**—The two polarities, though they may be molecularly decomposed, are not only naturally and essentially coexistent, and cannot be produced the one without the other; but they cannot be separated, or comprised separately, in any single detached magnetic mass; and wherever they *appear* to be separated, as in the two ends of the same magnetic bar, the separation is but conditional and relative—being dependent on the preservation of the bar in its original unbroken continuity.

11. **Experimental Verification.**—This proposition may, conclusively, be verified by a very easy experiment. Take a plate of steel, a compass-needle for instance, or a steel wire, or a narrow busk-steel plate, and when in a proper magnetic state it will be found, by the effect on a small compass, or on a common sewing needle (if magnetised and suspended by a hair or fine thread), that the whole of each half is of distinctive polarity—that is, one half wholly of northern and the other wholly of southern. Now here, *if it were possible*, we could easily obtain and secure each polarity separately by breaking the plate or needle in the middle. On trying the experiment, however, we shall find that each half, instead of exhibiting an uniform polarity, has suddenly changed to a distinct magnet with its separate and proportional force of polarity at the ends,—the end of the southern portion which was originally neutral, being now a strong north pole, and the original neutral place of the other section being a south pole. A similar result will be obtained if the plate be divided into several pieces, each piece, as soon as separated from the parent mass, becoming a true and independent magnet, but necessarily, because of the loss of material, of diminished power.

12. **The Quantity of each Polarity always equal.**—As the two polarities cannot be separated, nor either of them produced or

elicited separately; neither can they be produced in unequal measure or quantity.   Hence, if by any magnetising process we should produce a *northern* polarity in any piece or body of magnetisable substance, that development will produce spontaneously, and in precisely corresponding quantity, the *southern* polarity in another part of the same substance.   In one part, indeed, especially where polarity is given only to the extremity of a bar or plate of hard iron or of steel, the primary polarity may be strong and energetic, and in the opposite end or side weak, or scarcely apparent; but, however more extensively diffused, the relative quantities of the northern and southern polarities must be equal.

13. **Consequential Pole and its Power of Propagation.**—In soft malleable iron, as in a key for instance, when put in contact with, or suspended from, the pole of a magnet, the second or *consequential* pole will be always found at the lower extremity, or the end most distant from the magnet, and of the same denomination as that of the magnet's attracting pole. Thus the north pole of the magnet will produce southern polarity in the nearest end of the key, and that polarity the northern, consequentially, in the other end.   And as this lower polarity, in the case of soft iron, will have a degree of energy in some measure correspondent with the attracting force above, it will be endued with a propagating power of the nature of that of the original magnet, so as, in its turn, to elicit polarity, and then to attract it, in a second key; and that, if the primary power be sufficiently great, to do the same with a third key, and so on.   The pole of a very powerful straight-bar magnet may, in this way, be made to sustain a chain of keys, three or four in continuity, each key being, in itself, a separate magnet, whilst so situated, with its southern pole above, (the north end of the magnet being employed,) and its northern pole below; whilst in the employment of smaller substances, such as small polished bits of iron wire, sprig nails, or small balls of soft iron, the series may sometimes be run to an extent of ten or fifteen or even more in one chain.

CHAPTER II.

On the Inductive or Magnetic developing Power of
the Magnet. —Magnetisation, with new and powerful Processes.

I. THE INDUCTIVE POWER OF THE MAGNET.

14. **General Character or Definition of Induction.**—The well
known property of the magnet of producing or eliciting the
magnetic condition in other pieces of steel or iron, is the
property popularly distinguished by the term INDUCTION :
a property on which the attraction of iron, and the magnetising processes effected by the aid of magnetical instruments,
or other apparatus having magnetic polarity, are dependent.

15. **Each Pole of the Magnet induces converse Polarity.**—As
the two opposite polarities attract each other, and the same
kinds repel; so each polarity always tends to develope or
elicit the opposite kind of magnetism in any piece of iron or
steel to which the magnetic pole may be applied or brought
near.

16. **Principle of Attraction by the Magnet of previously
unmagnetic Substances.**—In this case of *attraction*, the inductive power of either pole of the magnet, it is plain, first
elicits a polarity opposite in kind to its own and then
attracts it. Hence, it follows, that attraction betwixt a
magnet and an *unmagnetic* substance, or a substance devoid,
at the time, of any sensible magnetism, can only take place
when the magnet by its own peculiar virtue, is able to elicit
the magnetic condition in such substance, and that with
a polarity contrary to its own.

17. **Two Effects of simple Induction.**—The induction of simple
contact, or of close proximity, *without motion*, or manipulation, is of two kinds, as to the ultimate effect of the
influence exerted.—the one transient, disappearing as in the

case of pure soft iron with the removal of the inducing magnet; the other remaining, with more or less permanency (as to a portion of the induced development), being sustained by the coercive force of the material acted on.

All kinds of iron and steel, in the metallic state, are susceptible of simple induction of the first or *transient quality*, from the softest and purest iron, where it is greatest, to the hardest steel and cast iron, where it is comparatively small; and all the harder kinds of the metal derive from contact with a powerful magnet, some *residual influence*, in the proportion, inversely, of the respective rigidity or coercive force of each particular specimen.

18. **Experiments on the two Effects of Induction.**—These effects may be thus characteristically exemplified :—

(1). *The transient Induction.*—Attach a short bar of iron, a key, or other piece of soft malleable iron, to the pole of a magnet, or place it with one of its extremities near to such pole, and the iron, as we have seen, will become a true magnet, and it will remain such as long as its proximity with the permanent magnet is maintained. But, instantly, on removing it away from the magnet, the polarity of the iron, to any obvious extent, will disappear.

(2). *Residual Polarity.*—If now, instead of a piece of *soft iron*, we apply a small bar or needle of steel, of cast iron, or of hard hammered malleable iron, to the pole of a powerful magnet, this will also become inductively magnetic, though not so strongly as the soft iron ; and, when removed from the magnet, a portion, probably of but feeble force, of the polarity will be retained. In this case, the inherent coercive quality of the material resists the return, in the magnetical elements, to a state of neutrality, on the same principle as the polarity of regularly constructed magnets is sustained and, in degree, made permanent.

19. **Magnetic coercive Force.**—The coercive force in the constitution of metallic iron and its several kindred qualities, is the property by which facility of magnetisation is opposed and permanency of the magnetic condition secured. It differs, in degree, in every variety of the metal and its ores. It is

the least (ordinarily not observable) in soft pure iron, becomes very sensible in hard, or crystalline, or hammered malleable iron, and increases in the several qualities of *cast iron* up to a degree which renders, as I found, this comparatively inexpensive material adequate for the construction of powerful permanent magnets. In steel, the coercive force has the greatest range, being very feeble in soft and bad qualities, and extremely strong and enduring in steel of the best qualities (that is, in steel converted out of the highest qualities of iron), and the most so in thin plates of the greatest degree of hardness.* And, finally, in certain magnetic iron-ores, such as the *loadstone*, the rigidity or power of retention of the magnetic condition, is, in the best qualities of the mineral, greatest of all.

20. **Relation of facility of Induction and Permanency.**— These follow a converse rule. Hence in the magnetising effects of simple induction, the kinds of iron or steel which realise the attractive force most easily and fully, or have the smallest coercive force, generally lose the magnetic condition with proportional facility,—pure soft iron, as we have shewn, instantly and entirely as the actuating magnet is withdrawn, whilst hard iron and steel in any of their kinds or qualities usually retain a portion of the magnetic development.

21. **Contact, or very close proximity not essential for Inductive Action.**—A remarkable law of induction is the power of acting magnetically at considerable distances, and with a degree of energy entirely unaffected by the interposition of the most solid non-magnetic substances. Nor is the power of magnetic induction prevented, though it is necessarily modified in its action, even by the interposition of iron, steel, or actual magnets. By contact, the most powerful inductive effects are of course produced; but at small or moderate dis-

---

* The relative susceptibilities for the magnetic condition, with their respective coercive qualities and adaptation for the construction of magnets, compass-needles, etc. of the different forms of iron and of steel—and of this latter material in a most extensive variety of quality, temperament, proportions and mass—are shewn in an elaborate series of researches discussed in " Magnetical Investigations," vol. i. Longman & Co. 1852.

tances from the extremity, or pole, for instance, of a powerful
straight-bar magnet, all the phenomena elicited by contact
(though in a rapidly diminishing ratio) may be produced.

22. Illustrative Examples of powerful Induction without contact.—If we place a strong bar magnet on a pedestal or support, we may attach to each of its poles, as we have seen,
keys, or other pieces of polished iron, and under the power
of propagating the magnetic condition from piece to piece,
we may attach a second, third, etc. in considerable series,
according to the power of the magnet. But the same thing,
though to an inferior extent, may be effected *without contact*,
so that, placing the flat of the hand, or a piece of brass, wood,
or glass, betwixt the pole of the magnet and the primary key,
it may be made to suspend (in the employment of a very
powerful magnet) a second, or even a third, notwithstanding
the separation, and entirely uninfluenced by the quality of
the substances interposed.

23. Ratio of Diminution of the Power of Induction in respect
of Distance.—The ratio in which the inductive force of either
pole of a magnet acts on a piece of iron when separated
from contact and tried at different distances, follows the law,
as to diminution of the power, of the square of the distance,
inversely,—a law and progress of diminution, which, in simplest form of expression amounts to this, that, if at the distance of a given space, say an inch, from the extremity or pole
of a long straight magnet, the magnetic or inductive force
be called 1, then, at the distance of 2 inches, the force will
be reduced in the inverse proportion of the squares of the
two distances, or as the square of 2=4 to the square of 1=1,
or as 4 to 1. In other words; if the magnetic force at 1 inch
distance be called 1, that at 2 inches would be as the square
inversely, of 2, equal to $\frac{1}{4}$th, at 3 inches to $\frac{1}{9}$th, at 4 inches to
$\frac{1}{16}$th, etc.

24. Absolute Reduction of Influence by Distance still more
rapid. —The proportions of magnetic influence with respect
of distance, as just stated, if a magnet could be had with but
one pole, would be found nearly correct at all distances ; but,
when the distances are increased to anything like the length

of the inducing magnet, and much more beyond it, the progress of diminution in the power is found, by experiment, to be much more rapid. This arises from the converse action of the opposite pole of the magnet, in counteracting, in its due proportion with respect of distance, the primary inductive force. Thus, taking the distances of one and two lengths of the magnet, we find the proportions of influence to run thus :—at one length distance, calling the force of the nearest pole of the bar 1, we should have this reduced, by the square of the distance of the counteracting pole, inversely, or $\frac{1}{4}$, giving $1-\frac{1}{4}=\frac{3}{4}$, for the correct expression of the power of the whole bar. In the case of two lengths distance, then, the power of the nearest end as represented by the square of 2, inversely, or expressed fractionally, equal to $\frac{1}{4}$, would be diminished by a force equal to the square of 3, inversely, or $\frac{1}{9}$, representing the whole power at this distance by the fraction $(\frac{1}{4}-\frac{1}{9}=)$ $\frac{5}{36}$,—thus giving the respective powers of the magnet, at 1 and 2 lengths, the proportion of $\frac{3}{4}$ to $\frac{5}{36}$, or $5\cdot36$ to 1, nearly. Thus, calculated in the same way, and tested by the deviations of a delicate compass-needle, the powers of a straight-bar magnet, acting at various distances and reckoned in measures (from 1 to 10) of the length of the bar, come out in this series,—1, $\frac{5}{27}$, $\frac{7}{108}$, $\frac{9}{300}$, $\frac{11}{675}$, $\frac{13}{1323}$, $\frac{15}{2352}$, $\frac{17}{3888}$, $\frac{19}{6075}$, $\frac{21}{9075}$,—instead of 1, $\frac{1}{4}$, $\frac{1}{9}$, $\frac{1}{16}$, $\frac{1}{25}$, $\frac{1}{36}$, $\frac{1}{49}$, $\frac{1}{64}$, $\frac{1}{81}$, $\frac{1}{100}$, which would represent the ratio of diminution of the power of the bar magnet, with the increase of the distance, if the magnet had but one pole, or were of unlimited length.[*]

II. MAGNETISATION.

25. **Magnetisation—Definition.**—Magnetisation, in practical

---

[*] This series of proportions requires, for strict accuracy, some correction, on account of the real actuating force of either end of a magnet not residing precisely at the extremities, but a little within.   The nature and quantity of the correction requisite has been shewn in a paper of mine in the " New Edinburgh Philosophical Journal," for April, 1832, where the *deviating* action of a bar-magnet, or set of magnets, on a compass at various and considerable distances, is adopted for the purpose of measuring distances through intervening masses of rock, and so applied to practical engineering for determining the thickness of rock or other material betwixt two opposite and approaching headways in tunnelling and mining.

science, is the producing of permanent polarity in substances capable of receiving and retaining the magnetic condition. This operation consists in the due application of extraneous magnetic force, for the decomposition of mutually attracting and combined qualities, or the development of properties or powers previously latent or dormant, in the metal acted on. The two polarities appear to reside in, or in connection with, each particle or molecule of the iron (as before noted), in respect of which they are susceptible of movement and arrangement, but from which, as belonging to the very constitution of matter, they cannot be separated. Magnetisation, therefore, may be considered as an arrangement, elementally, of the two polarities, giving the *effect* of a decomposition, in which an accumulation of energy is, in the most effective modes of the process, given to the two extremities of a steel bar in opposite denominations—the northern, at one end; the southern, at the other.

26. **Nothing added or infused by Magnetisation.**—That magnetising *adds nothing* to the metal operated on, but only develops a previously dormant power, is demonstrably shewn by the simple and well-known fact,—that a strong and rigidly permanent magnet is able to produce the like magnetic state in thousands of steel bars without material loss to itself; and that, by certain processes of manipulation, such a magnet is capable of giving power to bars or plates adapted for large and powerful combinations, to an extent incomparably beyond its own, whilst its original energy may be but little, or possibly not sensibly, reduced.

27. **Theory of the Magnetic Development.**—Hence, if these views be correct, it may be inferred (in respect to the theory of the magnetic condition), that the power or polarity developed is derived from the favourable *arrangement* accomplished by the inductive magnet in the molecular magnetisms of the iron or steel acted upon. And this view of the phenomenon admits of some striking illustrations:—

28. **Illustrations of Theory.**—We may suppose, for instance, the particles or molecules of the ferruginous metal to be represented by a series of small short magnets, or of magnet-

ised balls of hard steel, and the *mobility* of the molecules, or the magnetisms associated with them (as the theory implies) to be represented by suitable changes in the relative positions of the little magnets. Then we may illustrate what we suppose to be the two conditions of a steel bar—first, in its natural or neutral state; and secondly, in the magnetic state—in this manner :—

(1). *The Neutral State,*—Fig. 1.

N. $\overset{n}{\underset{s}{\text{O}}}$ $\overset{n}{\underset{s}{\text{O}}}$ $\overset{n}{\underset{s}{\text{O}}}$ $\overset{n}{\underset{s}{\text{O}}}$ $\overset{n}{\underset{s}{\text{O}}}$ $\overset{n}{\underset{s}{\text{O}}}$ **S.** $\updownarrow$ **C.**

Let the annexed diagram represent a series of such magnetic particles, as if constituting a slight bar of iron or steel, N. S., with their respective polarities *n s*, cast transversely across the bar, either with similar poles, in parallelism, or alternate poles opposite to each other, or in any other way irregularly across. Such a series of magnets, if tested by a small compass placed near the end, as at C., would exhibit no polarity whatever, and might fairly represent the unmagnetic state of a steel bar. But, if we merely alter the arrangement, and place the polar axes of each ball (or little bar magnet) in the longitudinal direction so as to run in a continuous line, and with all their northern polarities pointing one way and their southern the other, after the manner of this other series :—

(2). *The Magnetic Arrangement,*—Fig. 2.

N. $n\text{O}s$ $n\text{O}s$ $n\text{O}s$ $n\text{O}s$ $n\text{O}s$ $n\text{O}s$ **S.**   ↔   **C.**

Then we should have a representation, in the attraction of the north pole of the compass needle by the end S. of the series of balls, of the *magnetic* condition; and this attraction, on bringing the little magnets together with contrary poles in contact, would be found to be greatly increased, and the polarity at the extremities of considerable force.

29. **Magnetic Arrangement illustrated by the Voltaic Combination.**—Another illustration of the effect of arrangement, and of the increase of intensity with relation to the number of the particles brought into the arrangement, is very well supplied by the voltaic pile or the ordinary galvanic battery. Here the several magnetic molecules of a bar of iron or steel may be represented by the separate cups, or cells, or pairs of

plates, in the galvanic combination. These molecular galva-
nisms, so to speak, however numerous they may be, can have
no great external effects without particular and special
arrangement. A thousand of them uncombined, or combined
by wrong polarities, are all but powerless, and may thus
serve to represent the *unmagnetic* condition. Arranged,
however, with their respective metals, in proper alternate
connection, we then obtain from the extremities of the series,
energetic and powerful effects—effects generally proportionate
in energy to the number or quantity of the series brought
into connection. And here we have the representation of the
magnetic condition, as well as of the varieties of intensity
produced by difference of extent, or precision of adjustment,
brought under the magnetic arrangement.

30. **Requirements for obtaining a permanent Magnetic Development.**—Our remarks, here, as bearing on methods of magnetisation of the most ready and convenient application, will
have principal reference to the employment of permanent
artificial magnets, as the master-power, in the operation. The
requirement as to the *actuating force,* will hence be a good
and powerful compound magnet of the U or horse-shoe form,
or a pair of strong (single or compound) straight-bar magnets. The requirement for the *recipient energy* will be a bar
or bars of steel (hard cast-iron may do) of such rigidity,
or coercive force, of molecular condition (19), as may enable
the bar subjected to the magnetising process to retain the
polarity, at least in a large proportion, which may be
developed.

III. MAGNETISING PROCESSES.

31. **General Principle of the Magnetising Processes.**—The
object aimed at in these processes, is the application of the
scientific principles of induction, connected with judicious
forms of manipulation, etc., so as to elicit the highest intensity
of the magnetic development of which the material acted
upon is susceptible,—a condition which has been denominated
(though not very accordant with the strict meaning of the
expression) *magnetic saturation.*

32. Simple and elementary Processes of Magnetisation. — These familiar processes, having a certain measure of effectiveness adapted for the decomposition and arrangement of the magnetic elements in steel and other varieties of iron, are very various both in their modes of manipulation and their capabilities of development. Some of these it may be useful to notice.

(1). *The Magnetic Touch.*—If we merely cause one end of a steel bar or needle to touch the pole of a magnet, it will be found to have derived from the contact a slight magnetic development of the opposite denomination; and on touching the contrary pole of the magnet with the other end of the bar or needle, it will be found, however feeble in energy, to have become a true magnet. The magnetic development, indeed, will not, under this mode of operation, be either thorough and complete, or the arrangement of the molecular polarities regular and continuous; but the power developed in the ends will be sufficient, both for the attraction of iron filings or small bits of iron, and for enabling the bar or needle to take the direction of the compass-needle if suspended by a hair or thread. This, the simplest, perhaps, of magnetising processes, originated, no doubt, the use of the expression of the *magnetic touch.*

(2). *Manipulation by one Pole.*—If instead of mere contact of the end of a steel bar with the pole of a magnet, we draw the bar across the pole from end to end, we obtain a much greater development of the magnetic condition. In this way, indeed, though only one pole is employed, we may elicit both polarities at the extremities of the steel, though not in equable degrees of concentration. For magnetising small instruments, such as penknives, with a pretty strong magnet, all that is needful is thus effected,—the point of the knife, receiving the converse polarity from that of the magnet employed, and becoming strongly magnetic. Rubbing the blade backward and forward on the pole of the magnet, as the practice prevalently is, produces no benefit; it may be of much disadvantage. Two strokes on the pole of the magnet from heel to point, one on each side of the blade, drawing it quite off to a distance on both occasions, would generally be enough.

(3). *Apparent Discrepancy.*—In this process, whilst the heel of the blade, commenced with, ultimately obtains the *same*, not the converse, polarity as that of the magnet made use of, there is no contravening of the law of development of converse polarities. At the instant of commencement, the north pole of the magnet developes southern polarity at the heel of the knife, but, as the stroke towards the point takes place, each part in succession obtains the southern quality, which, as the two polarities must be equally existing at the same time [5 and 10] spontaneously inverts the original kind behind it, so as to give to the terminal point the proper polarity due to the action of the north pole of the magnet.

(4). *Simple Manipulation by both Poles alternately.*—A magnetising process, little less simple than those just described, which is of very convenient application, and will in many cases be found completely effective, is the following :—Take a strong bar-magnet, and place square on one of its poles, the plate or needle, by its middle, intended to be magnetised, as in the form of a T, the thick stem representing the bar-magnet, and the thin cross line the plate or needle. The arrangement,—N s being the magnet,

with its south pole, s, upward, and *s n* the steel plate or needle,—will stand as in the diagram. Now, slide the plate off, either way (say from s to *n*), keeping it in close contact till it leaves the bar and passes clear off at its extremity: then, inverting the magnet, so as to place its north pole upward, replace the steel plate by its middle and with same side in contact as before, and slide it off by the contrary end, *s*. Repeating the process on the other side of the plate, taking care, by having the ends of the plate marked, so as to be easily distinguishable, to draw the plate from the middle to the end *n*, when the south pole of the magnet is upward,

and the contrary way when the north is upward, and the
operation, for small plates of steel, or needles, will generally
be completely effective. A strong magnet of the horse-shoe,
or U, form, may be used, with a like degree of efficacy under
a similar method of manipulation.

33. **Superior and more elaborate Processes.**—The processes
just described, though in certain limited cases completely
effective, will only serve for the magnetising of bars or
needles of steel of small or moderate masses ; but where large
and massive bars, or even small bars of great hardness of
temper, are required to be thoroughly magnetised, or, as the
common expression is, magnetised to *saturation*, other and
more effective processes must be employed. Such magnetising processes, in considerable variety, will be found
described in all general treatises on magnetism,—the several
processes being usually discriminated by the name of the
discoverer,—such as the methods of *Dr. Knight, Du Hamel,
Michel, Canton, Æpinus, Coulomb, etc.* These methods, of
more or less efficacy and extent of applicability, are all
founded on scientific considerations, as to the adaptation of
the inductive power of a magnet to overcome the resisting
coercive force of the material to be acted on, so as to decom-
pose, and arrange, its inherent polarities, and thus to produce
the highest possible development of power.

IV. ORIGINAL PROCESSES OR IMPROVEMENTS
IN MAGNETISM.

34. **Improvements on certain popular Processes.**—The processes which constitute the bases of the improvements referred
to, are those popularly known by the names of Æpinus, Dr.
Knight, and Michel. They are described in " Magnetical
Investigations," vol. I. ch. ii. But, for the practical objects
herein contemplated, it will be sufficient to describe the
process founded on that of Dr. Knight. For this operation
—which is specially convenient for the magnetising of bars
or plates of the nature of compass needles, or for thin hard
plates of steel of much larger dimensions,—a pair of pretty
large straight bar-magnets of considerable power, either

single or compound, are requisite. Those I usually employ are 15 inches to 24 inches in length, an inch and a half broad, and about half an inch in thickness. They are constructed of the best steel and made quite hard. The manipulation is thus conducted :—

The two magnets being laid on a table in a straight line, with their opposite poles very near to each other, but not in contact,— the plate or bar to be magnetised is laid flat upon the magnets, ranging in the same line, so as to extend equally over the surface of both. The bars are then drawn asunder, till the plate just rests with its extremities in contact with the extreme poles of the two magnets, and then it is slid off sideways, and removed to some distance, preserving the parallelism of its position with that of the magnets till these are restored to the proximity with which the operation commenced. The process is repeated with the other side of the plate in contact with the magnets; and in the case of thin small plates,—such as I have adopted for the needles of sea-compasses,—the condition of saturation is usually found to be obtained. Generally, however, to secure the maximum more satisfactorily, the plates are subjected to four strokes of the magnets, two on each side; and in hard short bars, six or eight strokes are usually given, partly on the edges, as well as on the flat sides.

The method of magnetising steel plates or bars now described, admits of such celerity in performing the manipulations, that a dozen plates, bars, or needles, adapted for compasses, may be easily magnetised to saturation in four or five minutes; and plates of 16 inches to 2 feet in length, of various thicknesses, up to 0.1 or even 0.25 inch, may be thoroughly magnetised within a minute, the thinner plates, indeed, in half that time or less.

35. **General Principle of Scoresby's Method of Magnetisation.** —By ordinary usage, I am entitled, I believe, to designate thus an original discovery in magnetising processes. And the method may not be undeserving of particular description here, because of its superiority in effectiveness of magnetic development and rapidity of manipulation, over any of the

methods of magnetisation by permanent magnets heretofore referred to, or publicly known; and the more so on account of its convenient applicability to compass-needles, and the large bar-magnets employed for the adjustment of compasses in iron ships. The novelty of the process consists in the interposing of a thin bar or plate of steel or iron betwixt the actuating magnet and the bars or plates to be magnetised. Its most characteristic advantages are realized where the species of magnet employed in the manipulation is of the *horse-shoe* form. In the use of this instrument (the most convenient of all for the compass-maker, and, probably, too, for the manufacturer of " adjusting magnets ") the ordinary methods of magnetisation, by direct contact of the poles of the magnet with the surface of the bars acted on, all fail in producing a regular and continuous distribution of the magnetic elements throughout the extent of the bars, — and especially in thin bars or plates of very hard steel. The general cause of failure here, of the ordinary processes, seems to arise from the production of consecutive poles. The action of a powerful horse-shoe magnet applied directly to a hard thin plate or bar is *too local;* so that in the passage of its two closely contiguous poles a kind of magnetic wave is raised, highest under the magnet, which leaves behind it, probably, like the passage of a ship through the water, a series of other waves of diminishing altitude. By the interposition, however, of a thin bar of steel or plate of iron, the influence of the actuating magnetic polarity becomes diffused, whilst the more extended developing action serves, as it would seem, to yield an improved arrangement in the molecular magnetisms, and hence a higher and better result may be the general result.

This principle of magnetisation, it may be added, was first made known at the meetings of the British Association at York and Southampton, and published, in abstract—under the titles of "A new Process of Magnetic Manipulation," and " A new and powerful Mode of magnetising Steel" [and Cast Iron]—in the Reports of the Association for 1844 and 1846.

36. **Requirements for Scoresby's Process.**—The apparatus re-

quisite for effective magnetisation by Scoresby's method, are a good compound magnet, or pair of magnets, and a plate or thin bar of iron or steel, as the medium of the magnetic development. Here, the magnet may consist of two similar series or fasciculæ of straight bars fastened together in parallelism, with the opposite poles contiguous; or of a pair of such bundles of bars separate; or, what is most convenient and useful of all, of a strong compound magnet of the horse-shoe form. The plates for being interposed may consist of thin slips of polished soft iron of the width and general form of the bars or plates to be operated on. For magnetising compass-needles, the length of these plates may be 8 to 12 inches, and the thickness about that of a shilling. For large bars, thicker and longer slips of iron (such as iron-hooping) ground smooth on the flat sides will be needful. In the magnetising of rather thin and hard plates or bars of the horse-shoe form, a pair of bars of similar kind, of steel or iron, or a pair of the bars taken out of the general series designed for a compound magnet, may be employed.

37. **Scoresby's Process for the magnetising of Bars of the HORSE-SHOE Form.**—The application, here, is to sets of bars of tolerably similar width and thickness. A pair of these, arranged in the form, nearly, of the figure ∞, with converse poles in contact, are first placed on a table. Upon these are then laid a pair of interposing bars or plates of similar size, —a pair for instance, of the series belonging to the instrument under magnetisation. The process of manipulation, illustrated by the annexed figure, is then as follows :—

The compound or operating-magnet is first placed at the curvature of the end marked by an arrow, with its north pole towards the south, and its south pole towards the north, of the bars of the lower pair designed to be magnetised.

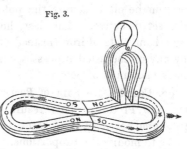

Fig. 3.

It is then slid gradually and smoothly forward (a slight oil-

ing being applied if needful), south pole in advance, or in the
direction indicated in the figure, towards the end designed for
the north pole of the lower bar, and continued across the junction of the bars in the course of the dotted line, keeping
the axis of the two poles in the central line of the bar till the
magnet comes round to the point at which the process commenced; it is then slid off in the direction of the arrow-
shaped mark. The upper pair of bars is then removed (keeping them together), and the lower pair turned over : and the
upper pair being also turned over is replaced on the top, and
the process of manipulation, changing also the direction of
the poles of the operating magnet, is repeated. Two complete circuits being thus made, will generally develope, in
the highest possible degree (if the operating magnet be sufficiently energetic), the magnetic power of the two *lower bars,*
whilst the upper pair, when steel instead of iron is used, is
found to be comparatively weak. Before separating the bars
of either pair, those above must be lifted together directly
upward; and then, if the highest capacity be designed to be
elicited, a separate conductor of soft iron should be laid across
the two poles of each of the lower magnets in order to sustain the power when they are separated. In this way, the
several bars of a compound horse-shoe magnet can be put in
a condition of superior, or rather, of the highest possible
energy, by the use only of the bars of its own series used in-
terposingly, except *one.* For, when all the bars have been
magnetised till the upper *pair* only is left, the same process
will serve for one of the remaining bars, provided a conductor
of iron be placed across the poles, so as to leave but one of
the set defective. This last, however, may be magnetised
by a bent plate of iron (instead of steel) being interposed, or,
by any of the usual processes, may be endued with the ordinary maximum of power.

38. **Comparative gain of Power in Horse-shoe Magnets by
the new Process.**—This result can only be decidedly represented for individual cases, as a variety of circumstances
serve to modify the proportions. But *the gain* I have always
found considerable, often astonishingly great. The case of a

small but powerful horse-shoe magnet of five bars of *fine steel,* may serve as an example. The bars weighed about 4000 grains each, and altogether 2·86 lbs. The magnet was constructed on my improved principles, referred to further on. Magnetised by the best of the processes of manipulation heretofore known, as improved by the figure of ∞ arrangement in single series, the total lifting power of the five bars, tried separately, was 37·1 lbs.; but the power obtained by "the double series process," gave 48·2 lbs. In this case, the lifting powers of the bars were, on an average, seventeen, and in one case above eighteen, times their own weight. When the five bars were combined (each sustained by separate conductors during the process of combination), this small magnet bore a load of 44·5 lbs., or about 16½ times its own weight. Of course, the sustaining power fell considerably after the first separation of the armature or conductor.— [*Report of British Association for* 1844, p. 102.]

39. **Gain of Power in Cast-Iron Bars.**—A set of five cast-iron bars of the horse-shoe form, cast very hard, and adapted for combination, derived even more advantage from this process. The bars measured 12 inches from the curved extremity to the poles, and weighed about 5·8 lbs. each. In this case, plates of soft iron of same form and size as the bars, and about one-fourth the thickness, were employed *inter-mediately.* By the best of other processes, four of these cast-iron bars obtain an average separate lifting power, *before* the separation of the armature, of 10·5 lbs., and a subsequent permanent power of 8·5 lbs. But, under *Scoresby's Process,* the primary power as increased to 19·5 lbs., *almost doubled,* and the permanent power extended to 11·8, being almost one-half gain.

40. **Scoresby's Process for the magnetising of large straight Bars of Steel in Series.**—For bars, such as those adapted for powerful magnetic machines, or for the adjustment of ships' compasses, the process is as follows :—If a large quantity be on hand for magnetisation, place as many as may conveniently lie on a kitchen table, or bench, in a line—say four or six in number—with a plate of iron hooping (made pretty

smooth on its surfaces) of a somewhat similar width to that
of the bars, of the full length of the whole series, or rather
more, and about the thickness of one-twentieth of an inch, or
somewhat less than that of a half-crown coin. The bars
having their ends marked, to distinguish the poles, must be
arranged with converse poles in connexion. At either extremity of the series, or a little beyond (the end of the iron
plate being supported by a lath of wood), commence the
manipulation by placing the horse-shoe permanent magnet
with its two poles ranging (one behind the other) in the direction of the length of the bars,—the one in advance being
of the same denomination as that of the polarity intended
be produced in the end of the bar commenced with, and in
the nearest end of each succeeding bar. Whilst the "master-
magnet" will never be found strongly adhering to the iron
plate, slide it smoothly, and *without any hitching*, along the
whole range of the bars, keeping always the same pole in advance, until it passes completely off at the farther end. A
little oil applied to the surface of the iron plate will serve for
preventing any hitching in the movement, which would mar
the perfection of the magnetisation. Removing now, by *sliding off sideways*, the iron plate, turn the series of bars over,
so as to bring their lower surfaces upward, being careful *not
to break the contact* of any two bars. The iron plate being
replaced on this new surface of the steel, repeat the manipulation, of one sliding stroke of the magnet, in the same way,
as to the direction of the poles and the plan of starting, as
before. If the bars be thick, such as to the extent of a
quarter of an inch, or more, or if they be very hard in temper, a third stroke, after again turning the series over, will
be requisite, and, indeed, may generally be desirable. With
good management, the whole series of bars will now be found
to be magnetised, not merely to what is usually considered as
saturation, but to an extent far beyond the power of the
coercive force of the steel bars to retain. At the same time,
because of the equable development of their magnetisms and
freedom from consecutive poles, the bars will be found endued
with a larger retentive energy than if magnetised by any of

the ordinary processes. If, however, it should happen, that by reason of defectiveness of the manipulation, or in the way of commencing the process, or the too small extension of the iron plate beyond the extremities of the line of bars, the extreme poles should prove weaker than the others, then the interior bars only need be taken as being strongly magnetised, and the defective bar replaced in a new series.

The effectiveness and rapidity of this process, it may be added, were well tested on occasion of the construction of "a large magnetic machine of enormous powers," (described in the Report of the British Association for 1845,) in which near 500 running feet of hard cast-steel bars were employed, —the bars of varying lengths, being an inch and a half in breadth, and a quarter inch in thickness. The amount of magnetisation, here, was found to be far greater than that produced by the same magnet by any of the forms of manipulation previously described; and so much more easy and rapid, that an operation which, under any of the former processes would have required, perhaps, two or three days' labour, could, under this new process, be accomplished and perfected in three or four hours !

41. For magnetising Plates or Needles for Ships' Compasses, by Scoresby's Process.—With a pair of powerful bar-magnets, results completely satisfactory are produced by the process referred to (35). But where a magnet of the horse-shoe form only is at hand, then, the process just described is found to be equally convenient—that is, wherever a considerable number of flat plates or needles are intended to be magnetised together. But, here, a long plate of iron is not found to be needful. It will generally suffice to employ a thin plate of 6 to 10 inches in length, and of about the thickness of a shilling.

42. Test Experiment resulting in extraordinary Powers.—A *Test Experiment*, exhibiting results in attractive force betwixt a set of twelve six bars of steel, which, I believe, had never been equalled or, probably, approached, was shown before the Physical Section of the British Association, at one of its meetings about ten or twelve years ago. The steel

bars subjected to the magnetising manipulation consisted of a
series which I had had constructed for determining the effects
of thickness and temper on the capacity of steel for the magnetic condition. Differing, therefore, as they did, in their
degrees of hardness, thickness, and mode of tempering,
they were by no means so favourable for my object as otherwise they might have been. They were of the uniform
length of six inches by about half an inch in breadth. Eight
of them were about a quarter of an inch in thickness, or
nearly so, the others were thinner, and of varying thicknesses. The magnet employed was a compound one, on my
own construction (described in " Magnetical Investigations,"
vol. I. part 2) of the horse-shoe form, weighing about eight
pounds. The intermediate iron plate, in this case, was only
about nine inches long, half an inch broad, and about the
thickness of a shilling.

For convenience of manipulation and exhibition, twelve of
these steel bars were placed on a piece of deal board about
7 feet in length and 6 inches in breadth, along which they
extended 6 feet. They were arranged in one line, commencing with the thickest bars, followed by the next in proportion of weight, and ending with the thinnest. Previous
to the application of the master magnet, the iron plate was
placed, by its middle, across the two poles, so that they respectively pointed, or were directed, towards the opposite
ends of the iron plate. An extremity of the plate, so combined with the magnet, was then placed partially on the first
steel bar in the series (the north pole of the magnet towards
the end designed for the north pole of the bars, and gradually made to slide smoothly along, with the magnet still
adhering to its middle, until (like the continuous stroke of a
plane along the edge of a deal board by the carpenter,)
both magnet and plate passed clear off the farther end of the
range of bars. The bars being now turned over whilst the
continuity of the series was carefully preserved, another
stroke of the magnet and iron plate, with the same pole as
before in advance, was made from end to end of the series,
and the magnetisation was complete.

The test of the extent of development of the polarities was the power of adhesion of this extraordinary series, weighing altogether about 37 ounces, and sustained by the attractive force of a primary bar of only 2150 grains. The second bar in the series, of about 1700 grains, here supported above seven times its own weight, and could have held considerably more. Whilst pressing with my thumb the first of the steel bars, with a force sufficient for the support of the weight of all the others, I placed the farther end of the board on the floor of the room, which was a foot or two below the level of the platform on which I stood, and then gradually elevating the other end which I was grasping, I brought it cautiously to a vertical position, so as to leave the twelve bars to their respective powers of magnetic attraction, and to enable me to allow them all to swing clear of the surface of the board (except the one in hand), for the support of the entire chain against the force of gravitation! Two or three years ago, I remember, whilst calling on a friend, a distingushed officer in the hydrographical department of the Admiralty, he mentioned to me, as a remarkable attainment in magnetisation, that an American gentleman, who had just preceded me in my call, had exhibited a set of small bar-magnets which were so strong that they actually supported one another to an extent of four in a chain. My friend, who had given some attention to such matters, was not a little surprised when I told him that I could show him a dozen similar bars sustained in one chain by a process of combination and magnetisation of my own!

Fig. 4.

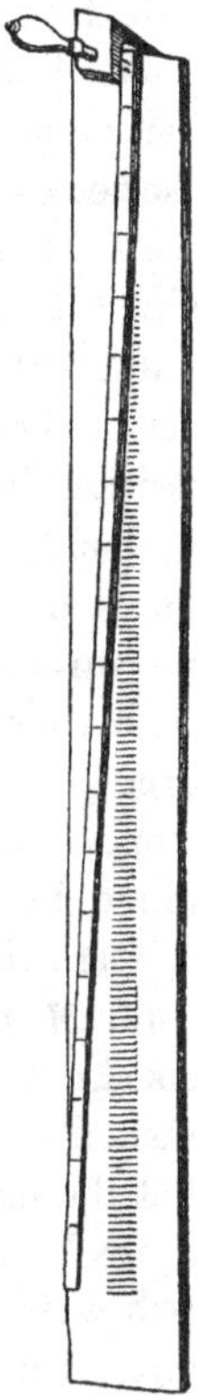

43. **A Chain of twenty Bars, of six inches in length each, sustained by Magnetic Attraction.**—Surprising as are the effects of the Test Experiment just described, the experiment itself is not at all a difficult one. For with a set of steel bars, very defective as regards degree of hardness and due graduation of the series in thickness, a chain of twelve bars, sustained

by the magnetic power developed by the new process of manipulation, may be combined magnetically with so much certainty, as to be unfailingly successful, even under the disadvantageous circumstances of a public lecture-room. But even with such defective quality of bars, the series can be carried greatly further. Thus, on more carefully proceeding with this experiment, with a further selection from my stock of six-inch steel plates, I was enabled to add no less than eight more to the magnetic chain—making, altogether, a series of *twenty* bars, extending to the entire length of *ten feet!* This astonishing chain, the plan and exhibition of which is represented in fig. 4, was freely sustained by the mutual attraction of the bars, when the board on which they were magnetised was raised, not only up to a perpendicular position, but so far beyond it that, under the unsteadiness of the hand, produced probably by pulsation, the lower part of the chain commenced a sort of undulatory motion clear of the surface of the board, which was allowed to go on for some time, without disturbing the adhesion of the magnetic force! To witness this extraordinary result of mutually supporting power, with the serpent-like movement, as if joined by hinges, of the lower portion of the chain, was very striking.

44. Maximum Capabilities of Scoresby's Process yet undetermined.—From the result of the experiment (43), accomplished without separation of the chain of bars, it is obvious that the maximum capabilities of this new process has by no means been reached. For the securing of the best results in such a test experiment, I should procure a set of at least a dozen bars of six inches in length, and half an inch, or five-eighths in breadth, and of graduated thicknesses, from three-eighths down to one-eighth of an inch. This set I should have made of the *best cast steel*, hardened thoroughly and equally from end to end, ground smooth on the surfaces, and the ends being square, carefully polished with a hone and buff-strop. A second set of about twenty bars or plates I should have prepared of best cast-steel, or shear-steel of the best quality, similar in length and breadth to the others,

and of graduated diminishing thicknesses, from an eighth to a twentieth of an inch. This set, ground and polished like the first, should, if constructed of *cast steel,* be reduced in temper, so as through the entire length of each plate to be of similar hardness to the best cutlery; but if made of shear-steel then each bar and plate should be fully hardened, according to the capabilities of this quality of metal, throughout its extent. Under such providings of bars and plates, it would hardly exceed my expectations if a chain of thirty, possibly three dozen, should be found capable of mutual suspension by this powerful process of magnetisation.

It must not be supposed, however, that the extraordinary polarity thus developed can possibly be sustained by the measure of coercive force or rigidity naturally inherent in the bars. On the contrary, the magnetic development *being excessive,* a great loss of energy must take place in the whole series of bars whenever the sustaining influence of the original contact is disturbed by the breaking of the connexion.

45. **Anomaly in Magnetisation.**—In certain magnetising processes, where magnets of the U form, or a pair of straight bar-magnets, with opposite poles combined in near proximity, are employed, I have often found that these magnets might have a damaging effect if *too powerful!* For all processes in which the poles of straight bar-magnets are used separately, or where the opposite poles are drawn asunder from the middle of the bar, under manipulation, to the ends, the *master magnets* cannot be too powerful. But for magnetising with a curved or combined magnet, a series of bars or needles, forming a continuous straight line, on the plan of Michel; or a pair of bars with iron armatures, forming a parallelogram, in the manner of Du Hamel, Dr. Knight, and Coulomb; or for augmenting the developing power by Scoresby's process,—the relative powers of the magnet and of the bars or plates acted on, require some adjustment. This fact, which I have abundantly proved, is not, as far as I am aware publicly known. An overwhelming power in the master magnet becomes thus, in certain cases, not only

deteriorating, but sometimes almost entirely defeats the aim of magnetic development.

Thus, in the magnetising of the series of bars employed in the Test Experiment (42), a single stroke of a horse-shoe magnet of less than 3 lbs. weight, will, with a thin slip of iron placed beneath it, produce a sufficient polarity for the sustaining of the twelve bars; but if a magnet twice as powerful be made use of, the thinner bars, below, will scarcely adhere. Hence, for the attainment of higher sustaining powers, I employ for the bars of three-eighths to one-eighth inch in thickness, the more powerful magnet, with a tempered steel plate of about a twentieth of an inch interposed; but before pressing downward over the thinner plates, I slide the master magnet off sideways, at the middle of the last of the thicker bars, and substitute the smaller magnet for the rest of the stroke. In this way one complete stroke on each side of the bar easily yielded a sustaining power for a chain of fifteen! Had the larger magnet, in this case, been used for the manipulation throughout, it would have been needful to interpose a thicker plate of iron or steel.

46. Highest magnetising Powers attained by Electro-magnetic Arrangements.—Great as the magnetising power of induction in permanent magnets of steel is found to be, especially in those of the best construction, as combined in the *compound* form, there is yet another agency, of comparatively modern discovery, which possesses powers of magnetic development incomparably exceeding those of ordinary magnets. For the most powerful of all agencies hereto employed for the effective coercing and developing of the latent magnetisms of iron and steel, is derived from electrical sources, and specially from the influence of galvanic arrangements. A long wire of copper, or other unmagnetic metal, covered with thread or silk, like bonnet-wire, being wrapped round some hard smooth ruler or roller of a cylindrical form, and in close arrangement of the coils, so as to form, when the roller (or mold) is removed, a close helix or spiral, becomes magnetic, if the ends of the wire be connected, respectively, with the two poles of any voltaic arrangement when in action—the

power of the magnetic development increasing with the increase of the heliacal coil, and the energy of the galvanic battery employed. The direction of the magnetism developed in the copper wire is transverse to the length of the wire, and therefore in the case of a helix is in the direction of the axis of the helix. If a bar of iron or steel, then, be placed *within* the helix, whilst under the galvanic power, the bar being so wrapped in paper, or otherwise supported, as to be kept near the centre of the cylindrical space—its magnetism will obtain high and powerful development by means of the inductive or accumulated inductive energy of the several coils of the wire around it. In this way an unlimited amount of inductive force may be obtained, adequate to the complete magnetisation of the largest steel bars, and of the most resisting material or condition, required to be acted on; whilst in bars of soft iron, when formed into temporary magnets by similar arrangements and agency, an amount of attraction can be easily developed of very astonishing power. Admirably adapted, however, as this mighty agent of magnetic development is for some particular purposes, it is far less convenient for the magnetising of plates and bars of steel ordinarily required for practical purposes, than the induction of good compound magnets which can be brought into use in a moment.

CHAPTER III.

MAGNETICAL APPARATUS.—IMPROVEMENTS IN COMPOUND
MAGNETS.——MAGNETIC MACHINES, AND COMPASS
NEEDLES.

I. IMPROVED COMPOUND MAGNETS.

47. Compound artificial Magnets. — Compound magnets,
having more or less of fixity of polarity, are usually constructed of magnetised steel bars of similar shape and size,
(though sometimes made shorter, as by steps, at the sides,)
secured firmly together by screws or clamps, so as to produce
somewhat massive and powerful instruments. The two kinds
in ordinary use are those of the horse-shoe, or U form, and of
the straight-bar form. These combinations possess differences
in amount of power, arising, not merely from the quantity of
steel employed in them, but from a variety of other circumstances, which we shall have occasion to notice. Besides the
convenience in bars of moderate substance or thickness, for
tempering and magnetising, other principles belonging to the
development of the magnetical condition, are in favour of
this system of combination in preference to the construction of
very massive solid magnets. But in any mode of construction,
combination as well as massiveness, becomes relatively deteriorating.

**48. Principles of Deterioration in the combining of Magnetic
Bars.**—Whilst the combining of bars or plates of steel in a
magnetised state, by their principal flat surfaces, and with
similar poles in contact, tends to yield a general result of
increasing power to the magnet as a whole; a general
damaging influence of the original powers of all the bars thus
situated necessarily takes place. For the series of bars,
requiring to be combined with the like polarities in contact,
necessarily tend to reduce, destroy, and invert each other's

power. And this tendency is found in practice to go to so great an extent that, in the instance of compound magnets constructed on the principles prevalent with manufacturers until a very recent period, the possibility of advantageous combination was found to be very limited. Thus with the ordinary stout bars of common steel designed for straight bar-magnets, the union of a second bar, instead of doubling the power of either of them, would only add, perhaps, a half; the union of a third bar, about a quarter more; of a fourth, an eighth more,—so that no extent of combination with such quality of magnets could, in those usually manufactured, effect more than the doubling of the power of a single bar! This fact as to compound straight bar-magnets of 12 to 15, or even 24 inches in length, with bars of about an inch in breadth, and a quarter of an inch, or less in thickness, I verified by the trial of a considerable variety of such combinations. And, further, I found that in attempting to increase the power of compound magnets of this kind, by other additions to their substance, sometimes the total power would be diminished; and, on separating the bundle of bars, some one, if not more, would be found to have altogether lost its polarity! Magnets of this kind, both of the straight-bar and of the horse-shoe form, and I have not met with anything superior, until after the publication of the researches referred to below, were all constructed on the same defective principles;—viz. hardened only at the ends of the bars, the rest being soft; made of those or similar massive proportions, and of the ordinary steel of commerce.

49. **Original Improvements in compound Magnets.**—Theoretic considerations, supported by a variety of experimental facts, led me to certain important principles of improvement, which were first published in the Report of the British Association for 1836, and subsequently, more in detail in parts I. and II. of " Magnetical Investigations," which were issued in 1839, and 1843.* These researches showed that great advances in

---

* It has been objected, as I am well aware, to the claim herein indicated for original improvements in magnetical instruments, that some of the principles insisted on are not new,—such as the hardening of the steel bars throughout,—the

magnetic energy in combined plates and bars could easily and with certainty be realized. Two leading principles,—to which others for *special* cases of form or massiveness were added,— were found to be of general applicability where powerful combinations are designed:—1st, The employment of the finest qualities of steel, such as that, for instance, converted out of the best charcoal manufactured iron of Sweden; and, 2ndly, The hardening of the bars or plates equally throughout their length. But modifications in the degree of hardness will be needful, where great masses are required to be combined, or, in certain cases, with reference to thickness in the separate bars, or to the shape into which the compound apparatus may be cast. These cases may claim some separate notices.

50. **Improvement in compound Magnets of the HORSE-SHOE Form.**—For compound magnets of this form, made of the best *cast steel*, I found, besides the application of the two general principles above noted, that a specific reduction of the general temperament, from that of very brittle hardness, was advantageous. The proper reduction was secured by placing the

---

same principle, it is said, being pointed to, or mentioned by, some of the magnet-icians of the last century. In two respects, however, they were demonstrably new; 1st, as the results of original investigations; and, 2ndly, as having, at all events, ceased to be acted on by modern manufacturers of magnetical apparatus. Thus, in a compound magnet in the Royal Institution, London, of the horse-shoe form, by one of the most eminent foreign magneticians; in another, in my own possession by a principal London instrument maker; in *intensity*, needles which I have examined, by a celebrated foreign philosopher, as well as in a great variety of other instruments, including a large number of compass needles,—not one was found which had the qualities essential for the best instruments, as derived from and demonstrated by my magnetical researches. So far otherwise, indeed, that on the first publication of my proposed improvements in compass needles before the British Association, at Bristol, it was objected by a gentleman of mathematical and scientific standing, that the principles were contrary to the results obtained by Capt. Kater, published in a paper in the "Philosophical Transactions," for which the Bakerian medal was awarded. Yet these very principles are now adopted for the naval department of the public service. Hence, if the principles of superiority contended for, had, any of them, been long ago published, they were not practically carried out, or scientifically received. The hardening of magnets, if recommended, had not been proved to be essential, and the principle having become obsolete, was practically repudiated or unknown.

whole of the series of bars for several minutes in a bath of oil heated to about 500 degrees. When the temperature of the oil was but 300 to 400 degrees, the capacity for magnetism, in *the cases tried*, were below a maximum ; and when the hardness was reduced in boiling linseed oil, there was also a loss. But in the employment of *shear-steel*, for this form of compound magnet, no reduction of the extreme hardness was needful. ["Magnetical Investigations," vol. i., 280]

Horse-shoe magnets of 5, 9, and 15 bars, being thus arranged and treated, exhibited very unwonted powers. One magnet, not of the best steel, which had originally been hardened only at the ends, was subjected to general tempering to the extent perhaps of a hardness about equal to that of ordinary carpenters' tools, and the simple effect was to advance the *prevalent* sustaining power of the armature from about 30 lbs. to upwards of 80 lbs. Other examples were but little less striking. ["Magnetical Investigations," ii., 172-3.]

The powers elicited in the bars of three or four compound magnets of this form were highly satisfactory,—many of the smaller bars, of half to three-quarters of a pound, being found capable of sustaining separately, and in the roughest manner of trial, 8 to 10 lbs. or more, or from 15 to near 20 times their own weight. If these, however, had been loaded in the careful progressive manner pointed out by M. Hæcker, they would probably have sustained a very large augmentation. The famous magnets by M. Logeman (produced many years after the publication of these researches), two of which were exhibited by Sir David Brewster, at the Meeting of the British Association, in Edinburgh, in 1850, derived their chief superiority from the adoption of these principles,—aided, not improbably, in the larger instrument, by magnetisation under the galvanic power of induction, which, in the case of massive bars, possesses a more powerful and *penetrative* action than can possibly be exerted by the mere passage of the poles of a permanent magnet across the surfaces. In the case of Logeman's horse-shoe shaped bars, however, of *light* or moderate masses,—such as half a pound to a pound in weight, some of the separate bars of one of my small compound magnets were proved to be considerably more powerful,proportionally, as tried by the simple test of sustaining power in weights hastily applied, than those exhibited in Edinburgh.

51. Scoresby's compound Plate Magnets.—This designation has been applied in the catalogues of Metropolitan instrument makers to straight bars or fasciculæ of hard thin plates of cast steel magnetically combined, constructed on principles described in " Magnetical Investigations," vol. i., p. 159, viz. The employment of thin cast-steel plates (such as the busk plates of commerce), hardened in oil pretty nearly to the brittleness of glass, magnetised separately, and, ultimately, formed into two fasciculæ,—the contrary polarities of the two compound bars being made by conductors of soft iron to be mutually sustaining during the process of combination and wrapping together.   The theory of this construction was deduced from results of experimental research, which went to prove :—1st, That, for moderately thick straight-bar magnets, and for all compound straight-bar magnets, superior cast steel was the best material. 2ndly, That for any considerable degree of massiveness, even such as a thickness of a fiftieth part of the length of the bar of ordinary proportions, the best temper was extreme hardness from end to end ; and 3rdly, that for all measures of massiveness or thicknesss, an increase of power was always obtained by dividing, so to speak, a single solid bar into slabs or plates, with a continual increase of the ultimate capacity with the increase in the number of the plates by which the dimensions and weight of the single solid bar might be made up.

52. Extraordinary inductive Powers of these thin Plate Magnets.—On these principles I constructed in the year 1838, a pair of straight bar-magnets, composed of 24 thin plates each ; the plates being of cast steel, 15 inches in length, an inch and a half in breadth, and of about a thirtieth of an inch in thickness. The magnetical powers of these bars, as tested by their deviating action on a compass, time of oscillation when suspended by a thread, and, above all, by their inductive action at a distance, were altogether unexampled.

Subsequently I made a much larger combination of similar

magnetised plates. Ninety-six plates were comprised in each of the fasciculæ (a slip of brown paper being interposed between every two plates), which were securely bound and kept firmly together by strong black linen tape. The ends of both magnets were ultimately ground flat and smooth. The entire bulk of each bar when finished measured 15 inches in length, by $1\frac{1}{2}$ inches in breadth, and 3 inches in thickness. For convenience of use in experiments, and for practical manipulation, I had them securely fixed in a strong frame of wood, adapted for the table, with a revolving axis, permitting their being placed, at pleasure, horizontally, vertically, or at any required angle of inclination.

The powers realized justified the most sanguine expectations. Their actual magnetic energy, as measured by the sure test of inductive action, was found to exceed that of any possible combination of straight bars of the ordinary or prevalent principles of construction, of like mass and dimensions (comparing them with bars, for instance, of a third or a quarter of an inch in thickness), just about *seven times*. Besides the capability of giving inductive force to a key of a pound weight, so that, at the distance of an inch or more, it could sustain two or three other large keys in diminishing series ; and besides the power of sustaining light iron or steel bodies (such as the larger kind of steel pens), against gravity with a thickness of half an inch or more of wood or glass interposed ; the two compound bars, when placed a little open, with opposite poles approximated, were found capable of supporting *ten thousand* polished half-inch sprigs !

II. MAGNETICAL MACHINES.

53. **Whether the Combination of artificial Magnets has, practically, any limit of Power?**—From certain well-known facts, as to magnets of small dimensions being more powerful than large ones, as to the loss of power, proportionally, (48), occasioned by combination, and as to all attempts to obtain anything like a due proportion of attractive force in large magnetic machines (until since the adoption of the principles of construction insisted on in art. 49) ; from these and other

cognate facts of experience, it might have been questioned whether some proximate limit, in the augmentation of mass or extent of combination, might not render the construction of huge machines practically nugatory? Magnetical investigations, in extensive series of experiments, pursued with this inquiry in view, enabled me to come to a satisfactory and decided conclusion, as forthwith will be seen.

54. **Determination of Principles necessary for unlimited Magnetical Powers.**—The experimental investigations referred to above, gave, in favour of the fact of the practicability of unlimited augmentation of power in magnetical machines, these several results: 1. That magnetic bars, designed for large combinations, may be conveniently, and often advantageously, constructed of various or numerous pieces. 2. That the separation of a long bar, say of 3 or 4 feet, into several portions, is not disadvantageous in regard to power, provided good contacts betwixt the ends of the pieces are secured. 3. That the resulting power is similar, whether, in the combination of several series of short bars, the elementary bars be of the same or of unequal lengths. 4. That the ultimate power of any given thickness of steel in combination is increased by the employment of thin bars or plates, the due hardening of which, too, is favoured by the thinness. 5. That although magnets of large dimensions are less powerful with respect to their masses, than small ones; yet in combinations of bars into magnets of *proportional* dimensions (the quality of the steel and degree of hardness being duly attended to,) an absolute increase of power, *ad infinitum*, may be attained. But for this result, let it be remembered, that the increase in the mass of the respective magnetic combinations must not be in any one of its dimensions, specially, but equably in all. For it has been shewn (art. 48) that the number of the bars of a compound magnet may be so multiplied in thickness as, ultimately, not only to become useless, but damaging. A proper regard, therefore, to suitable proportions of length and thickness, is essential to the attainment of the results contemplated in the present determinations. The practicability, then, of a beneficial enlargement of magnetic combinations to an

unlimited extent, must involve the condition of *enlargement in proportional dimensions.* Thus, if we fix on 2 feet long by 2 inches broad, and an inch thick, as favourable proportions, then good and effective magnets may be constructed of 10 times, 100 times, or 1000 times their several dimensions.

55. Construction of a large and powerful Magnetic Machine.— On these principles I constructed in the year 1842-3, a magnetic machine composed of two similar parts of about 4 feet in length, the centre row of bars projecting to 4 feet 4 inches, $4\frac{1}{2}$ inches in breadth, and 6 inches in depth. The two parts comprised 450 running feet of steel bars in about 380 separate pieces. Though the general arrangements and nature of the armaments were ill adapted for great lifting powers, yet 400 lbs. weight was easily carried by a simple plate of iron as a conductor. If, however, the same mass of magnetic materials had been cast into a ∪ shape, the sustaining power might probably have been extended to a ton weight. The inductive power of the machine realized every expectation. At six inches from the poles, a chain of six keys, generally of large size, was sustained, which could be then moved off to the distance of a foot without separation. With a board of nearly half an inch in thickness, interposed between the poles and the conductor, a weight of 15 lbs. was sustained. A bar of soft steel, six inches long, half an inch broad, and a sixteenth thick, was magnetised to saturation, by one transit or stroke across a single pole, *through a block of wood* two inches in thickness,—becoming so powerful that it carried large polished nails weighing from 600 to 822 grains. A compass was largely moved by the changing of the direction of the poles of the two magnets, through walls and other interposing substances, at the distance of 100 feet; and sensible movement might have been given 300 feet off. When fitted up with an armature and other apparatus, for electro-magnetic purposes, in which I was specially assisted by Mr. James P. Joule, F.R.S., it was found capable of decomposing water, and of electro-plating polished metal, *very rapidly;* and when worked by a galvanic battery of four tall cylinders, if exerted at one end only, a power of about a twelfth part of that of a horse.

But none of the capabilities of the machine have yet been fully developed.

From the results of this experiment, taken in connexion with the principles on which the machine is dependent for power, it is evident that a magnetic machine, or separate *compound beam* of magnetic force, might easily be constructed of the size of the main-mast of a line of battle ship,—the power of which, acting deflectingly on a compass, would be sufficient to produce a disturbance at the distance of a mile. It is, moreover, abundantly evident, that a monster machine, if required for any appropriate chemical or mechanical purpose might be constructed of hard plates of *cast iron,* so as, with the attainment of enormous powers, to come within no extravagant amount of cost.

III. COMPASS NEEDLES.

56. **Experimental Needle; original Improvements.**—Two distinctive qualities were obviously to be aimed at, for the obtaining of the most effective needles for the directing of ships' compasses : — 1st, The attainment of the greatest magnetic energy or *directive force* of which a needle of suitable size and weight might be susceptible; and, 2ndly, The securing of the greatest possible degree of fixity, or *permanency,* of the originally elicited power. The principles on which these desiderata depend, were largely investigated in experiments made more than twenty years ago. The first practical result of the researches was made public at the meeting of the British Association at Bristol, in the year 1836, by the exhibition and description of a new compound bar or needle, suspended on a pivot, and adapted, when constructed on a suitable scale, for the construction of very superior needles for ships' compasses.

This experimental bar, consisted of four thin plates of cast steel, tempered or hardened equably from end to end, which were placed nearly parallel to each other, rather closer at the ends, but not in contact,—spaces of about a quarter of an inch, by which a higher degree of power was secured, being produced by the interposition of discs of wood betwixt the several plates. For convenience of construction, the plates were pierced, correspondingly, with five small holes, one near each extremity, and two, an inch and a half apart, near the centre. Small brass pins passing through the several sets of holes and through the intermediate pieces of wood, afforded, by screw-nuts at the points, the means of ready and secure combination. The needle, as thus made up, was fitted with an agate and pivot, and also with the means of suspension by silk fibres, so as to traverse edge upward. The gain of power over any previous instrument of corresponding dimensions and weight was found to be great and decided, with a still more unusual permanency of the magnetic energy.

57. **General Character of Compass Needles previously in use.**— Previous to the time of the construction of the instrument just described, the needles in ordinary use, both in the Royal and Mercantile Navies, were excessively defective in qualities the most important for the real effectiveness and durability of action of the compass. They were generally weak and unenduring, whilst those supplied for the Royal Navy were, I believe, amongst the worst of their kind. I had tried and tested (about this period of time) a very considerable number of compass needles, by various makers; few were moderately good; not a few were *intolerably bad.* Nine needles were furnished me by the late Capt. Johnson, R.N., in 1839, as fair specimens of those then in use in the Navy; some of these as to mere directive force were tolerable of their kind; two out of four, by one of the accredited makers for the Admiralty, were utterly incapable of performing their intended function. Compared with needles constructed on the principles described above, all those which I had an opportunity of trying, were singularly inferior. Weight for weight (compared with needles of similar length), my compound edge-plate needles exceeded the others in the proportion of about $2\frac{1}{2}$ to 1, in the best of those supplied to me as specimens, and about 5 to 1, in the worst; whilst as to *fixity of power*, the superiority of mine, scarcely admitted of a comparison. For under the action of a very trying and damaging test (art. 60), which my needles could sustain with so trifling a loss, as to

leave them still in possession of above double the power of which any of the others were susceptible,—the others lost, without exception, the entirety of their original polarity.

58. **General Quality and Defects of the Needles at present supplied.**—The compasses now in prevalent use in the mercantile navy, whilst greatly improved in their general character, finish, and effectiveness, yet remain, so far as I may judge from a considerable number recently examined, *far behind* what they should be. In a recent trial of several compasses, accidentally gathered together, at one of our principal seaports, I found the old-fashioned principle of tempering the *ends* of the needles only, with no decided regard to quality or denomination of the steel, prevalent in the whole series, except in one by an Admiralty maker. They were all, of course, inferior to what they might be. The best had scarcely half the power of one of my four or six plate needles of corresponding weight, and one of the worst, *not a fourth*. And when subjected to the test for *permanency*, the best yielded up nearly two-thirds of its magnetism, and the others *the whole*. After the same severity of testing, one of my own needles, weighing only 600 grains, lifted a key of 6 to 7 ounces in weight!

It should be noted that the compasses of the Royal Navy, which, before the year 1836 were characteristically among the worst in use, are now as decidedly the best. The principles which I have described as working so admirably towards the improvement of magnets and magnetical combinations are now generally adopted in the construction of compasses for ships in the public service. The "Admiralty Compasses" have their directing magnets in the form of thin steel plates (two to four in combination) hardened throughout, and set on edge, and arranged in parallel series, generally two combinations of plates on each side of centre, I believe, beneath the card. The best description of these are powerful, effective, permanent, and admirable instruments, and care is now taken I doubt not that the whole of the varieties supplied for the public service are good instruments. Indeed the tests I have referred to (59) are so simple and so certain, and

COMPASS NEEDLES. 43

the principles on which the best capabilities are to be secured so specific, that there is no more excuse for sending out a bad compass-needle (as we shall immediately see), than for issuing a sovereign of inferior or bad gold from the Mint.*

59. **The testing of Compass Needles.**—The two principal qualities to be sought in the construction of compass needles, as I have stated, are *directive energy*, and *tenaciousness* or *permanency* in the communicated polarity. For each of these qualities, easy and satisfactory tests are suggested in "Magnetical Investigations," vol. i. chapters 3 and 4. These, for the convenience of compass-makers and other operative magneticians, may advantageously be mentioned here.

(1). *For the testing of the comparative Directive Powers.* —This, as elsewhere noted, is conclusively effected by observing their respective action on a delicate compass when they are placed, in succession, in the line of the east or west point, and at the distance of say, *two lengths* of the bars from the centre of the compass. The *tangents* of the *deviations* produced (subject to a slight correction for extreme accuracy) are proportional to the directive powers of the several bars. [If the deflecting bars are in the line at right angles to the deflected bars, then the directive powers are as the *sines* of the deflections produced.]

(2). *The testing for Permanency or Tenaciousness.*— For this object, I have suggested two easy and effective processes. As to one of these—designed for the testing of the permanency or tenaciousness of the several bars or plates of compound magnets, or of any series of magnetic bars of similar size and form, it may be sufficient, here, to refer to "Magnetical Investigations," vol. i. pp. 36 and 116, etc., where ample descriptions of the process and its results are given. But the other, as of much practical availableness for testing the quality of compass needles, may deserve more special consideration.

---

* These plans of improving and testing the needles designed for ships' compasses I exhibited before a Committee of the Navy Board, then engaged in the consideration of the special subject, at the Admiralty, in 1839.

60. **The Test Magnet for the Determination of Magnetic Permanency.**—The *test magnet* for this process should be constructed of somewhat similar dimensions, as to length and breadth, to the needles to be tested, and of the thickness of about half an inch. The best *cast steel* of cutlery is here required, with a hardening to its utmost capability, and a high degree of magnetisation. For needles of six inches in length, my test-bar consists of a square prism of six inches by half an inch. It is extremely hard and has a deviating action on the compass at two lengths, or 12 inches distance, of about 38 degrees. For longer needles, from 7 to 10 inches, whilst the length of the test magnet is made correspondent to that of the needles, or nearly so, the breadth adopted is usually three quarters of an inch, and the thickness half an inch.

61. **The Theory and Results of Action of the Test Magnet.**—The principle of the test-bar is founded on the general fact of the deteriorating influence of magnets placed with similar polarities in contact. Employing a test-bar of commanding power, and by reason of its extreme hardness, insusceptible of material reduction by the influence of comparatively feeble magnetic violence, the damage by unfavourable contact is sustained only (or mainly) by the needles subjected to the test. The testing of any needle or set of needles is thus effected:—1. After being fully magnetised, each needle (being marked or numbered) is placed in succession near the compass for the trial of its deviating action (59, 1), and a record taken of this its original directive power. It is then laid carefully down flat on the test-bar, so as to avoid any rubbing with like poles in contact, and moderately pressed down with the fingers. It is then lifted up, turned over, and replaced on the bar, pressed down as before, and being removed the operation is complete. The state of the various needles after the process is accurately indicated by their residual deviations on the compass; and their relative qualities as to fixity or tenaciousness of the magnetic power is conclusively shown. In this way a set of needles can be so satisfactorily tested, that the *best*, and the *next* in superiority, down

to the *worst*, can be picked out with unfailing accuracy. Of the needles in ordinary use a few years ago, hardly one that I ever met with could bear the power of the test-bar without almost an entire loss of polarity, while some actually suffered an inversion of the poles! Where, however, any or most of the following principles (62) have been adopted, the permanency is found correspondingly improved.

62. **Requirements for the best Needles.**—The improvements in the principle of construction of needles for the direction of ships' compasses, comprise, essentially, the following particulars:—1st, The employment of *cast steel,* "converted " out of the best qualities of foreign iron, or " *shear-steel*" of like quality if for light needles, may yield a stronger original power. 2nd, The adoption of the compound principle, in which 2, 4, or more thin plates are combined, the whole making up the weight considered desirable for the purpose designed. 3rd, The plates to be hardened equably throughout, and placed edgeways, in one or more sets, beneath the card, in true axial parallelism with the north and south line. 4th, In the use of a single set of plates, on my originally suggested plan, 2, 4, or 6 plates are placed in proximate parallelism, with moderate spaces betwixt them, and with a sufficient space in the middle for the pivot and its free action on the agate centre. 5th, On the plan as modified in the Admiralty compasses (as I have noted above) two to four sets of thin hard plates, each set having its plates in close contact, are employed.

Employing two separate combinations of plates, say, of two to four thin plates in each, and these being distributed after the manner of a two-bar compass card,—now extensively employed in the mercantile navy,—a very large augmentation of directive power is gained, with a permanency sufficient, with fair dealing, to last for years.

The needles which I first arranged on the compound plan, consisting of 2, 4, or 6 plates on edge, were spaced and combined by means of discs or blocks of wood, and thin screw-pins, exactly in the manner of the original experimental needle (56). Two of these, one of four and the

other of two plates, were transmitted to the " Compass Committee" of the Navy Board, in the year 1838 or 39. Besides
these a needle of two plates, which weighed, with its own
fittings, only about 400 grains, was adjusted to a *very light
card* of an azimuth compass sent down to me to Exeter
about the same period. It was designed for use under great
stillness, and its susceptibility of action was such that when
the instrument was placed on a table in *perfect repose* of
everything around, the card could be seen to oscillate down
to the small angular extent of the breadth of a horsehair.

63. **Permanency of Scoresby's Compound Needles tested.**—
Defect in permanency of these needles was intimated by
one of the Compass Committee, on an occasion when I was
present at a special meeting. It was stated that the needles
described above (62), and which had been in the hands of
the committee a few weeks, had been found rather weak, as
to permanency, a loss of ten per cent of the original power
having already taken place. Such a result, I confidently
affirmed, could not possibly have taken place unless the pair
of needles, which were packed with converse poles connected
by armatures of soft iron in the same box, had been put
into unfavourable contact or proximity with each other, or
some other magnet. But, at that time, I had no proof of
their tenaciousness of the magnetic condition except that of
the *test magnet*, which test, conclusive as on theoretic principles it seemed, might, without any fact of experience, be
questioned.

Since then the tenaciousness or fixity of these needles
has been amply proved. A thin *six*-plate needle, six inches
long, having about double the power of any needle I had
met with of similar length and weight, and a *four*-plate
needle, of 579 grains weight, much stronger than the other,
were put together in a small case in July, 1838. After about
16 years they were tested (meanwhile having been frequently
removed from the case and handled and examined separately), when the first was found to have lost only 6·6 per
cent. of its original power, and the other less than 6 per
cent.

Another needle of four plates, six inches in length, was magnetised, attached to a card in February, 1839, and, being tested as to its directive power, placed in a box *quite unprotected and unsustained.* This box, though very shallow and small, was taken no special care of, being put, from time to time, into different cabinets or other places, exposed to various magnetic influences from apparatus lying about, and occasionally removed from the box. In May, 1841, the needle, being examined, and its directive power carefully taken, was found singularly powerful. It was re-examined in the beginning of the year 1855—nearly *fourteen years* after the second and special testing—when it was found to have only lost, in that long interval of time, 7·5 per cent, or about a fifteenth part of its original power, and was yet of more than double the energy of the compass-needles in ordinary use! Several separate plates were also tested for the effects of time and position, and the results were equally satisfactory.

64. **On the storing or keeping of spare Compass Cards or Magnets.**—Familiar as the requirements for a proper keeping of spare compass cards or magnets now are, a cautionary remark may not be misplaced, especially as, when two or three spare cards happen to be taken out of their boxes together, the needles are very liable to get into damaging contact, whilst such contact, if but momentary, seems to be little thought of. But the deteriorating influence of similar poles, should never, in such cases, be lost sight of. For all magnets constructed of steel suffer deterioration, in however slight a degree, by contact, or even proximity, though but for an instant, of the similar poles of other magnets ; whilst contact or proximity of converse poles tends to sustain and increase the power of both magnets so situated. Hence we must be careful in the placing of magnets or magnetic needles together, not to allow the corresponding or like poles to come at all near to each other, much less to touch or lie in contact. And as the tendency in all cases must be to damage mutually the magnetic intensity in bars of similar size and kind, this tendency in case of one of the magnets

being much more powerful than the other, may possibly go to the entire destruction of the force of the weaker.

As to the keeping of compass needles constructed on the principles laid down above (62), indeed, the risk of material damage by proximity or contact with other needles is very small; but still, a judicious carefulness, especially where strong magnets are near, should always be practised.

65. Process for strengthening the Power of Compass Needles without removal.—In many cases where the needles of compasses have become weak, it is a matter of convenience to be able to restore their proper polarity whilst they remained affixed to their cards. With a pair of good straight-bar magnets, and by the process of magnetisation, art. 34, this, in the case of a single central needle, or a pair of needles considerably removed, or of a compound needle similarly placed, can be easily and effectually done. Thus placing the north end of one magnet against the central agate on the side towards the south pole of the needle, and the south end of the other on the opposite side, and drawing the magnets asunder till they come, respectively, to the extremities of the needle, and then sliding them off sideways, a high degree of energy may be developed. If the needles be on edge, two such strokes from the centre to the ends should be made on each side. If, however, there be two or more sets of plates or needles placed very near each other, say, within two or three inches, the process may not be relied on.

In the case of the needles of *pocket compasses*, where the glass is frequently plated down by the edge of the case, or of surveyors' instruments, where the needles may not be very accessible, I find that full magnetic powers may be easily and promptly restored, by making the passes of the magnet over the glass, and in the manner just described. For this very convenient process of magnetisation, a tolerably powerful pair of straight bar-magnets is requisite. With a pair of my busk-plate magnets (51) of about 24 hard plates of steel in each, the highest susceptible power of polarity may be elicited with certainty, and in a moment of time.

CHAPTER IV.

TERRESTRIAL MAGNETISM; ITS DIRECTIVE ACTION ON THE
COMPASS AND DIPPING NEEDLE.

I. DIRECTIVE ACTION OF TERRESTRIAL MAGNETISM ON THE COMPASS.

66. **The directive Property of the Compass**, with the cause
of it—familiar as the property is—may solicit some consideration. As a magnet can produce an influence developing
polarity, and exerting an attractive force on a piece of iron,
without being in contact or even in very close proximity, so
magnets, placed apart, exert an influence mutually on each
other. If placed very near to each other, with similar poles
contiguous, they will tend to invert each other's polarity,
and so to damage, in case of very high magnetisation, their
respective powers. But if one or both of them be supported
like a compass-needle on a point beneath, or suspended by
fibres of silk, or some other flexible material, from above,
they will exert, respectively, a sort of sympathetic influence,
tending when one is moved to put the other in motion, or
when left to their spontaneous action, to adjust themselves
with certain relations of conformity of direction with each
other.

67. **Illustrations of the Sympathies of suspended Magnets.**—
The mutual sympathies, or reciprocally directive tendencies
of magnets freely suspended, might be illustrated in great
variety of form and arrangement.

(1). Where *both* magnets are so suspended as to be *freely
moveable*, their respective sympathies may produce some
curious and interesting results. Thus, if we set a substantial and well-magnetised compass-needle on a point on
the table (the point of a *stout* sewing needle fixed point upward in a flat piece of wood answers very well), and hold

E

just over it a small compass, or a small separate magnetic needle suspended by fibres of silk, or otherwise, the upper needle will so sympathise in direction with that beneath it, and so follow its motions and oscillations, that if the larger one be set a whirling on its pivot, the smaller one, in obedience to this master-power, will follow it (unless incidentally checked) to the termination of its course. In this way, by employing a very strong bar-magnet, as the master influence, and placing around it (but beyond the reach of its extremities when in motion) several small compass needles mounted on separate points, the whole set will be put into sympathetic action, when the central magnet oscillates or is moved round.

(2). *For the case where one of the Magnets is fixed and the other moveable,* we may take this familiar example. If directly over the middle of a powerful straight bar-magnet laid flat, or horizontally, on a table, we hold rather near it a small compass, or a magnetised needle or a bit of magnetised steel wire suspended by a fine thread, the moveable needle will, after a few oscillations, adjust itself in parallelism with the axis of the magnet, with its north pole towards the south pole of the magnet. And there, whilst no further movement occurs, it will remain stationary. So if, whilst the bar-magnet is so laid on the table, the compass or needle is placed on the table in the line of the magnet, the needle, if sufficiently near, will turn into the same direction (so that the two will appear in a line), with its north pole pointing to the south pole of the magnet, or its south pole to the north pole of the magnet. In this experiment, however, the compass needle will require to be brought pretty near to the extremity of the magnet, or the exact line of uniform direction will not be taken, unless the direction of the magnet happen to be north and south (magnetic); the direction being due to another magnetic influence, that of the earth, of which we have now to speak.

68. The directive Property of the horizontal Magnetic Needle. —On these sympathetic influences, and on the power of a fixed and strongly magnetic body, to influence the direction

of a moveable magnetic needle depends the *directive* property of the compass. The earth, which is a powerfully magnetic body, here becomes the dominant or *master-magnet,* to whose influences the moveable or traversing needles, unless specially interfered with by other magnetic forces, yield a rigid obedience, pointing exactly as the earth's polarity solicits, and, when properly directed and undisturbed by extraneous influences, remaining stationary in such position.

69. **Apparent Discrepancy of the Law of Attraction of opposite Polarities.**—In this case, according to the universal law, the two poles of the needle must be respectively attracted by the *converse* polarities of the earth, so that the needle itself must stand in a contrary direction to that of the magnetic axis of the earth, considered as a magnet. But hence arises some confusion of expression when we speak of the north end of the needle pointing to (as if attracted by) the north of the earth. But, however we may designate the end of the needle pointing northward, the simple fact must be that the earth's arctic attractive force is, in reality, of the contrary kind. The discrepancy of expression has obviously arisen from calling the polarity of the end which points towards the north of the needle the *northern,* without reference to, or indeed knowledge of (at the time the name was given) the cause of that direction being taken. This discrepancy of expression has led to the adoption of a variety of denominations to express the two polarities. In what follows, I have deemed it the most convenient course to speak, generally, of the north and south poles of the needle, and of northern and southern polarity in their popular meaning ; but in speaking of *terrestrial magnetism,* where ambiguity might be caused by the use of these terms, the denominations *arctic* and *antarctic* (in preference to boreal and austral, which have, in reality, the same meaning as northern and southern) are used.

II. THE DIP OF THE NEEDLE.

70. **The Dip or Inclination of the Magnetic Needle.**— Though the compass-needle acted upon by the earth's direc-

tive force, moves in a *horizontal* plane, such position being required for the more convenient use of the instrument, this horizontal position does not represent the true direction of the earth's magnetic force. If a magnetic needle be so suspended as to be capable of freely assuming *any* position, it will generally be found to have an *inclination* to the horizon, one of its poles being depressed, or *dipping* below the horizontal position, by an amount which is called " dip " or " inclination." The *dipping-needle*—the instrument by which the true direction of the earth's magnetic force is measured—shows that in *our* hemisphere, generally, the *north* end of the needle dips, and that to an extent generally increasing from the equatorial regions northward, until, after entering the arctic circle, and in a position near to the wintering place of the late Sir John Ross, in his adventurous voyage in 1829-1833, the magnetic needle is found to stand in a vertical position.

71. **Terrestrial Magnetic Poles.**—The position,—where all the forces of terrestrial magnetism are so combined as to attract a correctly suspended and freely moving dipping-needle in a direction directly downward, and where the ordinary compass-needle would necessarily cease to traverse, —is considered as a *magnetic pole* of the earth. So, in correspondency with the terms in common use with respect to geographical latitude, the places where the dipping-needle is horizontal, having no inclination, are considered as being in the *magnetic equator;* and the degrees of the needle's inclination are taken as *degrees of magnetic latitude,* and denominated north or south, according as the northern or southern pole of the needle is depressed.

72. **Irregularity, geographically considered, of the terrestrial Magnetic Equator and Lines of equal Dip.**—The inclination, or dip, which, as a general rule, is northerly in the *arctic* hemisphere, and southerly in the *antarctic,* changes, with change of latitude, very irregularly, the magnetic equator forming a waving line intersecting and crossing the equator of the earth. Thus, whilst in the longitude of 30 degrees W the magnetic equator is in about 16 degrees *south*

latitude; on the western coast of Africa, close to the land, it rises, up to the Gulf of Guinea, to the northward of the line.

Like irregularities, comparing magnetic with geographical latitudes, are found to occur in the course of the lines of equal inclination. Thus, sailing from the Bristol Channel towards the S. W., or S. W. by W., we might have the *same* dip of about 70 degrees for 1800 miles. Or, still more remarkably, we should find that, pursuing a similar south-westerly course from the Cape of Good Hope, with a dip, at starting, of about 52 degrees, we might proceed for a distance of 2500 miles with but little alteration. But, on the other hand, sailing from Cape Horn, or from the southern part of New Zealand, on the same course, we might increase the southerly dip from 10 degrees to 12 degrees, in a distance of 1000 miles; or, in some positions and directions we might, within the same distance, find twice that amount of change, in the inclination of the needle. The observations of magneticians and navigators of the dip of the needle in various regions of the globe have been carefully collated by General Sabine, and, with his usual tact and ability, cast into curves of equal dip, or "isoclinal lines,"—a work, like that of his Variation Charts, comprising curves of equal "declination" of the magnetic needle, of much interest to men of science, and of great and admirable utility to navigators.*

III. VARIATION OF THE NEEDLE.

73. The Variation or Declination of the Compass.—There is another peculiarity in the direction taken by a freely suspended magnetic needle, which, however familiar to the sailor and others, it may not be unfitting to notice;—that is the "declination," or "variation of the compass." The direction of the compass-needle constrained to move in a

---

* A sheet of valuable charts of "Terrestrial Magnetism" has been recently published by Mr. Keith Johnstone, in his "Physical Atlas," of whom this set of charts, so valuable to navigators, may be had separately.

horizontal plane, as is well known by no means coincides with the true north on the surface of the earth; and this want of coincidence, or "variation" changes in amount and direction as we change our position on the earth's surface. In England the deflection is from 22 degrees to 28 degrees towards the west; in other countries and regions of the globe it varies, in some it being easterly, and in others westerly, to every possible extent. In General Sabine's Charts of the Magnetic Declination, and in recent charts published by the Admiralty,—especially those of the North and South Atlantic,—we have the variations of the needle, exhibited in curves of equal degrees, admirably adapted for the guidance and convenience of the navigator.

74. **Secular Changes of the Variation.**—The variation of the compass, fixed as for limited periods it may appear to be for any given places, is, in reality, subject to continued secular changes; not only to changes of a tolerably uniform character from year to year, but with a fluctuating motion perceptible with delicate instruments, from hour to hour in the day, and from month to month in the year. The *daily changes* are too minute for ordinary observation. The *annual changes*, too, are so small as not to produce any practical inconvenience to the navigator, except after a period of some years, and as the progress of change has, in very many cases, been pretty well ascertained, the careful navigator or scientific observer may easily apply the requisite correction.

IV. TERRESTRIAL MAGNETIC FORCE.

75. **Differences in the Terrestrial Force.**—The great magnetic body, the Earth—like other magnets—possesses different degrees of force or energy in different parts. Generally the total force is the weakest about, or near to, the region of the magnetic equator, increasing in power as we proceed northward or southward towards the places of the magnetic poles. The extreme differences are such that in some parts of the globe the total magnetic force has upwards of twice the intensity of the force in other parts; or, more accurately, perhaps, the greatest and least forces may

be represented by the proportions of about 10 to 21 or 22. This is the force which acts upon the dipping-needle.

76. **Processes for determining the Amount of the Earth's Magnetic Force.**—The relative amount of the earth's magnetic force in various regions may be determined by at least three or four methods of experiment.

(1). *By the Times of Oscillation of the Dipping-needle.*— The most familiar and easiest of these is by the oscillations of a dipping-needle, the oscillations being the quickest where the force is the greatest, and the force being proportionate to the square of the number of vibrations in a given time of the same needle, supposing its magnetic power to remain unchanged.

(2). *Other Methods of determining the Terrestrial Force.*— The differences in the magnetic force of the earth may be determined by other methods, such as by observing the deflections of a dipping-needle caused by a given small weight, or by observing the weight which will cause a given deflection (90 degrees) in a given dipping-needle; or by observing the times of oscillation of a horizontal needle in different regions where the dip is known. The same thing, too, might be equally well and much more rapidly done, by having a dipping-needle so constructed as to admit of a small permanent magnet (of highest fixity) being placed at right angles to the direction taken by the dipping-needle, and easily adjusted to the same precise distance from its centre or axis; by means of which, on observing the exact measure of the angular deflection from the proper inclination, we should have, in the sines of the deflections, a comparative measure of the terrestrial force in the different places where observations might be made.

77. **On certain important practical Consequences of the Varieties in the Dip and Force.**—On the *varieties in the dip and force,* just elucidated, are dependent many important consequences and phenomena in the action of the compass, and in the magnetism of iron ships, which, hereafter, we shall have occasion to bring before the reader, and of which it is desirable that persons engaged in the practical

duties of navigating iron ships should have a clear understanding. Here it is only necessary to explain the effect of
the terrestrial dip and force on the directive action of the
horizontal needle, or the needle of the mariner's compass.

The power by which the horizontal needle is directed into
its proper position must obviously have essential relation to
the directness or obliquity of action of the terrestrial attractions. Where there is *no dip* the directive influence of a
given magnetic force is the greatest, and where the dip is
90 degrees, or where the earth's force acts vertically, there
can be no directing influence at all. The great sluggishness
of action of the compass in high magnetic latitudes, generally,
is a fact with which every arctic navigator is familiar, whilst
the extreme case of *no* directive action under dips nearly
approaching to 90 degrees is a fact of observation which the
explorers of Barrow Strait, Regent's Inlet, and the adjacent
regions on the west of the land of Boothia of Sir John Ross,
have repeatedly observed and recorded.

78. **Principles regulating the directive Force acting on the
Compass Needle.**—The directive power of terrestrial magnetism on the horizontal needle, or the compass, in the
various regions of the globe, if the total magnetic force of
the earth were uniform, would be proportional to *the cosines
of the dip.*

But as the force of the earth's magnetism is not uniform,
but varies, as we have seen, in its probable extreme limits
in the proportion of about 1 to 2·2, the simple ratio of the
cosines of the dip requires to be combined with that of the
magnetic force in order to determine for any particular places
the real directive energy exerted on the needle of the compass. In other words the horizontal force of the earth's
magnetism at any place is proportional to the total force
multiplied by the cosine of the dip.

79. **Examples:—**

Place.	Dip.	Nat. Cosine.	Total Force.	Horizontal Force.
Liverpool	71°	0·326	15·5	5·05
Gibraltar .	60	0·500	14·	7·00
Bahia . . .	0	1·000	10·	10·00

CHAPTER V.

TERRESTRIAL MAGNETISM, AS TO ITS INDUCTIVE EFFECTS
ON IRON, WITH THE AUGMENTATION OF THE POLARITY
THUS ELICITED BY MECHANICAL ACTION.

80. **Inductive Action of Terrestrial Magnetism on Iron.**—
In the same way in which iron exposed to the influence of an
artificial magnet becomes magnetic by induction, so iron
exposed, as all iron must be, to the influence of the great
natural magnet, the earth, becomes magnetic by induction.

81. **Terrestrial Magnetic Induction may be thus exhibited.**
—Take a bar of malleable iron of two or three feet in
length—a full-sized *kitchen* poker (not a *steeled* one) will
answer very well—and, holding it upright, hold a small
magnetised needle, or pocket-compass, near its upper end.
In England the north pole of the compass will be strongly
attracted by it, and the south pole by its lower end. This
shows that the upper end of the iron bar has southern
polarity; the lower end, northern. Now, reverse the position of the poker, placing it upright as before, but with its
point upward. Being again tested by the compass (supposing the poker to be free from retentive magnetism in itself),
it will appear that the point of the poker now upward
has changed the character of its magnetism, and exhibits
southern polarity, and the knob end northern polarity.
These changes show that the magnetism of the poker was
not a special or abiding power in it, as in a permanent
magnet, but a transient influence, dependent on its position.
In pursuing the experiment we find that the magnetism developed in the poker or bar is greatest, not when the bar is
held vertically, but when it is held in the direction of the
dipping-needle. Its magnetism diminishes as its direction is
inclined to that of the dipping-needle, being proportionate
to the cosines of that inclination, and vanishes when its
direction is at right angles to that of the dipping-needle.

It will be easily seen to follow from this that when a horizontal bar is held in an east and west (magnetic) position, it will have no induced magnetism; that when held in a north and south magnetic position, it will have an amount of induced magnetism equal to that which it has when held in the direction of the dip, diminished in the ratio of the cosine of the dip : 1 . When the horizontal bar is held in any intermediate position its induced magnetism will be found by diminishing its magnetism when held north and south, in the proportion of the cosine of the angle between its direction and the magnetic meridian.

82. **Illustration of the Plane of Neutrality, or that of the Magnetic Equator.**—A clear understanding of the position and principle of adjustment of the neutral plane in the earth's inductive action being necessary for obtaining correct information from many of the matters investigated on my recent voyage, I shall transfer from the work before referred to* another illustration relating to the polar and equatorial directions of terrestrial magnetism for its inductive action.

For this illustration, a thin quarto or folio volume,—I prefer a music book,—two bits of wire, or wooden skewers, about $3\frac{1}{2}$ inches long, a straight rod or stick, and a common table, of a square or oblong form, constitute the requisite apparatus. First, place the table so as to bring one of its square sides in the line of the magnetic meridian. Then, on either of the sides running north and south, place the music

Fig 5.

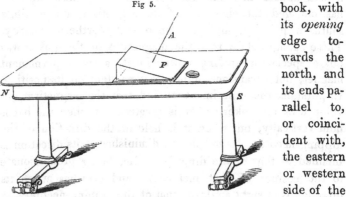

book, with its *opening* edge towards the north, and its ends parallel to, or coincident with, the eastern or western side of the

---

table. Raise now the upper board of the volume, and prop
it up by the two wires or skewers (which should be pointed)
set near the edge, or, to speak more accurately, at 10 inches
distance from the hinge of the book.

Thus arranged (as in fig. 5), the hinge and front edges
of the boards will lie horizontally in an east and west (magnetic) direction; and the plane of the upper board, P, will
lie at an inclination of about 20 degrees to the horizontal
level, rising towards the north, N, so as to represent the
plane of neutrality, or no-attraction. Parallel to this sloping
plane, therefore, or laid flat upon it, the iron bar, *if thoroughly demagnetised*, will exhibit a condition of neutrality,
and so have no effect, or no material effect, on a compass
placed near it, no matter in what direction as to the points
of the compass it may lie. This plane being in all regions
of the earth adjusted at right angles to the direction of the
dipping-needle, will show, in each case, *the plane of the
magnetic equator*, or rather the equatorial plane of the
terrestrial magnetism at the place of experiment.

83. **The Direction, in its Action, of Terrestrial Induction is
similar in all Forms of Iron.**—The laws of terrestrial induction, as regards the direction of the polar axis, and the position of the neutral or equatorial plane, apply, under necessary modifications, to all other forms of iron bodies, as well
as to bars   These modifications arise from the reflex inductive effects of the iron itself, when magnetised by the induction of the earth. To compute these accurately exceeds the
power of mathematical analysis, except in bodies of a very
simple form; but an estimate sufficiently near for practical
purposes can be made without any refined calculations.
Hence, under the guidance of the results obtained by experiments with iron bars, I had long ago anticipated all the leading facts elicited by subsequent experiments on the phenomena of terrestrial induction in *plates* of iron, *solid masses*,
and such huge and peculiarly constructed fabrics as those of
iron ships. The principles of inductive action, by the earth,
are found to be similar in all. Thus, if an *unmagnetic plate*
of iron be set upright, or nearly so, the upper *edge*, like the

upper *end* of the poker, will show southern polarity, and the *lower* northern, no matter which of the edges be placed upwards; and if it be placed at right angles to the line of dip, all its previously elicited magnetism will be found to have vanished.

84. **Augmentation of the Influence of Terrestrial Induction by mechanical Action.**—The foregoing expositions prepare us to advance another step in the consideration of the effects and phenomena of terrestrial magnetic induction. It fell to my lot, as a first-fruit of experiments undertaken in early life, to discover, in 1818—19, that the polarity terrestrially developed in iron and soft steel might be increased, controlled, inverted, or neutralized at will, by *mechanical violence,* such as hammering, bending, twisting, folding, etc.; or even by simple vibration.* This idea was suggested by the reflection, that if the simple influence of the earth's magnetism was so capable of eliciting the magnetic condition in iron, where there is always a *resistency* of the metal, especially in hard iron or steel, to such a state, that mechanical violence or vibration might assist in the overcoming of this resistency, and so render the iron susceptible of higher degrees of magnetic development. And this view, as I now proceed to show, was not a mistaken one.

85. **Experimental Elucidations of the Effects of Percussion.** —The general phenomena of percussion, or other mechanical action on iron, the laws of which have an essential and perpetual bearing on the magnetism and compass action of iron ships, may be exhibited by the following easily conducted experiments, for which no other apparatus is necessary than a pocket-compass, a bar of soft iron, or iron poker, and a hammer.

(1). *Experiment on the magnetising of an Iron Bar by a blow.*—Place the compass on a table and adjust it so that the needle may stand in the north and south line of the

---

* The entire investigation on the laws and phenomena of magnetism thus developed were originally published in the " Transactions of the Royal Society of Edinburgh," for 1821, vol. ix. p. 243; and in the " Philosophical Transactions," for 1822 and 1824.

instrument. Exactly in the line of the *west* or *east* point, and about three inches distant, place one end of an unmagnetic iron bar or poker, held vertically, and notice the number of degrees by which the south pole of the needle is deflected. Now strike the poker, whilst thus in an upright position, a smart blow with the hammer (or any part of it and in any direction), and, instantly, the needle will start aside, as if by magic, so as to treble or quadruple the original deviation. By repeatedly hammering or vibrating the iron, whilst still held in the same position, the power may be greatly augmented. If the bar be then turned into a horizontal position, east and west, where (art. 81) it was neutral, it will be found to be strongly magnetic, the end that was downward when it was struck, having northern polarity; and if it be inverted near the compass, with the contrary end downward, that end, which in the manner of the experiment referred to should have had northern polarity will still *repel* the north end of the needle. The poker, for the time, and for a long time afterwards,—if no further blow or vibration be given it,—will be found to be a true magnet. But this quality of magnetism is not, and cannot be, the same in fixity as that of *permanent* magnets, but a changeable kind, as the next experiment will show.

(2). *Reversing the Magnetism of the Bar.*—Restore the now magnetic poker to an upright position near the compass on the table, but with the other end (the knob) downward, when, as already intimated, it will attract the *north* end of the needle by its portion of retained magnetism; then, striking it a smart blow with the hammer, the needle will be seen to fly round, as if by magic, into the opposite side of deflection, the percussion having served, *in aid of the earth's influence*, to invert the polarity and render the bar again magnetic, with the polarity reversed, in consistency with the proper tendency of terrestrial induction.

(3). *The neutralizing of Magnetism in an Iron Bar by a blow.*—A third experiment with the same bar completes the series of mere characteristic phenomena. The object is to neutralize the polarity which has been elicited by the pre-

vious operations. For this purpose hold the bar in a horizontal position, and in the direction of the east and west
points of the compass, so that it may lie in the neutral or
equatorial plane. A blow or two with the hammer, when it
is so situated (the bar being assumed to be of soft iron),
will now be found to have reduced the previous polarity to
a small or inconsiderable quantity, which a little further
hammering (if needed) will generally render absolutely
neutral. The bar will have ceased to be a magnet.

86. **Scoresby's demagnetising Process for Bars or Plates
of Iron.**—On the last phenomenon, comprised among the
series of facts originally described in the "Transactions of
the Royal Society of Edinbugh," (vol. ix.), we obtain the
means of *demagnetising* iron bars or plates, which, accidentally, or otherwise, may have become magnetic, and
thus adapting them, by the simplest process, for the most
delicate magnetical experiments. Previous to the determination of this fact, and almost up to the present time, because of prevailing ignorance of the fact, it was usual for
experimenters, in demagnetising, to place the iron plate or
bar in a fire and heat it to redness, thus damaging it or destroying its polish, in order to rid it of the magnetic condition, which may in a moment be neutralized by the simple
process described.

Where the existing magnetism in an iron bar or plate
has been powerfully developed, its removal may require
a blow or two with the *north* pole of the bar or plate upward, or rising above the equatorial magnetic plane of the
earth.*

87. **Apparent Neutrality not necessarily perfect.**—It is difficult in a compendious description of magnetical principles
to exhibit with sufficient clearness some of the more abstruse
phenomena. Yet, because of its bearing on some peculiarities
in the magnetism of iron of ships, not hitherto explained, it
seems needful, here, briefly to notice this peculiar and little

* The niceties of the process for the most perfect neutralization are described
in "Magnetical Investigations," vol. ii. p. 218.

known fact. The phenomenon referred to is this:—that though a bar or sheet of iron may indicate no action (when in the equatorial plane) on a compass needle placed near it, it may yet, possibly, have its molecular magnetisms in a state, not of neutrality, but of constraint. This is a condition to which iron that has been subjected to long and severe hammering in a particular position is liable, as also bars of steel which have been long in a high magnetic condition without change of polarity. The fact which I first pointed out in the "Philosophical Transactions," for 1822-4, is indicated in various ways. Thus a bar-magnet of long standing and considerable power, having its polarity apparently neutralized by another magnet, will afterwards receive polarity more easily, and often much more strongly, in the original direction of the poles than in the contrary. Again, if such a magnet be but just neutralized, so as to have little or no action on a compass at a short distance whilst the bar is laid horizontally east and west, a few blows of a hammer on it will generally disturb its neutrality, and by the force of the molecular tension, elicit some of its original magnetic condition. If, again, a bar of iron which has been subjected to severe hammering with one of its ends always upward, be just neutralized by the process just described (art. 85, exp. 3), a blow or two whilst it is laid. exactly in the equatorial plane will restore some of its original polarity, though the terrestrial induction tends rather to neutralize it.

The general phenomenon is, that in iron, especially that which has had a high magnetic development under mechanical violence, there is a greater readiness (after apparent loss of polarity) to return to its original, than to assume a reverse polarity; so that in a bar of iron so circumstanced, whilst a single blow with the bar *wrong* way up may produce a very small measure of polarity, a similar blow, the *right* way up, may suffice to restore the characteristic polarity to three or four times that extent or more! In this peculiarity and principle we may, I hope, be able to elucidate certain phenomena in the changes in the magnetism of iron ships, hitherto passed over as inexplicable.

88. **Development and Change of Magnetic Condition by other processes of Mechanical Action.**—Results similar to those described in No. 81, as to the development of magnetism in malleable iron, cast iron, or soft steel, may be produced, though differing, probably, in intensity or degree, by any process of mechanical violence, such as bending, twisting, straining, or by mere *vibration*, of which the following experiments may serve for examples and illustration:—

(1). *To magnetise a Slip of Iron by Vibration or bending.*—Take a slip of thin plate-iron, about the thickness of a shilling, say 15 to 18 inches long and 2 inches broad; and to free it from accidental magnetism strike it a blow, or bend it slightly in the hand, whilst being held horizontally east and west. Now, whilst being held by one hand *in a vertical position*, strike it with the hand, or give it a moderate bend or twist, or shake it; then bringing it near a small pocket-compass or needle it will be found to have acquired considerable magnetic power; the upper end (which should be marked) attracting the north pole of the needle.

(2). *For the reversing of the Polarity.*—Strike or shake the plate again, but with the marked end downward, and the polarity will be reversed.

(3). *For neutralizing the Polarity.*—Finally, strike or vibrate the plate in a horizontal position, pointing east and west by compass, and the power will be destroyed. Few experiments in natural science, so simply managed as this is, are so startling, apparently mysterious, or interesting.

89. **Retentive Magnetism.**—The quality or character of the magnetism thus elicited by mechanical action, is, as it will be observed, very different from the qualities or denominations previously known or described—viz. *inductive* and *permanent*. The *inductive*, as we have seen, is transient, variable, or evanescent, with reference to change of position; the *permanent*, such as that in the loadstone and in hard steel magnets, is not affected, or only in a very slight degree,

by position, or ordinary mechanical action; but the other, whilst variable and changeable by mechanical action, under the fitting alterations of position, may be *retained*, if the bars so magnetised be kept in a quiescent state, for long periods of time, and even under great changes of position in relation to that of the polar force of the earth. Hence, for this quality of magnetism, I have suggested the name of *retentive*.

90. **Important bearing of the foregoing Experiments on Magnetism of Iron Ships.**— The discoveries thus far described, in respect to the phenomena of *retentive* magnetism as developed in iron by mechanical action, and which originally were looked upon by many as mere scientific curiosities, have, in recent times, turned out to be of vast importance in their bearing on the phenomena of magnetism in iron ships, and in pointing out the true theory and principles of its development, distribution, and changes. How they apply to this important subject, so essentially influencing the action of the compass on board such ships, and regulating its changes, will soon be seen; but before proceeding to that exposition it will be advantageous to our object to show the applicability of the principles so powerfully affecting the magnetism of iron bars, or slips of iron plates to sheets of iron, as well as to continuations of sheets of iron somewhat analogous to the "plating" of iron ships.

91. **The magnetical Results of terrestrial Induction with mechanical Action on Iron Plates.**—In the case of slender bars of iron we had to deal only with the character of the magnetism as *longitudinally* elicited, the whole of either end, from the middle, being endued, according to the position and inclination of the bar, with the same polarity. But in the case of sheets of iron, such as those used for iron ships, the effects of terrestrial induction, along with mechanical action, whilst following the same laws, are necessarily more complex in their phenomena. Yet with the help of the former diagrams, especially of fig. 5, these phenomena, I hope, as to their general nature, will be easily understood, and the more so,

F

perhaps, by the describing here of certain illustrative experiments.

(1). *Experimental Illustration of the Magnetic Distribution in Iron Plates.*—Take a plate of sheet iron of a square or oblong form, and of any convenient size and thickness, so as not to be smaller than a quarto or folio book board. A *strong* sheet of tinned iron, of the kind employed by tin-plate workers may do. But it must be free from actual magnetism, which will be shown by its having no action on a very small magnetic needle placed *near* to its several edges whilst the plate is laid in a position corresponding with that of the magnetic equatorialplane. (Fig. 5). If it be found to be magnetic my process of neutralizing such plates, as subsequently described (93), will easily correct it.

Place the plate upright, with any one of its edges resting on the table, and in the line or direction of the *magnetic meridian.* The upper edge, from end to end, will be found, on both sides of it, to attract the *north* pole of the needle, showing that it has received southern polarity inductively from the earth. The edge resting on the table will exhibit the converse, or northern polarity. Somewhere, then, intermediate betwixt the top and bottom, there must be a neutral position, the level of which will easily be found by gradually raising the compass from the table till, at an inch

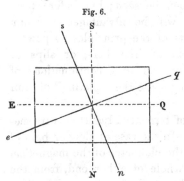

Fig. 6.

or two distance, the needle stands parallel with the side of the plate. But this neutral place will be found at a different level in the two ends of the plate, being low down at *e,* the southern end, and considerably higher at *q,* the northern. A straight line, E Q, betwixt these two points, running through, or near to, the middle of the plate, will mark the line of the magnetic equator ; and a line, *s n,* drawn at right angles to this, and also through the centre of the plate, (being in the same direction as that of the dipping-needle), will represent the *polar axis*. By reference to fig. 5, no. 82, we find that the line *e q* corresponds with that of the magnetic plane, and *n s* with that of the polar axis.

(2). *Direction of Polar Axis and Equatorial Plane, with the Plate east and west.*—For this determination, experimentally, place the plate again on edge, in an east and west line, inclining the upper edge backward towards the south, so that the sides may lean in the direction of the dipping-needle (or of the polar axis in fig. 5), and now, if totally free from acquired magnetism, the equatorial plane (though not so easy to be determined) will be found running horizontally through the centre of the plate in the direction of the dotted line E Q, the whole of the portion above it being of southern polarity, and below it northern.

(3). *Intermediate Directions of the Magnetic Lines.*—If the lower edge of the plate be placed in any line of direction intermediate betwixt that of the magnetic meridian, and that of the east and west points, and the plate be set upright, the equatorial line will, of course, run intermediately, as to inclination, betwixt the largest inclination of about 20 degrees due to a meridional position and the horizontal direction, as taken with plate in the east and west line.

(4). *Equatorial and neutral Position of the Plate.*—Returning the plate, finally, into the position in which its neutrality was first tested, that is, laying it flat in the plane of the magnetic equator (fig. 5), all the transiently induced magnetism, with its characteristic lines, will disappear.

But these simple and transient effects of simple induction, as we shall now see, may be, in a certain degree, fixed, and rendered *retentive* by percussion or other mechanical action.

92. **The Magnetic Distribution in Iron as augmented by Percussion or other mechanical Action.**—The effects of percussion, or other mechanical violence, in developing, controlling, or neutralizing *retentive magnetism* in iron plates, will be found to follow the same laws, as to the disposal of the polar axis and equatorial plane, as those just shown to

prevail with respect to simple territorial induction. But the results of the phenomena are more accurate and precise, especially as to the position of the lines of neutrality, when the plate is well hammered in various directions or inclinations. Hence, whilst the polarity of the plate becomes more energetic, by reason of the retentive magnetism developed, the general line or places of neutrality will be more definite, so that, should the plate, at the first, have been in any measure magnetic, the percussion in a position steadily maintained, will hardly fail to obliterate it, and to merge it in the new and more precise magnetic distribution, or adjustment of magnetic lines. Hammering the plate the contrary way up will, as in experiments with iron bars, reverse the polarities, and recast, into the proper direction, the equatorial or neutral line.

93. **Process for the Neutralization of any existing Magnetism in Plate Iron.**—The neutralizing in iron plates of any existing magnetism, either experimentally or accidentally acquired, may, as from analogy with previous experiments I had confidently inferred, be easily effected. For this demagnetisation the application of the same principles as those employed for the neutralizing of iron bars is found to be equally effective; that is, by hammering, or rapping the plate pretty extensively over the surface, whilst placed flat in the position of the magnetic equatorial plane. By this simple process, again, not only is the method formerly practised of heating the iron to redness dispensed with, but an incomparably more ready and perfect neutralization may be obtained.

94. **Application of the Laws of terrestrial Induction under mechanical Violence to the case of Iron Plates of peculiar form.**—An illustration, having reference to plates of a special form, will complete our designed series, and prepare the reader, I trust, for entering, with clear apprehensions of the effects of induction and mechanical action, on the fabric of iron ships.

For this I employ a long plate of sheet iron,—say, a foot broad and about 3 feet 6 inches in length. This may be

conveniently bent into the form of the midship section of an iron ship, as represented here in fig. 7. With an apparatus thus ar-
ranged and supported upright on a moveable piece of board, the foregoing principles may be extended still further ; for, here, the two sides of the bent iron (being in contact below, or at the place of the level) become mutually influential, so as, in most positions

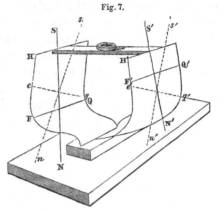

Fig. 7.

of the model, to produce a *different level* in the lines of no-attraction on the two sides.

(1). *Experimental Illustration : Head of the Model towards the north.*—Thus, if the model be placed with its keel in the line of magnetic meridian, and with the edge H, considered as lying towards the *head* of the ship, towards the north, then, after due hammering over the general surface of the iron plating, we find the neutral or equatorial lines running parallel to each other on the two sides, and rising towards the head in the direction of the dotted lines *e q ;* the polar axis, or place of greatest intensity at the upper edge inclining towards the stern, or in the direction *s n.*

(2). *Result with the Head eastward.*—Here we have a marked change. For if the keel of the model be placed in the line of the magnet east and west, the head H pointing *eastward,* and the surface be well tapped with a small hammer all over, we find, consistently with theory, that the lines of no-attraction now run horizontally on both sides, but at different levels,—that on the starboard side being low down in the position F Q, and that on the port side much higher, as represented by the line F' Q'.

(3). *Experimental Results, with the Head in intermediate directions.*—If, finally, the model be turned to a side from the direction of the magnetic meridian, with the keel in any intermediate position betwixt that and the east and west, and well tapped over the surface in the new position, to render its magnetic condition conformable to that of terrestrial action,—we again find an oblique equatorial line, becoming less and less inclined as the direction approaches the east and west, whilst consistently with this change, the line of no-attraction will be the highest on the side of the model lying the most towards the north, the difference of level being the greatest of all in an east and west direction.

CHAPTER VI.

On the Principles and Phenomena of the Magnetism in Iron Ships, as to its original Development, Direction of Polar Force and Equatorial Plane, and subsequent Changes.

95. **General Laws of the Magnetic Condition of Iron Ships.**—The general laws, already stated, may be extended to the more complex form and fabric of iron ships.

96. **Summary of leading Deductions on the Character and Distribution and liability to change of the Magnetism of Iron Ships.**

(1). *As to the Sources of the intense Magnetism of Iron Ships.*—Ships built of iron must not only be strongly magnetic, because of the vast body of this metal which is subjected to the action of terrestrial induction, but that, by reason of the elaborate system of hammering, as well as from the bending of plates and bars during the progress of construction, there must be *an extremely high* development of the quality of *retentive magnetism.*—[*Mag. Invest.* ii. 331.]

(2). *Effect of Position of the Ship when building.*—Each iron ship must have a special individuality of the magnetic distribution, depending *essentially* on the position of the keel and head whilst building, such distribution having, in each individual case, a polar axis and equatorial plane conformable to those of the earth at the place where the ship is built.—[*Mag. Invest.* ii. 332.]

(3). *Magnetic Lines of the inductive and retentive Magnetism the same.* —Whilst the spontaneous influence of simple induction must be to develope a transient magnetic condition, having, in each individual case, a polar axis and equatorial plane conformable to those of the earth at the place where the ship is built; so also the retentive magnetism developed during the building must have corresponding polar direction and distribution.

(4). *Liability of original Magnetic Distribution to change.* —The original distribution of the magnetism, or the casting of the magnetic lines, must be liable to change after the launching, under any violent mechanical action on the ship, when lying with her head in a new direction, or sailing in remote regions of the globe, having very different directions of the earth's magnetic force.

(5). *Sympathy of the Compass with a Ship's Magnetic Changes.*—All changes in a ship's magnetic condition must tend to produce disturbance in the action of compasses on or about the deck. And the effect must be, in however minute or insensible quantity in some particular cases, to change the amount of the original *deviations.*

How far these various inferences have been verified by recent observations and experiments will hereafter, and I hope satisfactorily, appear. Meanwhile, in our present descriptions and expositions, some verifications, of no inconsiderable importance, may be conveniently given.

97. As to the special Individuality of the Magnetic Distribution in Iron Ships.—Here, I proceed to show more distinctly, the characteristic form which, according to deductions from experiments described, the original polar axis and equatorial plane, with the otherwise general magnetic distribution of iron ships, should take as built with their heads in various directions. The diagrams annexed, which were cut in the wood in the year 1851, and published ["Magnetical Investigations," vol. ii. p. 269] in 1852, being some years before a single conclusive fact of experiment had been made—will serve to illustrate this important part of our subject.

(1). *The Range of different Cases considered.*—Suppose four vessels to be built of iron, with their heads, on the stocks, directed respectively (after the plan of the figures) to the four cardinal points of the compass, south, east, north, west.

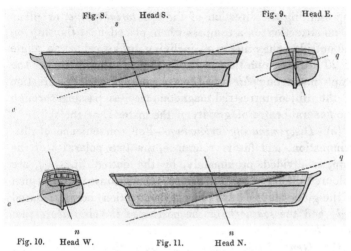

Fig. 8.   Head S.   Fig. 9.   Head E.

Fig. 10.   Head W.   Fig. 11.   Head N.

(2). *Case with the Head south*—In the case of the first of these vessels, fig. 8, with the head towards the *south*, the general magnetic distribution should cast the equatorial plane, or the plane dividing the northern and southern polarities, so as to rise from the keel, or fore-part of the ship, near the stem, up to the deck near the stern, somewhat in the manner of the dotted line *e*, in a fore and aft direction, this plane being at the same level on both sides of the ship.

(3). *Cases with the Head east and west, and north.*—Here, in cases of ships built with their heads towards the *east* and towards the *west*, the equatorial plane dividing the southern polarity above, from the northern below, should run obliquely across the hull, being highest on the port side with the head east, and highest on the starboard side with the head west. And, in the case of the ship's head towards the *north*, the equatorial plane should fall down to the keel abaft, rising up or above the deck forward.

(4). *Characteristic Differences.*—The characteristic differences in these four cases, as derived from theoretic considerations, are rendered obvious to the eye by the straight lines passing through or near to the centre of the ship in each of the diagrams; one line, *e q*, rising towards the right hand,

representing the direction of the *equatorial plane*, or plane
of no-attraction on a compass when placed near the ship on
the outside; the other *s n*, inclining backward, at an angle
of 20 degrees from the vertical, showing the direction of the
ship's magnetic *polar axis*, corresponding with the direction
of the dip, or terrestrial magnetic force, as passing through
the general centre of gravity of the material of the ship.

(5). *Discriminating colouring.*—For convenience of discrimination, and future reference, the two polarities of the
ship, as divided, proximately, by the dotted lines *e q*, are
coloured differently—the *northern*, comprising in our region
of the globe all *below* the line of no-attraction, being coloured
red; and the *southern*, or the portion of the ship *above* that
line, *blue.*

(6). *General Deductions on the Magnetic Development and
Intensity.*—On these principles and in such forms, varying
with every difference of the line of the keel when on the
stocks, should the magnetic polarities, as I considered, be
arranged in iron ships; and, by reason of the immense
amount of mechanical violence applied in the fixing of the
various masses of iron, and the hammering in of hundreds
of thousands of rivets, the *retentive magnetism* should be
developed in *the highest possible degree.*

98. The Relation betwixt the Ship's Magnetism and the Compass Disturbance.—Such relation or sympathy is a necessary
consequence of the laws of magnetic action we have described.   The precise quantity and form of disturbance con-
stitutes what has been termed the " deviation of the compass," a quality of error in the compass action on ship-board
common to all ships (however different in quality) whether
built of wood or iron, by reason of the magnetic condition
of the iron employed in the construction or fittings.   In
iron-built ships the disturbing action must necessarily be the
greatest, and the casting of the magnetic lines the most
varied, yet, in respect to the position of each ship whilst
building, the most specific.   So far as the general character
of the ship's magnetism goes—the ship being considered as
a single united magnetic body—theoretic views seemed to

guide to the conclusions just described ; and of the correctness of these views, subsequent experiments on the external magnetism and equatorial lines of a considerable number of ships, have afforded (as to the leading and characteristic features) abundant demonstration. But however specific and susceptible of prediction the general magnetic condition of any iron ship, externally examined may be, the extent and nature of the deviating influence of such magnetism within board, or on compasses placed on or near the deck, must necessarily be extremely various and uncertain. Indeed no two compasses, placed in different positions, can possibly be acted on with precisely the same measure of influence from the innumerable attracting forces of the surrounding iron; and, therefore, any compass, or rather any station or place adapted for receiving a compass, must have its own series of deviations.

99. **Determination of Compass Deviations by the Process of "swinging the Ship."**—The correct knowledge of the disturbing action of a ship on her compasses, by reason of the magnetic condition of the metallic iron used in her construction, is only to be obtained, as is well known to seamen and men of science, by the process technically designated *swinging the ship*. This process consists in determining by some convenient method of observation, the compass errors (for any compass or compasses to be tested) in the various directions of the ship's head whilst she is gradually turned round so as to make a complete circuit. For this purpose, where great precision is required, the ship is usually stopped in her progress, for a few moments, at each rhumb, or point of the compass, to which her head is successively directed, so as to obtain a series of 32 observations. These cast into a tabular form, or graphically into curves, show, for any particular course to be steered, the amount of the deviation, and the allowance, in way of correction to be made. But fewer observations of the compass errors, than for every point, will serve for all ordinary and practical purposes, such as observations for the 16 alternate points, or even for a still smaller number Nor is it absolutely necessary to have the ship's head precisely in

the direction of the rhumb to be observed, for the discrepancies can, by a practised observer, be easily allowed for.
The methods of determining the errors by the process of
swinging will hereafter come before us.

100. **Endless Varieties in the Deviations of Iron Ships.**—
The extent or quantity of the deviations in *iron ships,* with
certain qualities in their respective series, exhibit, within
noticeable limitations of analogy, an endless variety of differences. Ordinarily the errors are found to be easterly
(reckoning the *deviation* of the compass according to the
method employed for the *variation*) in one portion of the
circuit, and westerly in another; so that in the place of transition, or "points of change," the errors vanish; and, in these
two points, the compass points to the true magnetic meridian
of the earth. On each side of the points of change there is
a position of maximum error (sometimes a second and minor
maximum) ordinarily pretty well defined; but the two
maxima, though tending to similarity, do not necessarily
agree in quantity, nor do the sections of easterly and
westerly deviation in iron ships, necessarily take up equal
shares in the circuit. In ships built of wood, there is a general
tendency towards the adjustment of the points of change in
the north and south directions of the head, and to cast the
maximum errors on or near the east and west points. But
in iron ships, as theoretically we should gather from the
principles we have described of their magnetic distribution,
on such rule (whatever may be the shadowy tendencies)
prevails. Thus we find iron ships (on comparing an extensive series*) with *their points of change,* or no-deviation,
ranging throughout the compass circuit. As to *the proportion of easterly* and *westerly* deviations we find sometimes a
near equality; but, in other cases, 17 or 18 up to 20 points
or more in one way, against 15, 14, 12, or even less, the

---

* I have been indebted to Mr. Andrew Small, of Glasgow, and also to Mr.
Paul Cameron, of the same city, for a large number of deviation tables or curves
obtained in the course of their professional practice. From the former gentleman these extend to from 20 to 30, some of which (especially the comparative
tables) are specially instructive.

other way. As to the relation of the two maxima deviations
to each other we find a somewhat corresponding variety.
And as to *the quantity of the maximum deviations*, we find,
necessarily, every possible variety from a quantity so small,
as scarcely to exceed what may happen to occur in some
wooden ships, up to 8 or 10 points of error, if not more.
Cases, indeed, have been met with, from the indiscreet position selected for the steering compass, or the unfavourable
distribution of the disturbing influences around it, in which,
on the one side, the movement of the ship's head to an extent
of a point only has produced a change of three or four points
in the apparent direction as indicated by the compass; and,
on the other side, the change in the ship's head to an extent
of four points has been notified by the compass only as a
change of a point, or a little more!

101. **How far these diversified Deviations are endangering.**
—The question of the effect of these various and great irregularities in the compass action and its deviations in iron ships,
—irregularities arising out of the variously cast and intensely
disturbing influence of the ship's magnetical condition—is
one of vast importance in its bearing on practical navigation.
Summarily, and conditionally, we may reply, that no essential
difficulty would have been found as to the compass guidance,
and navigation of such ships, provided the magnetism so
powerfully elicited, and the disturbances so largely and
variously produced, were *permanent*, or *fixed*, like that in
good artificial magnets constructed of steel; for in the hands
of the scientific mathematician, the compass deviations for all
regions of the globe could have been pre-determined, or
the original errors of the compass corrected by magnetical
appliances so placed as to act antagonistically, and so cor-
rectively, against the disturbing force of the ship's iron.

102. **These diversified Deviations become endangering by
reason of their unfixed or changeable Character.**—For the
magnetism of iron ships, exerting an influence so powerful
and various on the compass, as we have shown, is not a
condition of fixity or permanency. And here, it may be
noted, that the principles and investigations described in the

foregoing pages, had long ago led me to the conclusion, that the *magnetism to be counteracted, the retentive*, which by some men most eminent in science amongst us had been deemed to be fixed or permanent, or otherwise, at least, analogous in fixity to that in ordinary steel magnets, must be liable to change, and that under much variety of conditions, and sometimes so as greatly to affect the compass, and, under certain circumstances, to change its direction very suddenly. And the conditions mainly necessary for such changes, it appeared to me, were only the application to the hull of the ship of violent or vibratory mechanical force, such as might be produced by strokes of the sea, or by hard pitching or straining in gales of wind, or even by the vibratory action of the engines in steamers, *when* the force of terrestrial magnetism might be acting in a direction different from that of the ship's then existing polar axis. The most sudden of such changes, and the largest in amount should (as pointed out by theoretic deduction) be expected to occur when the line of the dipping-needle passing through the centre of the ship might form the largest angle with the line of the ship's original polar axis, in a *new ship*, or with her existing or *normal* axis in a ship which has had much knocking about at sea.

102. Theory of sudden Compass Changes illustrated.—For an illustration of the nature and action of terrestrial induction in producing large or sudden compass changes, we will take the case of a new ship, built, for example, with her head towards the magnetic *east*, where, as in fig. 9, art. 97, the polar axis (representing the greatest intensity of magnetism, would lie over to the starboard side. Suppose, then, such a ship, soon after putting to sea, to encounter heavy weather, and now to have her head directed, as in figure 10, toward the west. Here, the new direction of the earth's polarity, now inclining to the *port* side, must obviously tend, especially under the violence of hard pitching, or straining, or heavy shocks from the waves, to alter the ship's magnetism and carry over the polar axis, *n s*, towards the port side. In such case, and more especially if the wind

should be on the port side, so as to produce a large angle
of heeling to starboard, for then the earth's influence would
be still more oblique, in respect to the ship's polarities, than
that of the dotted line, so as to influence the change in a
more forcible degree,—the shifting of the ship's polarities
might be such as to occasion large, sudden, and dangerous
errors in the indications of the compasses. And it is observable that these conditions were actually fulfilled in the
melancholy loss of the *Tayleur*, where, on evidence that could
not rationally be resisted, the compasses went astray to an
extent of from two to four points in the course of a day or
two of hard service only in battling with a southerly gale.

103. **Principles connecting the Changes in a Ship's Magnetism with Compass Disturbances.**—The little difficulty which,
in certain cases, has been experienced in the navigation of
iron ships—especially of ships regularly pursuing the same
voyage—had encouraged the doubt (elsewhere referred to)
that has prevailed in the minds of many practical men
and scientific investigators, as to whether any considerable
changes, especially sudden changes in the organic magnetism
of iron ships, could take place? But observation and experience—of which the voyage of the *Royal Charter* will,
of itself, afford very many examples—abundantly demon-
strate the fact of the prevalency of such changes in the magnetism of iron ships.

But how, it may be naturally enquired, may the changes
in the ship's magnetism affect the compass so as to explain
the immense variety in the results, when changes do take
place, which experience indicates? Here, theoretic principles can give us but very partial guidance. They will
show that all essential changes in the ship's magnetic condition (as already stated) must, unless in exceptional cases
of compasses peculiarly and favourably placed, produce
changes in the compass deviations, but not necessarily in
such quantity or amount as to be of practical consequence,
or, indeed, under ordinary inspection, observable Yet, it is
obvious that the greatest amount of change in the compasses
should not only be expected to attend the greatest magnetic

changes in the ship's polar axis and equatorial plane, but, generally, should be looked for in the greatest degree in compasses having originally the largest errors or deviations.

An illustration derived from further experiments with the model, fig. 7, page 69, described in the last chapter, may here avail us towards the explanation of a particular case, from the general principles, connecting the change in a ship's magnetism with compass disturbance. And here I need to refer only to the magnetism acting athwartship. Let the figure in question represent a section of an iron ship in the place, we will suppose, where one of the compasses has been fixed. H will mark the starboard, and H' the port sides. Suppose now, the ship's polar axis, as derived from being built in that position to lie far over towards the starboard side, H, or in other words, that this side has far stronger *southern* polarity, of the nature of *retentive* magnetism, than the other. The effect of this difference when the ship's head lies east or west, it is obvious, would not be perceptible, because of one or other of the poles of the needle pointing directly towards the place of the most powerful or dominant influence. But let the ship's head be turned away from this position,—say to the *north*, and, as now will be obvious, the two sides of the ship, possessing the same polarity, southern, will both be *attractive* of north pole of the compass so as, if both sides were alike in force, to neutralize one another, and leave the compass correct. But, it being understood that the retentive magnetism will be mainly unchanged in this new position of the ship's head; the assumed dominancy of the polarity of the starboard side H must also remain unchanged, the effect of which must be to draw the north pole of the needle towards that side, occasioning an *easterly deviation* with the ship's head north! Thus, let the attraction be to the extent of two points; then, with the compass showing N.N.W, we should have just N., on applying the deviation in the true magnetic direction.

104. **Experimental Verifications with the Model.**—The above deduction from theory obtained admirable verifications from a set of experiments with the model.

(1). *On the Dominancy of the upper Polarity of the Star-board side with the Head* H *directed towards the east.*—The model being placed in a direction at right angles to the magnetic meridian, with the edge H (the head) pointing east, and then tapped all over both sides with a small hammer, attained the condition described in art. 94, of horizontal equatorial lines on both sides, with that of the port side considerably the highest in level. And as the polar axis was thus thrown over towards the starboard side, the polarity of that side became, according to theory, more strongly magnetic than the other port side.

(2). *Effect of this Dominancy of Polarity on the Compass with the Keel of the Model in the line of the magnetic meridian.*—The supporting board of the model was, in this experiment, slewed into the meridional position with the head, H, towards the north. In this case the magnetism of simple induction, of course, changed in correspondency with that of the earth, whilst the more fixed polarity, *the retentive*, produced by hammering with the head *east*, remained. Hence the starboard side retained, in marked and obvious measure, its magnetic dominancy. This was beautifully shown by a small compass placed on the middle of a piece of deal board fixed directly across the model, as shown in the figure. The pointing of this compass, when the keel of the model was in the true magnetic meridian, would, it is obvious, have been exactly north, provided the magnetisms of the two sides had been the same. But, according to theoretic deductions, the southern polarity of the starboard side must attract the north of the needle more strongly than the same polarity of the port side, and so draw the north pole towards the east, occasioning an easterly deviation. And this was precisely the result and disturbance which the compass indicated. The north pole of the needle (on the result of various experiments) was always drawn to the starboard, to the extent of from 16 degrees to two points of easterly deviation.

The model after being again well tapped over the surface with the hammer, with the head lying towards the east, was,

G

on another trial, slewed into the direction of the magnetic south, when, as anticipated, deviations of the compass on the board pretty similar in quantity, but of contrary denomination, were obtained. In this case, the deviation was, of course, westerly.

(3). *Reverse Deviations on the Model being hammered with the Head towards the west.*—This reversal of the deviations produced by the dominancy of the retentive magnetism being transferred to the port side, in consequence of the model, when hammered, being turned in the contrary direction, was an inevitable result of magnetic principles. But the case, as proved by experiment, is here noticed as an element in the series of experiments bearing upon the principles, now being elucidated, on the relation in a ship's magnetism with compass disturbances.

105. **Change in the dominant Polarity (87) productive of Change in the Compass Deviations.**—Experiments with the model, fig. 7, abundantly prove, for the special case, this general proposition; and having already discussed the theoretic principle (96), we have now only to add the experimental elucidations. The experiments just described show how a *dominant* polarity, on one or other side of an iron ship, must necessarily be produced during the progress of construction, whenever the keel on the stocks happens to lie in an east and west direction, or, in strict science, in any direction to the eastward or westward of the magnetic meridian. The extreme case of east and west, however, serves the best for experiments.

But, proceeding with our illustrations, we will suppose that, from any of the causes which might change the ship's polarity, the dominant power of southern magnetism should be *diminished* on the starboard side, or *transferred* to the opposite side,—then the disturbing force originally drawing the north pole of the needle towards the starboard side would now either draw the less powerfully or solicit it to the opposite direction, and thus change the deviation,—in one case by *reducing* its quantity, and in the other by *reversing* its direction.

(1). *Experimental Illustration of new Deviations being obtained by changing the dominant Polarity of either side.* —For elucidating this important fact we may conveniently go back to the experiment with the model (94), where, by hammering it with the head *easterly*, we found, on turning the head to the north (2), a deviation of the compass of near two points easterly. We now slew the model with the head *westerly*, and tap it well over the two sides with a small hammer, and then slew it, as before, towards the north. We now find a very similar quantity of deviation, but in the reverse direction, being changed from easterly to westerly.

These experiments, I think, may serve clearly to explain something of the nature of the changes to which a ship's magnetic condition must be most specially liable, and to show that a change in the deviations of any compass lying nearly in the transverse line of the dominant and minimum action of the two sides of the ship, must be an inevitable consequence. On the same principle we might elucidate the natural tendency of changes in the more dominant polarity of the ship in the line of the keel, in producing changes in the deviations, more especially with the ship's head in and proximate to the east and west points.

(2). *General experimental Elucidation of Changes in the Deviations by mechanical action in a Model in new positions.* —This tendency to change in compass deviations, under the circumstances referred to, was illustrated by experiment at the meeting of the British Association, at Liverpool, in 1854, on a plan which, from the convenience of its arrangements, it may here be useful to describe. The experiments were made with a simple and easily obtained apparatus, a dish-cover of tinned iron, and a small compass; and they stand commended for the consistent and somewhat surprising character of the results. The dish-cover was placed concave side upwards, on a flat broad board, with radiating lines drawn upon its surface at angles corresponding with the eight principal points of the compass. It was attached by a pin through its centre to the middle of the board where the radiating lines met; and, whilst free to revolve on this centre,

was kept upright and steady in its progress by an intermediate piece of wood (through which the centre pin passed)
scooped out to fit its convexity below. The compass, a delicate one of three inches diameter, was fixed on a platform of
copper near to one end, so that whilst the cover might represent the model of a ship, it might have the position,
relatively, of the wheel-compass near the stern.

The experiment, which was designed to illustrate the
change in a ship's *original* magnetism, by straining with its
head in an opposite or different direction, was as follows:—
The model was first placed with its head in the direction of
one of the lines on the board—that of *north-east* was selected
—and in that position it was struck a number of moderate
strokes with a hammer on the upper sides and ends. The
retentive magnetism thus developed and arranged was then
tested by turning the head on the east and west line of the
board, and noting the deviation of the compass. The model
was then slewed into an opposite direction, head *south-west*,
and there it was strained by the hands merely, so as, within
the range of the elasticity of the metal, to press the sides,
momentarily, a little outward. On returning it to the testing
position, head east, the compass deviation was found to have
greatly changed! In various repetitions of the experiment,
with two different models of the same kind, the alteration
(always uniform in its direction) extended generally from 10
to 20 degrees, or, where the percussion and straining were
more elaborately carried out, sometimes to full 30 degrees!
It is difficult to imagine anything in the manner of an illustration more impressively conclusive than this experiment.

106. **Tendency of Sea Service to produce a normal Direction
of the Ship's polar Axis and equatorial Plane.**—Theoretic deductions from the magnetic principles we have hitherto been
attempting to establish and elucidate are calculated to carry
us an important step further in our anticipations of the
changes to which the original magnetic distribution in iron
ships, after considerable sea-service, should be liable. The
general effect of a ship's being well knocked about at sea—
that is, whilst voyaging only within the same hemisphere of

the globe—should be the production of a somewhat settled medium distribution of her magnetic lines. Thus, theoretically considered, it is obvious that, if the magnetic lines have a tendency, under the force of terrestrial action, to change into conformity with such action in all new positions of the ship's head,—then assuming a due vibratory or straining action of the sea to have taken place with the ship's head on *all* the principal points of the compass,—the ultimate result should be the production of a *horizontal* equatorial plane, and a *vertical* magnetic axis.

A large variety of facts of experience go towards the verification of this theoretic case of the *normal* magnetic distribution in iron ships,—such as the gradual reduction of the ship's original deviations, the diminished liability to large or sudden changes in her magnetic condition, the improved action and reliability of the compasses, besides a variety of facts of experiment and experience.*

But whilst asserting these demonstrable *tendencies towards a normal result*, it is by no means assumed that the true normal condition is necessarily attained; on the contrary, as certain peculiarities in the observed phenomena appear to indicate, it is more than probable that something of the original organic magnetism is of long, possibly of perpetual, continuance, and that there is a particular *tendency* in the normal approximation to a disturbing reaction, in each case, towards the *original* oblique magnetic lines.

107. **Magnetic Condition of Iron Ships, if built in high southerly Dip, must have converse Polarities.**

(1). The establishment of the theory of the original magnetism of iron ships built in England and other northern regions, as made powerful and retentive by mechanical action in process of construction,—must, of course, render the principle applicable, with differences corresponding with the changes of dip and magnetic force, to similar structures raised in all other regions. The necessary result must be, that iron ships, if built at Melbourne or in Port Philip, or in any part of Tasmania, where the magnetic dip is about

---

* " Letter to Liverpool Underwriters."

the converse of what it is in the Thames and more southern
parts of England, must have magnetic lines running in corresponding directions, but in reverse positions of the ships'
heads, with opposite polarities—the northern here being above,
and the southern below the equatorial or *no-attraction* plane.

(2). In like manner, if our deductions (106) as to the tendency in iron ships of the original oblique lines to fall into a
medium or normal condition be correct, such iron ships, if
built in Australia or Tasmania, and navigating only the seas
of the southern hemisphere, should be expected to show the
reverse polarities of their *normal condition.*

Thus, to repeat our illustration, we may represent such
antarctic normal condition, in the form annexed.

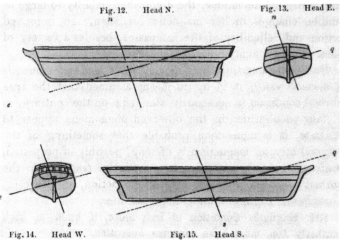

Fig. 12.   Head N.

Fig. 13.   Head E.

Fig. 14.   Head W.

Fig. 15.   Head S.

As the magnetic lines in ships built, say, at Liverpool and
at Melbourne, or about Port Philip, would, in their original
casting, be symmetrical (that is if built with their heads in
contrary directions), so that *normal* condition which much
knocking about at sea must tend to produce, would be symmetrical as to its lines, whilst the polarities would be necessarily the reverse of each other, the upper half of the ships
which in Liverpool would have southern polarity, having in
Australia, as indicated by the *red* colouring in the diagrams,
northern.

CHAPTER VII.

OBJECTIONS SUGGESTED AGAINST THE FOREGOING THEO-
RETIC PRINCIPLES.—REFUTATION, BY FACTS AND EXPERI-
MENTS, OF MANY OF THE OBJECTIONS, AND PROPOSAL OF
A PLAN FOR THE TESTING OF THE ENTIRE THEORY.

108. **Introductory Notice of Promulgation of Theory and
general Character of Objections.**—The views and principles
which have been compendiously stated, being deductions
from experiments, mainly on the Phenomena of Magnetism,
as developed in its *retentive* quality, by mechanical action in
iron, seemed to me when, many years ago, considering the
condition and difficulties about the compass in iron ships, to
point out the true theory of the ship's disturbing and varying influence, and to show something of the requirements
for a sound and safe remedy   They were first publicly promulgated, and elucidated by some experiments, at the meeting of the British Association, at Oxford, in the year 1847,
in a communication " On the Defects and Dangers arising
from the use of corrective Magnets for Local Attraction in
the Compasses of Iron-built Vessels ; " and shortly afterwards, in October of the same year, these views were more
comprehensively set forth in a public lecture at Glasgow.

The principles, however, which, on these occasions, I had
set forth, and still more elaborately in vol. ii. of "Magnetical
Investigations," published in 1852 ; and, furthermore, in
various expositions at the meetings of the British Association, at Liverpool and Glasgow, in 1854 and 1855, etc., were,
in various forms of publication, and from different quarters, objected to, denied, or even attempted to be scoffed
at. For it had been previously assumed, by some other
investigators, that the more intense portion of an iron ship's

magnetism, which I have designated *retentive*, was of the *permanent* quality, and so not subject to any rapid alterations; that the essential condition for change was *time*, so that changes which might not improbably occur, would require long periods for their being effected; and that, under certain arrangements and adjustments, the compasses might be made to act as correctly, at least, in iron ships, as in others.

109. **Varieties of Objections suggested to the Principles here contended for.**—Out of assumptions just glanced at, as well as no small variety of others, numerous objections to my views and expositions sprung, which, from time to time have been promulgated in public journals or elsewhere.

It was stated in objection, for instance (1), that experiments with mere bars or slips of iron could not be adapted for being applied, as I had supposed, to the case of such a huge and elaborate structure, and of a form altogether so peculiar, as that of an iron ship; or (2), that the facts of experience with regard to changes in a ship's organic magnetism, would not bear out my deductions. It was denied (3), on the ground of want of agreement in "compass deviations" of any two iron ships, that the distribution of magnetism elicited on the stocks could, as I had inferred, be definite and capable of anticipation; or (4), that such applications of mechanical force as ships might be exposed to at sea, would be likely to produce the effects exhibited in my experiments. It was affirmed in objecting to my views (5), that they were those of an alarmist and theorist, and calculated to produce unnecessary distrust, if not damage, to the property of iron ships. It was denied (6), that we had any facts of experience sufficient to authorise the statement that the compass on board iron ships, at sea, might undergo rapid or sudden and considerable changes; and (7), that the action of the waves could give such decided shocks or blows to a ship as the requirements for magnetical changes seemed to demand. It was asserted, in opposition to what I had suggested, and as a supposed result of experiment (8), that the magnetism of a series of iron plates rivetted together, as in ships, was not subject to the same influences as in single plates or bars. It was affirmed, in the way of refutation of my views concerning the great changes and probable inversion of the polarities of iron ships trading into high southern latitudes, or into regions of great *southern* dip of the needle (9), that the deviations of the compasses of such ships, however they might alter, still continued of the same character as they primarily were, and so, indicated the general permanency of the original organic magnetism. In opposition, too, to my plan of *a reference compass aloft*, for the daily or constant regulation of the course at the helm, it was publicly set forth (10), that the action of such elevated compass would be so disturbed by oscillations at sea as to render it useless; (11), that the points of suspension of such compass would be rapidly damaged by the agate of the needle, so as in brief space to spoil the performance of the instrument; and (12), that, on the trial of the plan, it had, in many cases, been found to fail; and, finally (13), that Mr. Airy's adjustments could be relied on in all parts of the world.

110. **Reply to, and, in many cases, Refutation of, suggested Objections.**—It would not be convenient here—as it would extend much too far this introductory part of my publication—to enter, in detail, into a refutation of the various objections which have been noticed in the preceding article. But I may refer, by the way, to papers of mine in certain public journals, etc., for a record of facts, applying, as I apprehend, conclusively, to some of these, and dwell, afterwards, more descriptively, on certain characteristic observations and experiments made on board iron ships, in verification of the deductions derived from experimental investigations on the phenomena of retentive magnetism, and the principles of its changes, in iron plates and bars. Taking the guidance of the numerical affixes to the series of objections noticed above, to which the reader, who may feel interest in the subject, will be pleased to refer, we reply:—

(1). That the objection to the applicability of the results of my experiments on iron plates and bars to the elucidation of the magnetical phenomena in such an elaborate structure as

that of an iron ship—will be found to be answered by the
facts and results, as to actual ships, recorded further on.   An
iron ship, though of such huge and elaborate formation, is
an *unity*, magnetically considered, and the terrestrial magnetism, combined with mechanical violence, will be found to
act on such a fabric on the grand system of natural laws.

(2). As to the fact of changes, sometimes rapid and considerable, in a ship's organic magnetism, and, sympatheti-
cally, in her compasses, I may refer again to the case of the
Tayleur, involving, by the action of heavy seas, a rapid,
almost sudden change in the compasses of two to four points;
that of the *Ottawa*, of Liverpool, where an instantaneous
change of two points was produced by the stroke of a sea;
that of the *Tiber*, where a similar sudden change was produced in the compass by a collision; besides the cases of the
Canadian, involving a still greater change in the steering
compass, very quickly or suddenly produced, the *City of
Philadelphia*, and others.

(3). As to the determination, theoretically, of the original
character and lines of the magnetism of iron ships whilst
building, to which different writers (including one of the
highest celebrity in science) have objected,—we have now
such an overwhelming measure of evidence, as demonstrably
to establish the proposition attempted to be denied.   But
this great fact will claim further and more particular consideration.

(4). The objection to the adequacy of the mechanical
force or violence to which a ship is subjected at sea for producing rapid changes in a ship's magnetism and compass
direction has already been met by the facts of the cases of
the *Ottawa* and *Tayleur* (2), and will hereafter obtain out of
the records of the voyage of the *Royal Charter* additional
facts in reply.

(5). As to the objection to my deductions on the instability
of the magnetism, and the liability to compass changes in iron
ships, as being those of an alarmist and theorist, it is only
needful to reply that these deductions as to the circumstances
and conditions referred to will now be seen to be demon-

strable facts.   And if the facts be so, the supposed alarmist
doctrines should result in salutary cautions and improved
safety to the navigation of this class of vessels.

(6). As to the denial of there being any facts of experience
to warrant the statement of occasional, rapid, or sudden
changes, of any important extent, in the compass deviation
in iron ships, it should be sufficient merely to refer back to
our reply to objection (2).

(7). As to the denial of the capability of the waves to
give violent shocks to a ship of the nature of blows, and
such as to be adequate in mechanical violence to the probable production of magnetic changes, a sufficient answer
(in addition to recollections by many readers of facts of experience), will, I think, be found in the subsequent records
of the effects of the seas of the Indian Ocean.

(8). The objection to the applicability of my experiments
on simple plates to the case of series of iron plates rivetted
together, on the ground of a different result from the actual
trial, is founded in mistake.   For, on combining numerous
pieces of plate-iron, by rivetting, into a slip   inches long,
the same general phenomena as to the production of retentive magnetism by percussion, and its inversion by proper
change of position when hammered, as resulted from similar
mechanical action on simple continuous plates, were obtained.

(9). The consideration of the objection to the general
principles herein set forth of the great changes, to the extent, possibly, of the inversion of the polarities, to which,
under certain extreme conditions, the organic magnetism of
iron ships must be liable, may be best referred to the actual
facts in the magnetic changes of the *Royal Charter*, as determined at Melbourne.

(10, 11, 12). The several objections made against the
desirableness of the plan of a compass aloft, for reference
and general security against misguidance, to the navigator, on
occasion of changes or sudden disturbances in the steering-
compass,—will hereafter come more specially under consideration.   In this place it may suffice to say that so far as the
experience of the voyage of the *Royal Charter* may afford

guidance, these different forms of objection were completely and severally negatived. The plan, though by no means well carried out, was yet found entirely reliable and successful.

(13). Compasses adjusted by magnets cannot be relied on on a change of magnetic latitude.

111. **Verification, by Experiment, of the general Theory of Magnetic Distribution in Iron Ships.**—The important bearing of this great fact,—viz., the dependency and essential relation of the original polarities and magnetic distribution in iron ships to their respective position on the stocks, on the general phenomena of the magnetism in such ships individually, its changes, and the consequent compass disturbances, claims for the particular proposition, deduced from theory, a special demonstration by reference to observation and experiment.

The correctness of the theory in the case of ships yet on the stocks, I had an opportunity, first of all, of satisfactorily determining by experiment on the *Elizabeth Harrison*, of 1400 tons (now the *Ellen Stewart*), of Liverpool, in October, 1854. The head of this ship pointed E.N.E., a position not very different from that of our figure 9, art. 97, showing, theoretically, the course of the magnetic axis, or polar plane, and of the equatorial or neutral plane, with the head at *east*. Being provided with a ship's steering-compass, I attached it to a piece of board about three feet in length, so that the distance of its centre from one end of the board was just two feet. The board was carefully slung with cords so as to be easily let down or adjusted to any level or distance from the top of the iron plating above, and the end intended to rest against the ship's side being cut square, kept the compass (in all part of the ship's main breadth or straight side) with its lubber points in the line of the keel, and gave, without other adjustment, an uniform distance from the iron plating of two feet.

Professor Traill, the Rev. John Moss, and Mr. Towson, were witnesses of the experiments here recorded, and the

descriptions are taken, in substance, from notes made by
Professor Traill.   Our first station was near the middle and
mainbreadth section of the ship.   Near the gunwale, on both
sides, the *north* pole of the compass was vigorously attracted.
It was easy to perceive the progress of diminution of the
force of attraction as the compass was gradually let down to
the level of the plane of no-attraction at which the compass
pointed correctly, and the change to the attraction of the
south pole, and its rate of increase, as the compass was still
further let down.

The diagram, fig. 16, gives a stern elevation of the ship,
and shows the course, theoretically, of the magnetic lines,
n s being the polar axis, and *e q* the equatorial plane.
Here on the actual trial, the level of the equatorial plane
e q, as taken on both sides, was found to be 11 feet 6 inches
lower down on the starboard side than on the port side,
whilst the difference of level by calculation, as previously
announced to the friends who witnessed the experiment,
was 11 feet.   The quality of the polarity below this plane
(tinted red) was, in accordance with theory, all northern,
and that above it (coloured blue) southern.   The ship's
main breadth, it should be noted, was 36 feet 6 inches.

The annexed diagram, representing the stern elevation of
an iron ship, will serve to convey more impressively to the
eye the manner in which her magnetic lines were found to
be cast.

Here, the section below the equatorial lines, or lines of
no-attraction, coloured red, had northern polarity from stem
to stern, and the portion above, coloured blue, southern.
The line *s n*, running obliquely towards the starboard side
represents the magnetic axis of the main breadth section, in
parallelism with the direction of the earth's magnetic force,
or the position assumed by the dipping-needle.

In the case of a ship built with her head towards the
west, the plane of no-attraction would run in the direction
e′ q′, and the red and the blue polarities be separated by
that line.

Fig. 16.

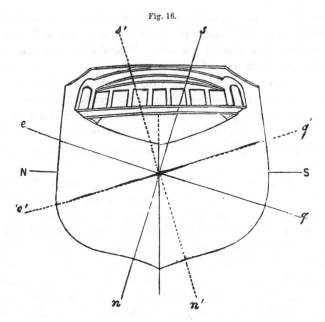

112. **Further Verification of the Theory of the Magnetic Condition of Iron Ships.**—In the instance just described, where the correctness of the theory received such beautiful verification, the trial of the ship's magnetism was only made in a few positions near the middle of the ship, so as not fully to verify previous anticipations in the rest of the ship's hull.

This lack, however, was soon supplied by further experiments, which I suggested to friends concerned in the building of iron ships, by which every prediction from theory was fulfilled, even to the *exceptions* or *deviations* from regularity of the lines of no-attraction which I had suggested might probably occur where there were large or particular masses of iron inside, such as stringers, bulk-heads, beams, etc. For these confirmations I was indebted chiefly to Mr. James R. Napier, of Glasgow, in experiments made on the *Fiery Cross*, the *Persia*, etc.; to Mr. R. Newell, of Newcastle, for the determination, whilst I was present, of the magnetic lines of the *Elba*, built at Jarrow, on the Tyne, with her head lying nearly *south;* and to Mr. George Palmer, for the magnetic lines of another ship built on the same site as the *Elba*. These experiments, to which others might now be added, sufficiently established the principles elucidated in articles 96, 97, etc., so as to leave no reasonable ground for persistency in objections against this important portion of the theory affirmed.

There were other results of inductive research on magnetical phenomena in iron,—some of the greatest possible importance in their bearing on the great practical question of safe compass guidance in iron ships, which neither home experiments, nor the information hitherto obtained respecting magnetic changes in such ships abroad, were sufficient to reach, so as to establish or refute them. Nor could the extent of the operation of influences tending to the inversion of a ship's magnetism in far southern latitudes, or the extent of compass changes from circumstances adapted for producing them, be gathered from the deductions of science. If, indeed, it had been practicable to put an iron ship to the test of experiment, as in the case of iron bars and plates (arts. 85, 91), so as to turn it upside down, and in that position subject it to a thorough hammering, or other mechanical action, we should then have had the asserted principles of the complete inversion of the ship's polarities by the reverse operation of terrestrial induction under mechanical action put to the test, and the truth or error of the theory in that particular satisfactorily determined.

But such, with various other forms of experiment and research, on organic changes in the magnetism of iron ships, which at home are impracticable, might be virtually made, it is obvious, by observing the effects of powerful mechanical force in straining in heavy seas, or from shocks from the waves in regions of the globe where, magnetically regarded, the ship would actually be inverted in position. The series of home-built ships which we have figured in art. 97, with northern polarities below and southern above, would, if built in Tasmania, or in the southern part of Australia, where the south end of the needle dips just about as much as the north end does in England, have necessarily the reverse magnetic condition, with the southern polarities below and the northern above.

[Here Dr. Scoresby's MS. abruptly terminates. The Editor, without attempting to complete what he left imperfect, will endeavour to point out the results which from the principles laid down by Dr. Scoresby might be expected on a ship going from the northern to the southern hemisphere.

In the case of a ship built in England, with the head east, Dr. Scoresby has shown that there will be a general deviation of the north end of the compass to the starboard side; because, although both top sides will attract the north end of the needle, the starboard side will do so most powerfully.

On the same ship going to the southern hemisphere each of the topsides will probably have its magnetism inverted, and each will repel the north end of the needle, but the starboard side from retaining some of its original magnetism will do so least; and the result will, therefore, be as before, a deviation of the north end of the needle to the starboard side. This will generally be the result when the deviation is caused, not by a want of symmetry in the arrangement of the iron on each side of the compass, but by one portion of the iron having, in building, acquired an advantage over the symmetrical portion on the other side.

But when the deviation is due to a want of symmetry, as a deviation towards the bow or stern of a ship built east and west, this deviation will generally be inverted on the ship going to a southern latitude.

The plan proposed and advocated warmly by Dr. Scoresby for preventing the deviation of the compass was to fix a compass aloft on the mizen-mast far removed from the influence of iron, and to compare that with the steering compass on every course steered. This plan is described by him fully in his "Letter to the Members of the Underwriters' Association, at Liverpool."]

NARRATIVE OF THE VOYAGE.

CHAPTER I.

THE large and increasing use of iron in ship-building, the magnitude of the commercial and other enterprises carried on in iron ships, and the extent to which they are made use of for the conveyance of passengers, have given to questions regarding the efficiency and safety of such ships a degree of importance that may be fairly designated as *national*. Among the questions respecting the navigation of iron ships, where every portion of the main fabric is so powerfully magnetic, that of correct compass guidance must, from its bearing on the safety of life and property, be felt to be of the very highest importance. For whilst a hundred other circumstances go to swell the sad list of maritime calamities, a delusive confidence in the compass, liable as that instrument has been shown to be to great and sometimes sudden changes, must necessarily occupy a position of deep consideration in the sources of danger.

To obtain a correct and useful knowledge of the nature of the disturbing influences on the compass, and to form a due estimate of the plans devised for their neutralisation, or of other modes of remedying the evil, the first grand requirement obviously must be the attainment of sound views on the theory or principles regulating the magnetical phenomena exhibited in iron ships. The theory and principles which I had suggested, and have now in the foregoing pages described and expanded, appear to me to be scientifically grounded and true. But they were not generally admitted; on the contrary, many of the principles lying at the very root of the leading theory were, as I have intimated, opposed or

B

denied. And whilst the very facts as to compass errors, and the liability of the original or prevailing deviations to change, were so resisted—the necessity for more than existing appliances and the availableness of the plan I had so long been urging, of a compass aloft for reference, could not be duly felt or recognised by those most directly interested in the question—the owners and navigators of our iron shipping.

Under a strong conviction of the importance of testing the theory of compass disturbances and the remedy contemplated, herein propounded and elucidated, the idea of a voyage of research to Australia originally suggested itself; and it was the anxiety to establish by demonstration what might, under the blessing of Providence, be useful in the cause of humanity as well as commerce, which ultimately led to the voyage, the particulars or results of which it is the leading purport of this volume to place before the public.

The objects more prominently before me when first I began to move in the undertaking (in which I was the more encouraged as a voyage of this kind might possibly be beneficial to my health, which was by no means strong,) may perhaps be most conveniently and comprehensively exhibited by transferring to these pages (as far as my rough copy enables me) the substance of a letter to a friend in Liverpool, which led to the arrangements ultimately made. It ran thus:—

"GROSMONT, NEAR YORK,
"*August 25th*, 1855.

" DEAR MR. RATHBONE,—As I have already received important aid in my investigations about the *magnetism of iron ships* from you and Mr. Philip Rathbone, I am led to trouble you further on the same subject, with a project in which I feel sure you will aid me if you can.

" However my expositions may at first have occasioned distrust, objection, or alarm, the results I have now every reason to know have already been largely beneficial. I believe no intelligent captain will now be found, especially in a ship not well tried, who will trust his compasses in an iron ship as being safely to be relied on; nor will it be long before the compass aloft will become a general arrangement for reference, the Astronomer-Royal himself having now proposed this very plan (suggested in my voyage to Greenland in 1822) for the readjustment of the magnets on his method of correction ; and Lieut. Maury, U. S. Navy, having, on reading the discussions at Liverpool, on the subject of my paper to the British Association, at its meeting there, recommended the same thing when writing to the underwriters of New York. My position has obtained the highest supports, and can, as I am prepared to show, be demonstrably vindicated; and so that a compass duly placed aloft may be safely referred to, and (if corrected by a table of deviations where needful) safely relied on. It can be proved that such a compass will be free from two grand sources of error,—that by heeling, which all admit, and that from sudden changes in the ship's magnetic condition. The more the subject has been investigated, the more is the theory I have so long sought to elucidate and apply to practical navigation confirmed. I have now a paper in the hands of the editor of a London periodical (the *Mercantile Marine Magazine*), in which some thirty propositions previously drawn from theoretic and experimental grounds, and for the most part published several years ago, have obtained the most conclusive and beautiful verifications. As to changes I have always felt sure I could not be mistaken; but as to the quantity of such changes, or the frequency they may be dangerous, I have been greatly mistaken and misrepresented by others. But when distinguished men (for science) deny such changes to be possible, with crowds of facts against them, it is impossible to meet them without appearing to give too much prominency to the actual cases and facts which have already come to light. These indeed have greatly multiplied on my hands by visits to Newcastle and Plymouth, where the subject was cordially entered into by builders and owners of iron ships, and every information frankly given.

" The above statement it may perhaps be desirable to confirm by an extract from my ' Magnetical Investigations,' vol. ii. pp. 289–290, as that portion of the work was printed off I believe in 1851, having been published in 1852. ' How far these changes,' in the ship's magetism, ' under the more influential forms of mechanical violence may be thorough and complete, or to what extent they may act on the compass deviations, or be discernible by ordinary observations, it would be presumptuous and vain to attempt to predict. For the tendency to change, and for the presumption of considerable and important changes, we mainly contend. The quantity of approximation in the ship's magnetisms to the earth's inductive action, under any of the conditions referred to' (that is, change of latitude and hemisphere, change of course and voyage, and encountering of gales of wind, etc.) 'is a question which observation and experiment on the actual cases can alone determine.'

" Now it is for the determination of these actual cases, under the changes of condition above indicated, that I am induced to trouble you with this long letter, and to submit to you the project I have for some time had in contemplation.

" My project is, if I can find a suitable iron ship of a first-rate or good class, either sailing or steaming, voyaging to Australia (or New Zealand), and if with the owners of such ship I could made a favourable arrangement so as to put me to little cost (for I have already spent many hundreds in this branch of the public service), I am disposed, God willing, to undertake such a voyage, in order to verify and complete the investigations I have been so long pursuing, on the important national object of the compass-action in iron ships.

" No scientific investigations from these regions, as far as I am aware, have yet come before the public which can serve any useful end, but only enough to show that the freaks of the compass in high southern latitudes are as perplexing as they are dangerous. A voyage to Australia, where the earth's magnetic force is *inverted*, and where the S. end of the needle *dips* about the same extent as the N. end with us, would no doubt afford the means of making the most valuable experiments and observations, and gaining a species of information which could not fail to place our iron shipping in a very superior condition as to safety and public confidence. This service I am disposed to undertake, whatever sacrifice of private comfort and domestic privileges might necessarily attach to it.

" You know me too well to suppose I would in any way attempt to interfere with captain or officers, but only to have it arranged that facilities should be given for experiments and observations whenever it might be consistent, or not interfere with the ship's duties and the progress or despatch of the voyage. And on my part it would be my special object and desire to give information to the captain and officers on subjects most deeply concerning their profession, as far as opportunity or disposition on their part might admit.

" The requirements I contemplate are,—a large or first-class iron ship for Australia, to return without much delay ; without iron masts ; with some reasonable arrangement for compass aloft ; and with such appliances of azimuth compass, and previous swinging of the ship for determining the state of the compasses, as a prudent management might demand. The time of sailing should be October, November, or December, and I should require the use of a small separate berth, and the consent and approval of the captain.

" Will you, as early as you can, obtain for me the information,— 1st, Whether any such ship as I require may be had? 2nd, Whether a notion could be obtained as to the probability of the plan being made available. "Very faithfully yours,

"W. S."

In the absence from home of the friend to whom this letter was addressed, it was put into the hands of S. R. Graves, Esq., Chairman of the Liverpool Shipowners' Association, a gentleman taking much interest in the researches in which I had been engaged, and having a lively appreciation of their importance.

My demands for the contemplated voyage were so special, and so many conditions were required to meet in the same ship, that I felt it exceedingly uncertain whether the undertaking could be carried out. A letter, however, received from Mr. Graves on the 30th of October indicated a fair prospect of practicability :—

" Since receipt of your note of the 5th September," he writes, " I have been looking about for a vessel suitable to your purposes, and owned by parties willing to aid you in your valuable researches; such a vessel I have found in the *Royal Charter*, 3000 tons, iron screw steamer, sailing from this port for Australia—wooden masts ; but how the Company (the Liverpool and Australian Steam Navigation Company) may enter into your views is to be tried. I have put the matter in writing to be laid before the Directors. The setting aside of a room for you would no doubt form an item of some moment in the consideration of the proposition."

Under Mr. Graves' friendly influence the matters of arrangement gradually but slowly progressed. On the 20th of November, in a letter reporting progress, Mr. Graves states, "I believe the owners of the *Royal Charter* will at once meet your views." This opinion was to a certain extent confirmed in an obliging and cordial communication from

Messrs. Gibbs, Bright, & Co., Agents for the Company, dated November 24th. A free passage out and home, on the ground of their conviction of the importance of my scientific researches, was frankly conceded to me; but from the press of first-class passengers, and the value of a whole cabin (estimated at 240*l.*), it was inquired whether a single berth might not meet my views.

It may here suffice briefly to state, that as I declined sharing a room with another party, and the Company, after their liberal offer, feeling that they had done their fair share for the promotion of such an undertaking, the treaty seemed likely to fail, had not Mr. Graves, with a most obliging promptness and zeal, taken the matter in hand, and formed the plan of raising the sum needful to obtain the second berth through the medium of different public bodies representing the interests of shipping in Liverpool. His application was so generously and promptly responded to by the " Liverpool Compass Committee," the " Shipowners' Association," and the " Underwriters' Association," that in the course of a day or two all that was requisite was attained. In communicating the result, in a letter dated December 5th, Mr. Graves says—

" By this post you will receive communications from the Compass Committee and Shipowners' Association; the opinion of the Underwriters' Association is expressed in the enclosed letter from the Secretary. Throughout the whole community of Liverpool there is but one feeling, and that is to give you every aid. I cannot help expressing my satisfaction in having given the various Associations the opportunity of recording their appreciation of your spirited project, rather than, as at first intended, accepting the liberality of a few private individuals, and," he adds truly, " I am sure you will be gratified with the mode adopted."

As it is equally creditable to the right and liberal feeling of the gentlemen representing the several Associations referred to, as it was gratifying to myself, I shall perhaps be excused giving the substance of the communications respectively made to me by their Secretaries. They were as follow:—

No. I.

From the Secretary of the Liverpool Compass Committee.

"LIVERPOOL, *December 4th*, 1855.

" Rev. Dr. SCORESBY, Torquay.

" SIR,—I am directed to forward to you the subjoined copy of the Resolutions, which have been unanimously adopted at a general meeting of the Liverpool Compass Committee, held this day in the rooms of the Underwriters' Association. I am further instructed to convey to you the expression of the great admiration felt by the members of the Committee, at the new instance of your devotion to magnetic science, afforded by the contemplated sacrifice of home comforts and other personal considerations, to investigate the causes of the compass errors which are experienced in southern latitudes.

" The Committee the more appreciate the importance of your intended researches, from having already felt the necessity for, and the great difficulty of, obtaining accurate and trustworthy data respecting this portion of the compass inquiry.

" It will afford the Committee much gratification to learn, that their co-operation can in any way facilitate your arrangements in Liverpool preparatory to the voyage, and their services are placed at your disposal for any preliminary experiments on board the *Royal Charter* which you may think necessary.

" I have the honour to be, etc.

" W. W. RUNDELL, *Secretary.*"

" Resolved,—That the Rev. Dr. Scoresby be elected an honorary member of the Liverpool Compass Committee.

" Resolved,—That the members of this Committee, having been informed of Dr. Scoresby's wish to personally investigate the nature and extent of the compass deviations, observed on board iron ships while traversing southern latitudes, desire to express to him the high sense which they entertain of his noble self-devotion to the elucidation of this department of science, and their desire to co-operate with him in carrying out his projected voyage to Australia in the *Royal Charter*. To this end, the sub-committee are requested to place themselves in communication with Dr. Scoresby.

" Resolved,—That this Committee undertake to raise the sum of 50*l.*, in aid of the fund for presenting the Rev. Dr. Scoresby with a free passage out and home, with suitable accommodation."

No. II.

" At a special meeting of the Liverpool Shipowners' Association, held the 5th day of December, 1855, convened to consider, ' A recommendation of the Committee, that a grant be made from the funds of the Association towards providing a free passage for the Rev. Dr. Scoresby in the *Royal Charter*, to Australia, his object being to investigate and determine the changes which take place in the needle in iron ships in high southern latitudes.'

" PRESENT,—S. R. GRAVES, Esq. in the Chair, etc. etc. etc.

" The Chairman submitted to the consideration of the meeting an offer made by the Rev. Dr. Scoresby, to proceed in the *Royal Charter* to Australia, with the view of investigating and determining the changes which take place in the needle in iron ships in high southern latitudes, and the recommendation of the Committee at their meeting on the 3rd instant, that a sum be voted out of the funds of the Association towards Dr. Scoresby's laudable object.

" Resolved,—That this meeting is desirous of expressing its appreciation of the disinterested zeal of Dr. Scoresby in the promotion of the security of navigation by his contemplated voyage and scientific observations, and in order to testify the sense of the importance this meeting entertains of Dr. Scoresby's proposal, the sum of 30*l.* be contributed out of the funds of the Association towards providing for him a free passage in the *Royal Charter*.

" That a copy of the foregoing Resolution be transmitted by the Chairman to Dr. Scoresby.

" Extracted from the proceedings, and signed

" E. CARSON, *Secretary.*"

No. III.

" LIVERPOOL, *December 4th*, 1855,
" *Underwriters' Rooms.*

" To S. R. GRAVES, Esq., Liverpool.

" DEAR SIR,—Your letter of yesterday to the Chairman of the Association mentioning the wish of the Rev. Dr. Scoresby to ascertain from personal observation the changes which take place in the magnetic condition of iron ships when in high southern latitudes, and that for this object it is necessary to raise a sum of 100*l.* in addition to that of 120*l.* subscribed by the owners of the *Royal Charter*, was duly laid before the Committee at their meeting yester--

day, and I am desired to acquaint you in reply, that, highly appreciating this further instance of the Rev. Dr. Scoresby's active exertions in the cause of a science calculated to increase the safety of navigation, they have unanimously voted 20*l.* from the funds of the Association in aid of the object contemplated.

"I remain, etc.

"THOMAS COURT, *Secretary.*"

Matters being thus far arranged, with a provision considerately made by Mr. Graves, but unknown to me at the time, for the appropriation of my second berth, if wished, to the use of`Mrs. Scoresby, I forthwith proceeded with inquiries and applications as to the instruments, etc., I should require for the most effectual carrying out of my objects; and these objects being of a public nature, I felt I might reasonably apply in official quarters.

From each department, comprising the Admiralty, the Compass Department at Woolwich, and the Colonial Office, the communications with the last of which were kindly made for me by His Grace the Duke of Argyll, my several applications for furtherance or aid in my objects, it is gratifying to me to have occasion to mention, were promptly and liberally responded to. The instruments which were entrusted to my care by the Lords Commissioners of the Admiralty comprised one of the best azimuth compasses, pocket chronometer, Fox's dipping-needle and gimbal table, with an ample selection of charts, etc., applying to the tracks in which we were expected to sail; and by General Sabine, Captain Becher, R.N., and Commander Edge, I was furnished with instruments of their respective invention. To Captain Washington, R.N., Hydrographer to the Admiralty, I was indebted for special and continuous furtherance in the completing of my stock of experimental apparatus; and for ready response in matters I had occasion to write upon, by Mr. Archibald Smith and Mr. Airy.

Other instruments were also sent to me by their respective inventors, for trial on the voyage, chiefly of the nature of compasses for the correction or determination of the ship's local attraction. These comprised the Polerus of Lieutenant

Friend, R.N., for observing the Pole Star; the Patent Compass of Captain Andrew Small, compass adjuster, of Glasgow; and the Dial Compass of Mr. Paul Cameron, compass maker and adjuster, in the same city.

It was now needful to ascertain how far the arrangements about the compasses on board the *Royal Charter* were adapted for my contemplated researches. On inquiry I found that there was a compass aloft, on the plan I had long been urging, attached to the mizen mast, well elevated, and clear of all specially disturbing influences from proximate ironwork. Two other compasses had also been fixed upon deck, *adjusted* by magnets on Mr. Airy's principle. But, to my great dismay, I ascertained that when the ship was swung for the placing of the magnets for these compasses no table of original deviations had been taken, nor was there any unadjusted compass on the deck! As the swinging of a large ship like the *Royal Charter* is an affair of considerable difficulty and expense, as well as loss of time, I became deeply anxious about the practicability of obtaining a repetition of the operation; yet without this, some leading purposes contemplated in the voyage would be greatly embarrassed and abridged, if not defeated. Blinded as the ship's disturbing influence on the compasses on deck was by the antagonistic forces excited by adjusting magnets, I could know little of the work they were doing, nor anything of the original measure or peculiarities of the ship's magnetic distribution; nor, consequently, could I obtain any just appreciation of the nature of the changes taking place in ship and compasses. Had there been one unadjusted compass, with a table of errors taken when swinging, groundwork would have been secured for any subsequent determination and comparison; but, as matters now stood, I must grope my way altogether in the dark as to my position at starting, and be content with a very limited and imperfect knowledge of the deviations of some other compass incidentally determined.

My difficulties, strenuously and repeatedly written about, were explained to the Company's Agents, and I was not slack in urging the immense importance to navigation and science of a repetition of the operation of swinging. The time, however, in reserve for fitting out the ship and taking in the cargo had now become so limited, and the difficulties were so great, that considering everything had been done which was deemed needful for guidance and safety in navigation, little encouragement could be given to me, beyond the obvious desire of furthering my views, as far as they could possibly be made consistent with the chief interests of the concern, for expecting the realization of this difficult attainment. Letter after letter passed, but unfavourable weather and accumulating difficulties seemed more and more to discourage the residue of hope.

My relief, and I may say joy, were in consequence greatly enhanced when a telegraphic message reached me at Torquay, in the evening of Monday, December 31st, to the effect—
" Will swing the ship, if possible, on Wednesday, the 2nd; if we fail we cannot try again." The next evening I was in Liverpool. Disappointment, however, for the time awaited us; it was neap tides, the water was *unusually* low in the docks, and the *Royal Charter* had grounded. This unexpected hindrance was the more felt as the day designed for the swinging (Wednesday) was fine, and the wind moderate. Again our much raised hopes were cast down. The ship, it was calculated, would not float till Friday, the 4th; and as (on account of the tides) she must ultimately leave the dock on the 15th, in order to sail, as advertised, on the 17th, there were even now only ten clear days for taking in the cargo and completing an almost overwhelming quantity of work. The case was one of great difficulty, but the wish of the owners to assist my objects as far as possible prevented any summary determination. The Liverpool public, too, showed an interest in the matter; and the carrying it out, in accordance with my views, was urged by the local press. Happily the conclusion conceded was that we should try again.

Meanwhile, I was enabled to make better preparations for the various observations contemplated at the swinging, and particularly for determining the ship's local attraction in the position which had been decided on for a " standard compass."

This position was on the deck-house, ranging on a level with the poop deck, and not very far from the centre of the ship, being 181 feet from the stern. The compass, a beautiful and complete instrument, with tripod for its temporary support, was taken along with me to the ship early on Friday morning, the time fixed on for our operations; but we lost more than half the day before the Wellington Dock, in which the *Royal Charter* lay, and where it was determined she should be swung, could be sufficiently cleared of encumbering shipping, and other arrangements not duly anticipated could be made.

With two magnetic needles on shore, which were shifted about the deck side or precincts as the obtaining of a clear view from the standard compass required, reciprocal bearings were taken, and those from the shore, assumed to be true, were notified, either by the voice of an assistant or by large figures chalked on a board. Assisted by Mr. Rundell, Mr. Towson, and another gentleman, members of the Compass Committee, on board, we commenced operations about 1 P.M., with, unfortunately, only about three hours of good daylight at command; and as the ship, from her great length, nearness to the ground, and the resistance of a fresh breeze of wind, was hove round with extreme slowness; we could only advance through a portion of the circuit. As far as we went, however, that is, by intervals of two points from N. W., east about to S., the work was satisfactorily done. We obtained for each station generally two readings of all the fixed compasses on board, as well as the standard; and thus obtained the deviations of the *adjusted* compass, I believe a rare acquisition with the navigator in aid of future comparisons.

So much having been done, the next forenoon was also conceded to us for the completion of our operations. The day soon became wet, and rendered our proceedings uncomfortable, but we were enabled to go on to a satisfactory conclusion.

As these details of a very common operation, that of swinging a ship for the determination of the compass errors, may possibly seem to some readers unnecessarily particular, I would simply excuse myself on the ground that the operation thus accomplished, and which had been so long in abeyance as of doubtful practicability, constituted in my particular instance, and for the objects of my adventure, by far the most important of my preparations; the very basis and ground-work on which any satisfactory determinations on compass action and changes in iron ships must rest.

The following table gives the results of the observations:—

Course, reckoned from N. by E. round to N.	DEVIATIONS. { E denoting an Easterly deviation. W denoting a Westerly deviation.							
Correct Magnetic Course.	Admiralty Compass.	Mast-head Compass.	Steering Compass.	Companion Compass.	After-mast Compass.	Gray's Azimuth Compass.	Dipping Needle Compass.	Fore Compass.
° '	° '	° '	° '	° '	° '	° '	° '	° '
0·30	W 17·	W 3·30	E 0·30	W 3·	. .	W 24·	W 19·30	W 26·30
9·	W 13·30	E 1·	E 4·	W 0·30	. .	W 21·	W 14·30	W 22·30
36·40	W 8·20	W 0·35	E 4·40	W 0· 5	. .	W 18· 5	W 11·50	W 16·35
62·45	W 4·45	E 0·45	E 4·45	E 0·45	. .	W 16·15	W 10·30	W 12·15
85·30	W 4·30	W 1·30	E 0·30	W 1·30	. .	W 15·30	W 12·	W 10·
112·15	W 0·15	W 1·15	E 2·15	E 0·45	. .	W 8·15	W 6·45	W 3·45
138·30	E 3·30	E 1·	W 0·30	E 0·45	. .	W 2·30	W 1·30	E 5·
138·	E 3·	E 0·30	W 1·30	W 0·15	. .	W 3·	W 3·	E 4·30
168·15	E 10·45	W 0·45	W 3·45	W 0·15	. .	E 7·15	E 6·15	E 12.15
194·45	E 17·30	E 0·45	W 0·15	W 3·45	. .	E 16·45	E 16·45	E 19·45
198·	E 18·	E 1·	W 1·30	. .	E 19·	E 19·	E 16·	E 23·
225.30	E 23·	E 5·30	E 0·30	E 3·	E 35·30	E 29·	E 25·30	E 28·30
248·	E 23·	E 3·	E 2·45	E 4·30	E 45·	E 33·30	E 27·	E 31·
260·	E 13·30	W 2·30	E 0·30	E 1·15	E 51·	E 30·	E 20·	E 26·
274·30	E 4·30	. .	E 0·30·	E 2·30	E 52·30	E 26·	E 15·30	E 19·30
286·	W 6·30	W 4·	E 0·30	E 0·30	E 53·	E 16·	E 7·	E 2·30
296·10	W 17·10	W 7·20	. .	W 1·50	. .	. .	W 5·50	. .
299·	W 16·	W 7·	. .	W 0·45	E 50·	E 2·	W 6·	W 10·
305·30	W 23·	. .	. .	W 3·30	. .	. .	W 14·45	. .
335·30	W 23.50	W 7.38	. .	W 6·	. .	. .	W 22·30	. .

The general results afforded by swinging the *Royal Charter* gave for the standard compass *maximum deviations* of about 24° W. and 23° E.; for the steering compass, about 4½° E.; the companion compass, 4½° E.; and the compass aloft (evidently in excess by reason of defects of reading, for want of proper graduation), about 7° W. Adjusted by the method of Mr. Archibald Smith, these deviations came out, for the standard, 23° 22' W., and 22° 19' E.; and the others, respectively, 4° 2' for the steering; 4° 47' for the companion; and 5° 25' for the compass aloft.

These results indicate a maximum error in each of the three fixed compasses of about half a point, and in the standard, not being adjusted or compensated, of nearly two points. Inspection of the table, moreover, shows that in all directions of the ship's head, from N., east about to S. or S.W., the compass aloft and the companion compass were, as to every practical purpose, quite correct. The extent of disturbance in the elevated compass, in the north-easterly quarter, were probably due to some local influence from the ironwork about the mast.

Up to the time of my first sight of the *Royal Charter*, the question of whether my wife should accompany me had never been raised. A deep feeling of responsibility, as to her undertaking such a long voyage, and the risks of health, life, and safety in passing through every variety of climate, and some of the most stormy regions of the globe, prevented my making any suggestion on the subject, much as my personal comfort was concerned.

Feelings of deep anxiety, however, on her mind, as to an adventurous undertaking on my part, in which several months would probably elapse before any communication could reach her, served to overcome an inherent dread of a long sea voyage, and induced her to urge that she might accompany me; and so, happily for both, it was eventually arranged that she and her maid should accompany me.

As the *Royal Charter's* day of sailing continued fixed for the 17th of January, we left our home at Torquay on the 15th, arriving safely at Liverpool the same evening. The ship we found was already in the river. The next day we spent on shore, in completing our outfit and stock of incidental articles for the voyage.

Before concluding this chapter, it may be a convenient course to introduce here some description of the ship in which we were about to embark, accompanied by such details of dimensions and equipment as may serve for reference in the subsequent pages.

The *Royal Charter* is the property of the Liverpool and Australian Steam Navigation Company. Messrs. Gibbs,

Bright & Co., are the agents in Liverpool for the Company. She was built at Sandycroft, Hawarden, Flintshire, under the superintendence of Mr. W. Patterson, of Bristol, the builder of the *Great Britain.*

The following are the *Royal Charter's* dimensions:— Length over all, 326 feet; breadth of beam, 41 feet 6 inches; depth of hold amidships, 26 feet 6 inches; area of the midship section, 605 feet; the entire length of the 'tween decks, near 320 feet; height of 'tween decks, clear of the beams, 8 feet 4 inches; of the orlop deck, 7 feet 6 inches; and of the saloon the same. The length of the saloon or poop cabin is 100 feet; of the deck-house, 50 feet; of the forecastle, 62 feet. Burthen, 3000 tons. The vessel is of great strength. A box kelson of large dimensions runs from end to end of the ship inside, besides two other kelsons, one on each side. The space within the body of the ship is divided, transversely, by bulkheads, and comprises altogether seven watertight compartments. There is a strong room for gold and other treasure. Her water tanks have the united capacity of about 64,000 gallons. The coal bunkers are calculated for 600 tons of fuel, of which, the anthracite kind being provided for the avoidance of smoke, the quantity of 20 tons is considered as sufficient, and proved on trial more than sufficient, for a day's consumption, with full steam in operation.

The ship was launched in September 1855, and then towed round to Liverpool, where she had her engines fixed, and was rigged, fitted out, and finally loaded.

The steam arrangements comprised an auxiliary propeller, and a pair of direct acting horizontal trunk-engines, nominally of 200 horse-power, or about one-third horse-power for every square foot of the midship section. The cylinders are 50 inches in diameter (less 21 inches trunk); stroke, 2 feet 3 inches; proposed number of revolutions, 75. The propeller, which is arranged for being detached and raised up clear of the water, is 14 feet in diameter, and 18 feet pitch. The boilers, which are placed abaft the engines, have 12 furnaces. A "donkey engine," a small detached apparatus, is fixed on light cast-iron brackets against the coal bunker. The funnel is 44 inches in diameter, appearing but small and short. Messrs. Penn & Son, of Greenwich, were the constructors of the engines.

The *Royal Charter* is three masted, and full rigged, with double topsails, on the American plan, on each mast; the lower topsails being similar to close reefed sails, but with the yards *fixed* at the lower mast heads, so as merely to traverse by the braces like the lower yards. The upper topsails, comprising one reef, are hoisted up by the yards in the usual way, and if suddenly required to be taken in, may be dropped behind the lower sails without further immediate attention being required. The total quantity of canvass capable of being spread is about 15,000 yards.

In dimensions, the masts and yards are very stout and heavily rigged. The main mast is 95 feet long by 42 inches in diameter; the fore mast, 90 feet by 42 inches; the mizen mast, 84 feet by 32 inches. The principal lower yards are 95 feet long by 26 inches in diameter; the lower topsail yards, 85 feet by 20 inches; the upper topsail yards, 76 feet by 18 inches; the top-gallant yards, 56 feet by 14 inches; the royal yards, 42 feet by 11 inches diameter.

The poop cabin, which is elegantly fitted up, has a range of tables calculated for dining upwards of 60 persons. Fourteen state rooms, of 8 to 10 feet in width by 6 feet 4 inches fore and aft, run along each side, mostly fitted with two berths; besides a considerable number of additional rooms of still more roomy construction below. The saloon is lighted by spacious skylights, and entered by a chief companion and stair near the fore part of the poop, and two smaller on the sides near the stern. A ladies' boudoir runs across the stern in the poop, lighted by a wide range of stern windows. The berths, besides deck lights above, are lighted and ventilated by circular port windows through the ship's sides, of which there are about 55 in each side, betwixt the stern and the stem. The entire accommodation for three classes of passengers extends to about 500, besides for the crew, about 85 in number. The hatchways and companions afford abundant facilities for access for people and cargo below; and the shafts for windsails, skylights, and port windows, afford excellent arrangements for ventilation.

In the *deckhouse*, on the middle of the main deck, are the rooms and appliances for cooking, officers' berths and mess-room, purser's storerooms, butcher's shop, etc. Besides the separate accommodation in the forecastle for the crew, other distinctive arrangements in the 'tween decks, in accordance with the improved regulations now happily required and effectually carried out under the Board of Trade, are carefully provided, with a view to separate the several classes of females and children, of married persons and of single men.

In respect to the adjuncts, furniture, etc. of the ship, we may just mention the provision of eight boats, including four fine life-boats, an armament of several guns (24 pounders), with an ample supply of fire-arms.

The following particulars respecting the ship and compasses are here placed for convenience of reference.

The *Royal Charter* was built with her head in the direction of N. 50° W. magnetic.

The keel when on the stocks, inclined 1 in 24, or in an angle of 2° 23' from the horizontal. The direction of her original magnetic lines, as calculated from the above data, assuming the dip to be about 70°, would proximately be—

Equatorial plane, transversely, inclined at an angle of 15° 11', and calculated as rising on the *starboard* side, at the midship section, about 11 feet higher than the line of no-attraction on the *port* side.

Magnetic central polar axis, inclined from the horizontal in an angle of about 74° 49', and leaning above or at the upper end towards the *port side*, in an angle, reckoned from the vertical, of about 15° 11'.

Inclination of the magnetic plane, longitudinally, or fore and aft, whilst on the stocks, should be at an angle of about 12° 42', horizontally reckoned, or deducting 2° 23', for the angle of rise in the keel, about 10° 19' in and near the main body of the ship, this plane rising from aft towards the stern.

The fixed compasses (three in number) were, as to their description and position, as follow :—

c

1. *A compass aloft,* unadjusted. This, the ship's reference and standard compass, was fixed to the mizen mast, a little above the spanker gaff, by strong non-magnetic metallic bars, projecting $7\frac{1}{2}$ feet aft, or towards the stern, and at the height of 42 feet above the poop deck.

2. An adjusted *steering* or *wheel compass.* The binnacle was 68 feet from the taffrail. The compass, on the floating principle, had its needle 3 feet 3 inches above the deck, and 3 feet above the level of the top plating, and was adjusted by two sets of magnets, moveable for readjustment, and boxes of soft iron chain. One set consisted of two cases of single bar magnets, one on either side of the compass, placed flat on the seat-board of the skylight, two feet from the deck; the magnets were ranged fore and aft, 18 to 22 inches from the compass, with their north ends towards the stern. The other corrector was in a case placed vertically and athwart ship, abaft the compass, 18 inches from it, and rising about $2\frac{1}{2}$ feet above the deck.

3. *An adjusted compass,* near the main companion, and hence designated the "companion compass." Its distance from the taffrail was 89 feet, and the height of the needle, 3 feet above the deck. It was also on the floating principle, and adjusted by two magnets, with boxes of soft iron on the sides of the instrument. One of the adjusting magnets, ranging fore and aft, was fixed on the deck, abreast of the compass, 2 feet 6 inches towards the port side of the ship, with pole towards the stern. The other, laid flat on the seat of the skylight adjoining the compass, nearly 4 feet abaft its centre, and 8 inches below the level of the card ranged athwart ship, with its south pole towards the starboard side.

These instruments were placed and adjusted by Mr. John Gray, maker and adjuster of compasses, under the general direction of Captain Martin, the Managers' Superintendent, and Mr. W. Rundell, Secretary to the Liverpool Compass Committee. The arrangements for the fourth, the Admiralty compass, were made by myself.

4. The *standard compass,* not adjusted, on the deck-house, 181 feet from the stern, with the needle 4 feet 6 inches above

the spar deck. This compass, lent me by the Admiralty, was one of their best and most complete instruments,-beautifully constructed, and with great and enduring magnetic power. The superior power is derived from compound plates of hard steel, set on edge on the leading principles suggested when I first brought forward my improved compass-needle, at the meeting of the British Association, at Bristol, in the year 1837.

Four sets of compound plates, distributed in parallel series beneath the card, give great and superior directive energy, whilst two cards, of very different weights, provide for the different conditions of steadiness and liveliness, under considerable motion at sea, in the quietness of port or very still water.

CHAPTER II.

On Thursday, January 17, 1856, at 2 P.M., we went on board the *Royal Charter*, then anchored in the Mersey, in the steamer carrying the mails. The ship, which had received on board near 400 souls, viz., crew, 106 persons, and the passengers, near 300, presented a scene of much confusion.

We had expected to have sailed immediately, but the afternoon being thick and rainy, the Captain did not think it safe to sail on the near approach of night. But the next morning, Friday the 18th, as the day began to break, the steam being already up, the anchor was weighed; the wind was light, seeming at first to draw from the northward, leaving us to our auxiliary steam-power for our progress. Subsequently, we had it calm, and then got a southerly breeze, but we proceeded for Holyhead under steam alone.

From the period of our starting, it was an object of interest to me to watch the practical operations of the compasses. The agreement with our system of deviations determined at the swinging proved highly satisfactory. Noting the three fixed compasses (the pedestal for the standard not being yet in its place), viz., the mast-head compass, the steering and companion compasses, the last was observed to point true as we steamed N. ½ W. out of the Channel; the second, N. by W.; and the elevated one, N. by W., agreeing very closely with their relative deviations. When our course was N. by W. ½ W. the two adjusted compasses on the deck differed from 6° to 7°, in general conformity with their known errors. When we proceeded W. by N., after getting clear out of the Channel, all the three compasses agreed. This course was close upon their almost common point of change, and therefore the agreement, as well as the truth of the general direction, was a consequence, affording at the same time, as far as the particular case went, a beautiful evidence of the accuracy of our previous observations. The pilot having remarked on the exceeding precision of his course as taken by the compasses, I informed him of the occasion of the special agreement, and ventured to predict, that on our change of course more southerly, the precision of agreement with the compasses would be proved to have ceased. To test the prediction, we compared the compasses again, as we proceeded on a S. W. ½ W. course past Holyhead, and then found a difference in the deck compasses, small indeed, but accordant with their deviations, of 2° to a quarter of a point.

On reaching Holyhead at 5 P.M., where the pilot should have been landed, we had a fresh breeze from the S. S. E., which, with the running of a strong tide, produced such a turbulence in the sea that, getting dark as it was, the Captain thought it unsafe to send a boat on shore. Whilst, therefore, a course was shaped S. W. ½ W., as was supposed sufficient to give a good berth to the Tusker, we took the pilot on, with the hope of landing him somewhere near Cork. But his destiny, so to speak, went otherwise. We were now under sail and steam, making about 7½ knots. The barometer, which had stood during the day at about 29·30, began to fall in the evening, and in the morning of Saturday, 19th, early, had sunk below 29 inches.

Jan. 19.—I went upon deck at 7·30 A.M. (ship's time). The morning was hazy, wind, a fresh breeze about south. The ship going a little off the wind, the pilot being anxious for a position where he might go on shore, and supposing his position to be 20 miles clear of the Tusker. But the mistake was soon made apparent by the unexpected discovery of breakers within a mile of us, probably three-quarters, to leeward! This, with the wind, not blowing hard indeed but dead on shore, was a disagreeable incident. It was but just full daylight, and the gloom of a dense hazy atmosphere gave an almost phosphorescent glare to the white threatening rollers. The ship was of course immediately hauled to the wind, and put in stays, but from her extreme length, notwithstanding the headway partially maintained by the screw, she came round very slowly, and for a while, perhaps two minutes, anxiously to those who like myself watched with a sailor's gaze, seemed to hang head to wind. At length she began to "pay off," but owing to the mainyard hanging square, the stiffness of the ropes, and the newness of the ship to the hands, the tacking occupied from eight to nine minutes. Lying off on the starboard tack, about E. to E. by N., we soon lost sight of the unpleasant shoals, the Lucifer Shoals, as they proved to be, lying about eight miles to the northward of the Tusker Rock, the lighthouse on which was dimly seen, soon after we tacked, on our weather beam.

This was an annoying and anxious incident. Had it occurred in the night, or had the screw failed in maintaining some headway, the consequences might have been dangerous.

The cause of the mistake became a subject of inquiry of interest and some importance. Had the ship gone to leeward of her supposed course? Was there an inset of current produced by late easterly winds on this shore? Or had the ship's magnetism changed?

This latter supposition at first prevailed, though I felt a difficulty in realising the conviction that either the course we had been steering, or the action of the very moderate sea, could have sufficed to produce sensible change. For, on the principles already referred to, applied to the special case of the magnetic condition of the *Royal Charter*, having regard to the direction of her head whilst building, I had confidently inferred that no material change in the ship's original magnetism, even under the most violent action of the sea or extreme vibrating influence, could be produced on a northwesterly or proximate course ; and theoretic induction led me to the further inference that little change in the action of the compasses, if any, was to be expected whilst the course of the ship should lie within six or even eight points of her direction on the stocks, that is, from S. W. round by N. W. to N. E. Hence the apparent compass misguidance by which it might be supposed we had been led so close upon the Lucifer Shoals, was to me a new fact, if it were established, for which I was by no means prepared.

Happily our former experiments on the compass deviations enabled me to determine the negative of this supposition.

This was accomplished by comparing the two adjusted compasses with that aloft and with each other whilst the ship was on different tacks.

The results were as follow :—

Time.	Compass aloft.	Steering Compass.	Companion Compass.
10 A.M.	E. $\frac{3}{4}$ N.	E. $\frac{1}{2}$ N.	E. $\frac{3}{4}$ N.
,,	E. $\frac{1}{4}$ N.	E. $\frac{1}{4}$ N.	E. $\frac{1}{2}$ N.
,,	E. $\frac{3}{4}$ N.	E. 7° $\frac{1}{2}$ N.	E. 10° N.
11 A.M.	S.W. by W. $\frac{1}{2}$ W.	S.W. by W. $\frac{1}{2}$ W.	S.W. by W. $\frac{1}{2}$ W.
,,	S.W. by W. $\frac{1}{4}$ W.	S.W. by W. $\frac{1}{2}$ W.	S.W. by W. $\frac{1}{2}$ W.

Here the agreement of the compasses was so near, and the small differences so accordant with former experiments, that the conclusion was certain that no change, as yet, affecting the compass deviation, had taken place in the ship's magnetism. On the latter course, it is observable, the tabulated differences of the two adjusted compasses was about 3° or 4°; by observation somewhat less. But our observed deviations could not be relied on to a degree.

The barometer was in the morning as low as 28·80, yet no gale. Baffling breezes and occasional calms prevailed most of the day. With our auxiliary steam we generally made five or six knots, sometimes seven. A ship which crossed us about three miles distant to windward as we stood off, we weathered above a mile on standing in. At 2·30 P.M. we tacked. At 3 P.M. the rain, previously all but constant, cleared off, and the ship standing to the westward, soon sighted the Tusker. At 4 P.M it bore N.W. by N. From hence, under a brisk breeze at S.W. by S., we proceeded on the port tack, somewhat closely hugging these dangerous shores.

Seldom, in my own experience, had I felt more anxiety under circumstances of no immediate danger, than in our progress during this day. A gale, probably one of most formidable fierceness, threatened us, and had threatened us for a considerable period; and here, as indicated by a heavy swell from the W. S.W., and sometimes southerly swells, the gale might catch us at any moment in our most critical position, critical specially from the slowness and difficulty of our evolutions (the tacking having occupied nine, seven, and twelve minutes on different occasions); and wearing, if pressed on by a gale of wind, being likely to require a space of a mile or miles. Truly grateful was the feeling that the storm as yet was stayed. Tacking in the evening, after just passing the Saltees, in dusk and haze, we stood all night on southerly courses, as the varying wind allowed us, ranging from S. W. to S. by E.

Sunday, Jan. 20.—Having attained as much sea room as the Channel in our position permitted, we tacked at 8 A.M., heading W. N. W. to W., having sometimes a fresh breeze, but not unfrequently light airs or a calm, the engine then accomplishing generally about five knots.

The ship being yet in considerable disarrangement, and the Captain, from over exertion in his efforts to get ready for sea, and cold, being ill in his room, but imperfect arrangements could be made for the due sanctifying, by Divine Service, this sacred day. However, having assembled some 60 to 80 of the passengers, I performed the usual morning service, with an extemporaneous exposition of Mark vi. 43-51, the interesting story of our Lord's walking on the sea, of His pitiful concern for His toiling disciples, and consoling salutation, " Be of good cheer : it is I ; be not afraid."

The following set of compass comparisons was made at 10.30 A.M. :—

Compass aloft.	Steering Compass.	Companion Compass.
N. W. $\frac{1}{2}$ W.	N. W. $\frac{1}{2}$ W.	N. W. $\frac{1}{2}$ W.
N. W. $\frac{3}{4}$ W.	N. W. 8° W.	N. W. 6° W.

In the afternoon and evening we had still light variable winds, hazy, or wet weather. The barometer, which had risen about one-tenth of an inch during the previous night, now sank back again to 28·86. This circumstance, so ominous of a heavy gale, connected with a swell, sometimes very considerable, from the W. S. W., plainly indicated an aërial conflict—violent atmospheric currents setting antagonistically, and in our position producing a sort of aërial node. It seemed probable that in mid-England and on proximate Europe south-easterly or north-easterly winds might prevail, whilst in the Atlantic it was certain, from the westerly swell experienced by us during great part of two days, westerly gales must have been blowing hard; and I could not but suppose that this wind would ultimately reach us. Though our position was much improved from what it was, yet, in case of a hard westerly gale, it was by no means a comfortable one.

At noon, Cork harbour, supposed to lie about N.W. by W. of us by compass, distance some 80 miles.

The motion of the ship being disordering, and the weather wet and unpleasant, the majority of the passengers were unable to leave their berths.

Jan. 21.—The wind during the night generally enabled us to lie well about to windward of the southern coast of Ireland ; but calms and variable breezes were not unfrequent. The steam power, however, kept us in good headway, so that on the morning of the 21st, at daylight, we considered ourselves off Cape Clear, perhaps 20 to 25 miles. About this time, after an instant of calm, a breeze sprung up at N. E. This, rapidly freshening, soon rendered the steam power more than useless. A ship at this time, too, heaving in sight, and reaching towards us by the wind, we immediately hauled to, for the twofold purpose of disengaging and raising up the screw and of sending off the pilot. The latter object, however, failed, by reason of the considerable sea rendering the lowering of a boat difficult and somewhat hazardous.

With an exchange of numbers, therefore, we parted. The screw, from too close fitting, or rusting, gave considerable trouble, being removed with unwonted difficulty, and a loss of an hour and a half of time.

The following comparisons of the compasses were made during our lying to, and after making sail :—

Mast-head Compass.	Steering Compass.	Companion Compass.
N. N. W. ¼ W.	N. N. W. ½ W.	N. N. W. ¼ W.
S. by W. ¼ W.	S. by W. ¾ W.	S. by W. ½ W.
S. by W. ¼ W.	S. S. W.	S. by W. ½ W.

Ship heeling about 4° to port.

The north-easterly wind, which promised to aid us so much on a S. W. course, and which freshened to the extent of enabling us to make at one time 10 to 11 knots, soon began to show signs of want of predominance as to the antagonistic aërial currents, and by 8 P.M. had subsided into a light breeze. This long continuance of such variable and moderate winds, with the barometer *below* 29 inches, a continuance now extended to some 50 hours, I have rarely observed; but that we should altogether escape the gale I could hardly believe.

Jan. 22.—Light and variable winds prevailed during the night, and in the morning, about 6 A.M., a breeze, freshening to a strong gale, sprung up at W. by S., varying from W. to W. S. W., which soon put us under single topsails and fore sail, and led us in a direction on the starboard tack two or three points to the eastward of our course.

By this time, or before this time, we had obtained considerable knowledge of the capabilities and qualities (from which so much had been anticipated) of the *Royal Charter*. But, unfortunately, the superior capabilities of the ship were found not only to be marred, but in many respects completely destroyed, by her quantity and manner of loading and great depth in the water. Not expecting a full cargo, and fearing that the ship, possessing a very small breadth of beam in comparison of her length, might be defective in stability, a large quantity of stone ballast (several hundreds of tons) was distributed from end to end in the hold. Cargo, however, offered to an unexpected and superabundant extent during the progress of the lading; and, when it was too late to remedy the mistake, the ship was found to be too deeply laden by more than two feet—her draught of water, which ought to have been under 20 feet, being no less than 22 feet 6 inches. This actually left in this large and splendid vessel *less* than 6 feet of clear side above the main deck. The effect of this was not only to deteriorate, for the time, some of her best qualities as a "sea boat," but to embarrass the steering by the partial and occasional submersion of the screw, and greatly to retard the ship in her proper speed. Confident, therefore, are we in the assumption that great disappointment will be experienced by the owners and managers in the rate of speed attainable, and ultimate period requisite for making her passage. In other respects her good qualities triumphed over all these disadvantages.

At noon, the summing up of the ship's log gave the latitude 47° 17' N., longitude 10° 27' W. But it was obvious, where no observation or sight of land had been obtained for three days, whilst the ship had been sailing on the majority of points of the compass and cast about by very variable winds, there could be no pretension to accuracy, nor even to a moderately proximate result from the reckoning. No navigator, under the circumstances, unless he had been always on deck watching the compass and log, could tell the position of the ship to perhaps 60 miles or more. The hazy and obscure atmosphere, indeed, was very unusual; for no sight of the sun calculated for determining our latitude had been obtained since the day of sailing.

Just before dark I compared the three fixed compasses. The ship lying about S. by W., close hauled, heeling perhaps 10° or 12°, and rolling not a little. The adjusted compasses agreed within ½ point, but the compass aloft was (according to the reading when oscillating about two points) perhaps ¾ of a point more easterly, giving about south as the real direction of the head. And here it was very gratifying to find that instrument practically available in a strong or head gale of wind and a corresponding high sea. Before night the weather moderated, but the wind remained in the same quarter.

Wednesday, Jan. 23 (W. S. W.)—The day broke finer

than we had yet had the weather, and the barometer was found to have risen to 29·30, and continued to rise to 29·56. In the forenoon, happily, we got good sights of the sun (at 9ʜ. 31ᴍ. 40s. apparent time), which gave the then longitude, as derived from three good chronometers, of 8° 33′ 30″ W.; whilst good altitudes at noon gave the latitude 46° 28′ 52″, a position as had been assumed probable of not less than 70 miles in the rear and eastward of the reckoning.

We continued on the port tack with various progress from 3½ to 11 knots, making little better than a S. E. ½ S. (true) course, and getting rapidly to the eastward of Cape Finisterre. At 3·30 ᴘ.ᴍ. tacked, and lay nearly up to N. W. (compass), making little headway, under a low sail and heavy westerly sea, and much "leeway." The wind increased in the evening, with rain and lightning. During the night it blew tremendously in the squalls. The compasses were particularly compared on both tacks. With the ship's head on S. ½ E. by compass aloft, the wheel compass indicated S. 2° E.; companion compass, S. ¼ E. On the port tack, lying N. W. ½ N., the differences were greater. The compass aloft acted so well that, though there was sometimes a considerable oscillation from the heavy sea, its average indication was very nearly ascertained. The height of the waves I found, in respect to perhaps two out of ten, was, in the evening, about 21 to 22 feet.

Thursday, Jan. 24 (W. to W. N. W.)—The gale blew all night, and, with intervals of more moderate weather, with rain, and, in the afternoon, thunder and lightning. Sea high and cross; waves generally 20 feet; some waves, perhaps one in ten, rose to the height (as determined by the crest intercepting the horizon, and the estimation of the height of the eye,) of about 24, occasionally, 26 feet; regular crests without breakers.

We stood on the port tack making very little way till 8 ᴀ.ᴍ., wind W. to W. by N., and then attempted to wear round. But with favourable and predominant head sail and suitable dispositions elsewhere, she paid little attention to the helm, alternately falling off and coming to for an hour, that is, with the wind then a moderate gale, this evolution occupied an hour, and a distance of some six miles !

This unmanageableness of the ship, with the screw elevated (but only rising about 18 inches above the ordinary level of the water), was a circumstance of most serious alarm, affording us no chance of escape, by wearing in heavy weather, from any suddenly appearing danger. Nor did it seem likely that the evil (owing chiefly to the heaviness of the ship's lading) could materially abate. What, with a prudent Captain,— prudent, and yet considerate of his owners' interests,—was to be done ? Happily for facilitating and clearing the Captain's embarassment, other circumstances, of a different but also of a serious nature, and those equally irremediable at sea as the other, were concurring. This was the state of the ship in the 'tween decks, which had been flooded with water (entering mainly as it seemed by the covering boards) for above two days. From the day after our sailing, indeed, much wet had passed through the seams of the deck, by the skylights and bull's-eye lights, screw bolts, etc. This becoming daily and constantly, after the sea became considerable, more and more distressing to the passengers, and threatening damage to the cargo. But after the commencement of the westerly gale, the water had flowed down in such quantities, that all the third class, and most of the second class, and some of the first class, were more or less flooded. This had been before reported to the Captain, still confined to his room by sickness, but what could be done ?

He hesitated on the question, on my visiting him in the morning of this day, of the justifiableness of putting back. My own mind, as an old sailor, was well made up on the point; but fearing in any way to influence him in a step which might seriously compromise the owners' interests, and damage the character of the ship, I told him I should volunteer no opinion until his own decision was taken ; but, meanwhile, I would go below into all departments of the passengers, and then report to him the result of those impressions.

The examination occupied about an hour, an hour of most

pitiful solicitude and sympathy with the wretched condition
of the passengers. Everything presented a picture of wretched-
ness and damage. Dirt and water defiled, especially in the
third-class department, every cabin and berth. The deck
was in any case saturated with wet, and generally defiled by
dirt. Boards, lids of boxes, and cabin doors, were laid down
to tread on. And in despite of these, the water was in
several cabins overflowing all; and at every roll of the ship
dashing from side to side in a depth of several inches of
water. It was most pitiable to see the prevalent suffering.
Wives, children, and young women, wet throughout their
dress up to the knees, without possibility of drying their
petticoats, or benefit from shifting them. One very interesting
looking young person, speaking in an accent and in language
indicative of superiority of original position to those with
whom she was associated, was in this very condition. She
spoke feelingly of the impracticability of improving herself
in comfort. She was reclining on her bed, a few feet from
the deck, whilst a middle-aged woman, without shoes or
stockings, and wet up to the knees, was baling out the water
as it ran from side to side on the floor of the berth. In
many cases, the upper berths near the ship's side had been
deserted, and the bedding removed half saturated with wet.
Bags, boxes, articles of culinary or domestic character, were
scattered about, or heaped together, drenched in water, or
with the object of preserving them as far as possible from
damage or destruction. It was lamentable to witness the
small property of the poorer emigrants thus damaged or
destroyed, a loss or injury to them, in some cases, of their
"little all." Having visited the majority of berths, or sections
of berths, in this *third*-class department, I then proceeded to
that of the *second* class, which, if not so badly drenched, or
the furniture and clothing damaged, was sufficiently wretched.
Here also, in some of the berths, the water had washed about
to a depth of some inches *below*, whilst in numerous places
above, droppings, and almost streams of water had poured
down from bolts and seams upon their sleeping places. Persons of respectable classes and habits—men, women, and chil-

dren—were here exposed to these discomforting and damaging evils, and some sickly females were suffering deplorably.

Reporting to the Captain the result of my observations, which, whilst more in detail than the official notifications, were in perfect agreement, he told me he had made up his mind immediately to put about, the low barometer, and the bad weather looking sky, giving no token of amendment, and make for Plymouth, then within a day's good run. I had now no hesitation in expressing my decided conviction, that this was the only course he could wisely and safely pursue. A sensitive feeling, indeed, of the heavy disappointment this step would prove to the company and managers of the concern; a clear perception of the damaging influence, in the first instance, which might result from the putting back of the *Royal Charter*, might, if an onward course could safely or by possibility be pursued, have counteracted this judicious resolve; but, on the other hand, the grave and perilous prospects of an endeavour to persevere in the voyage gave abundant support to the necessity for returning to some British port. As to the prospective evils we had, in the wet state of the 'tween decks and the condition of the passengers there, an almost certain source of disease, when on coming into a warm climate the evaporation from the saturated wood-work and drenched bedding and clothing of the people must be expected to produce a pernicious malaria. In regard to the cargo, too, comprising a large quantity of valuable goods, there was more than a mere risk; there was all but positive certainty of damage by the escape downward of some of the vast quantities of water which in rough weather were perpetually flowing into the 'tween decks. And lastly, in regard to the safety of the ship, with life and property embarked, there was enormous risk in pursuing so distant a voyage from the unmanageableness of the ship in blowing weather from the defect in performing whilst so deeply loaded the ordinary evolution of " wearing;" for our experience had already shown that should we come unexpectedly near to shore, or other danger, in a gale, there would be no chance of getting the ship round under a run of some miles, if at all! On the

causes contributing to this latter danger we got further
information during the day.

At 12 noon, (lat. 46° 32', long. 7° 59' W.), having a commanding sail on the ship and a "lull" in the gale, we tacked to
the northward, and then shaped a course N. E. by E. for the
Lizard. This proceeding, as might be expected, produced a
great diversity of feeling among the passengers—disappointment, annoyance at delay, vexation at so much endurance of
evil thus turning to no account; but a sentiment of approval
at the judiciousness of or necessity for the proceeding was
all but universal I believe (if not entirely so) throughout the
ship, whilst numbers of suffering ones rejoiced greatly at the
prospect of a speedy restoration of comfort and of future
benefit in the correction of the experienced evils. The wind,
in the showers, blowing still hard, the ship was steered with
difficulty; at last, about 4 P.M., during a heavy squall, and
when going under low sail about 11 knots, she "broached to,"
and for four hours remained the master of the navigators, in
refusing obedience to helm and sails in the endeavour to
"pay her off." As the most fitting course, pointed out by
experience, Captain Boyce determined to "ship" the screw,
and put on the steam power. This difficult operation in a
strong gale and heavy sea was well and effectively accomplished in about an hour, when happily we found the ship
became subject to the helm and was soon got upon the prescribed course.

Whilst pursuing our way N. E. by E., at the rate of 10 to 8½
knots, I observed, before retiring to rest, the direction of the
Pole Star by the companion compass, then agreeing pretty
nearly with the steering compass. The bearing was about
N. by E. ½ E., indicating a point of easterly deviation, which
was found nearly correspondent with the pointing of the
compass aloft, which gave ¾ of a point more easterly than
the course steered. The ship was rolling perhaps 20° to 25°
to starboard, with a prevalent heeling of about 10° to 15°.

Friday, Jan. 25 (W.)—We proceeded in this way till 8 A.M.,
the ship overrunning the oblique progress of the screw, and
in her heavy rolling producing a great vibration on the shaft

and connection with the engine, when the engineer reported
to the Captain that if sail were not taken in and speed reduced
(though then not exceeding eight knots) he could not answer
for the safety of the machinery. Here, then, was a new and
unexpected source of antagonism in our case betwixt the
sailing and steam propelling power. It had been supposed
that the provision for allowing the screw to run correspondently with the ship's speed, when the revolution of the engine
should fall short, would be adequate for yielding accordance
of action; but it lamentably failed us, so that during some
hours the ship's way was reduced to about six knots. Afterwards, the disconnecting of the shaft from the engine was
tried, but the progress in our course was much retarded.
During the previous night (24th—25th), the wind occasionally blew very violently, and seas now and then rose to the
height of 26 feet measured from the floating line of the ship.

At noon, the latitude by observation was 49° 7′, being
many leagues short of that given by the reckoning, and the
longitude reduced from sights at 8·30 A.M. was 5° 52′ W.

The Lizard, reckoned at noon as bearing N. N. E. ¼ E.
(true) or N. E. ½ E. (magnetic), 56 miles distant, a course of
N. E. ½ E. to 6 P.M. and N. E. by E. afterwards was adopted,
calculated as was supposed to lead us very near to it. My
own conviction, however, was that a change had taken place
in the adjusted compasses by the mechanical action of the sea
on the ship's polarity whilst steering north-easterly. This
conviction, derived from the direction of the Pole star and
compass aloft the preceding evening, was strongly supported
by the Pole star observed again about 6 or 7 P.M., the bearing
by the companion compass then being N. by E. or N. ¾ E.,
instead of N. N. E. ¼ E., indicating a deviation in that
compass (which on this point, as observed at Liverpool, was
correct) of 1¼ points or more. Applying this difference or
error to the time when the longitude was obtained (8·30 A.M.),
and a distance from thence to the Lizard of about 80 miles,
would give an error, after running that distance of 19 to 20
miles. No light appeared in sight when we ought by the
reckoning to have been pretty near the Lizard; and when at

length the Lizard light was seen, it was found bearing north-westerly about 20 miles off, exactly agreeing with the deviated position! And at 10 P.M. the Start light bore N. E. by E. (ship's head N. N. W.) about 12 miles off, being nearer us by 10 miles than by the undeviated reckoning.

Hence no doubt could be entertained, in a case where no material error could well arise from the fairly balanced tides, that the adjusted compasses on a north-easterly course were in error to the extent fully of a point.

The weather was now (in the evening) fine and clear, and the wind being at west, enabled us first to haul up on a course N. N. W., which about 2 A.M. (Saturday, Jan. 26th) brought us, aided by the screw propeller, into Plymouth Sound, within the Breakwater, where we safely, after no small measure of difficulty and anxiety, came, thank God, to an anchor.

The suspended storm was renewed in the morning, and all day we had strong gales, with hard squalls and heavy showers of rain. A fine Aberdeen clipper ship was found to have preceded us to this anchorage with loss of spars.

Soon after daylight, Captain Boyce had telegraphed the managers, Messrs. Gibbs, Bright, & Co., Liverpool, communicating the painful and disappointing intelligence of his having been necessitated to put back, by reason of the wet and wretched condition of the passengers, and under the heavy gales encountered, the unexpectedly bad effects of overweighting of the ship, noting the necessity of lightening her of some of the dead weight in the form of ballast. A gratifying return message, as to the liberal terms in which it was expressed, was received early in the day, directing the Captain to adopt such measures as he might deem needful, and referring him to Messrs. Fox, Son, & Co., ship agents, for the requisite assistance.

After dinner, Mrs. Scoresby and myself accompanied Captain Boyce on shore, in the midst of a heavy squall of rain, which we passed through without damage. Proceeding to the railway station, I telegraphed to Mrs. Ker, Torquay, this message :—

"*Royal Charter, Plymouth Sound, Jan. 26th.*—Put back from bad weather. All safe, thank God. Georgina quite well. Expect us this evening at the Castle."

This afforded a timely indication of what would otherwise have been a startling appearance of those whom they expected to be far off upon the wide sea. About nine o'clock we had the pleasure of embracing our domestic family circle.

It may here be noted, in respect to our disappointment and difficulties with the defects in the condition of the ship, that there is always a special risk in a large new ship of inconveniences and defects, which can only come out to observation as developed by the actual circumstances of storm and bad weather; for defects, in any case, may exist, which it is impossible otherwise to test. As I remarked to some of the passengers who were complaining that in all their experience they had never met with any such difficulties or evils: "The first putting to sea in such a ship must always be an experiment, and the adventurers on a first voyage bear the burden of the trial and testing of the various qualities and matters affecting personal comfort in a ship, of which subsequent passengers reap the benefit!"

On Monday, the 28th, I returned to the *Royal Charter*, and remained on board until the afternoon of the following day. Meanwhile, a survey was made on the condition of the ship, about the gunwale, decks, and upper works, by Plymouth officers, but without arriving at any decided result as to the causes of leakage therein. One of the principals of the firm of Gibbs, Bright, & Co. (Mr. Tyndall Bright), with Captain Martin, their shipping manager, arrived; and the removal of the ballast, a difficult and tedious operation, was forthwith commenced and perseveringly carried on, by relays of labourers, night and day. The ship being built in compartments, separated by water-tight transverse bulkheads, and having, moreover, a spacious orlop deck, the process could be carried on without the discharge of cargo, beyond what the main deck and 'tween decks might for the time receive.

Thus, by the removal only of the portion of the cargo

lying beneath one of the hatchways, the ballast below was soon reached; and, as it could be raised to the level of the deck or bulwark, was discharged into small craft moored close alongside. Fortunately, fine weather set in and continued for several days, so as to allow this part of the alteration in the ship's condition to be carried steadily and satisfactorily forward; and the effect was soon apparent, for on Thursday the ship had risen in the water, forward, near two feet.

The following compass comparison was made on board, on Tuesday the 29th; ship upright:—

Mast-head.	Steering.	Companion.	Admiralty.
N. E. ¾ E.	N. E. ¼ E.	N. E. ½ E.	
N. E. 7° E.	N. E. 2° E.	N. E. ½ E.	
... ...	N. 35½° E.	... ...	N. 43½° E.
... ...	N. 35½° E.	... ...	N. 44° E.

On Thursday, the 31st, I again visited the *Royal Charter*, and remained some time on board whilst a second survey by officers from Liverpool, and Mr. Paterson, the builder of the ship, was going on. The upper overlapping plate (next the gunwale outside) was found to have admitted water, from the upward force of the sea, in various places—a defect happily considered to be of easy reparation; much water too, it was ascertained, had run below by the "companions" of the second and third-class compartments, as also by the hatchways, skylights, hause-holes, etc.; but the faults in the gunwales and decks seemed by no means to be very apparent. The next day, however (Feb. 1), I received a telegraphic message: "Survey completed; we hope to be ready to sail on Wednesday."

The ship's head on the 31st January having changed in direction to a position where originally there was considerable deviation in the azimuth compass, I made the following comparisons:—

Compass aloft.	Steering.	Companion.	Azimuth Standard.
N. 9° W.	N. by W. ½ W.	N. by W. ¾ W.	
N. by W.	N. by W. ½ W.	N. by W. ¾ W.	
N. by W.	N. by W. ¾ W.	... ...	N. 3° W.
N. by W.	N. by W. ¾ W.	... ...	N. 2° W.

On this visit (Feb. 1) to the *Royal Charter*, now shifted as to her berth nearer to shore, I found the work of unballasting actively going on. The ship's draught of water, originally 22 feet 2 inches abaft, and 21 feet 7 inches forward, being now about 21½ feet abaft, and 18 feet forward, but not finally adjusted. The work outside the plating had been finished, and the caulking of several seams of the deck, about and near to the waterways, was just drawing to completion, and it was thought that the ship might be ready to sail on Saturday.

A heavy gale, with almost constant rain, from S. to S.W., set in on Wednesday, 5th, and continued the two following days, which considerably retarded the work on board the *Royal Charter*. The continued delay was a great disappointment to all parties interested in the ship and voyage. Yet with all the delay, say a fortnight, or even more, my personal conviction is that, please God, we have a prosperous voyage, with ordinarily favourable winds and weather, the *Royal Charter* will lose nothing in time, and will gain much as to the development of her real capabilities, in which she must have greatly suffered by a long, bad, and difficult passage. Hence her putting back, disappointing or vexatious to many as the incident has been, will, through the good effects of her being lightened of much dead weight and put into good and effective trim, prove ultimately very beneficial to the enlarged and admirable enterprise of the " Australian Navigation Company." These fine ships, it should be observed, adapted for the fastest sailing, suffer more than any other sea-going craft from being overloaded or out of trim! With ships of ordinary build a cargo somewhat overweighted is seldom destructive of their evolutionary properties, and not often dangerous,

but with fine clipper ships of large tonnage the bad consequences are incomparably more serious.

Intimation having been received that the ship was now
about ready for sea, and would probably sail on Thursday,
we (myself, Mrs. Scoresby, and maid,) took leave of our
friends at Torquay, on Wednesday morning, February 13th,
and arrived at the Barbican, Plymouth, with a view to embarkation, at 2 P.M., but the wind blowing very strong from
the south-westward, and the preparations not being expected
to be completed for another day, we took up our quarters,
with much comfort, at the Royal Hotel, where some 15 or 20
of the saloon passengers were yet remaining; the large proportion having resumed their quarters on board.

On Saturday, the 16th, as we were about to sit down to
breakfast, it was announced that the wind was fair (the weather being also fine), and that we must be at the Barbican
pier at 10 A.M. for embarkation. A little before 11 A.M. the
majority of the party with whom for two days we had been
associated went on board.

We now found the ship (with the exception of some confusion inseparable from the re-embarkation of the passengers)
in high order. Above 400 tons of ballast had been removed
from the hold, and the trim of the ship greatly changed and
improved. From drawing about 22 feet abaft and 21 feet
forward, she was now 21 feet abaft and 19 feet forward, looking gracefully buoyant, and "bravely" lifting up her head.
We resumed our places on board with far different feelings
and prospects, under the Divine furtherance, than what prevailed on our original outset. There was an accession, too,
to our numbers in the saloon: Lieut. Chimmo, an officer who
had done good work in surveying the Southern hemisphere,
having, under Admiralty orders, taken passage in the *Royal
Charter* "on particular service."

The day of our sailing was beautiful; the wind moderate
from south-eastward, which pleasantly contrasting with the
previous stormy weather, was very cheering. At 2 P.M. we
steamed slowly out of the Sound, and soon passed fairly out
into the open Channel.

After the pilot left us, about 3 P.M., we steered W. S. W. with a fresh breeze at S.E., and as the ship outran the screw it was at first disconnected, and at 4·30 P.M., whilst we partly lay to, taken up. At 5 P.M. the Eddystone light bore N. by E. ½ E., and at 8 P.M. the Lizard lights were at N.W., about 11 miles distant.

Sunday, Feb. 17 (S. S. E., S. to E., W. S. W., N. W.)— During the night of Saturday—Sunday we continued to steer W. S. W. (by compass), going from 9½ down to 5½ knots; the wind being moderate, sometimes a brisk breeze, light, with frequent showers of rain. A high swell from the north-westward constantly prevailed.

At 10·30 A.M., a portion of the passengers, with the Captain, etc., being assembled in the saloon, I performed Divine Service, with an extemporaneous exposition and discourse on Genesis xxvii. from the first lesson for the day. Visiting the second and third-class departments in the afternoon, I offered, if it were wished, to perform a short service in the mess-room of the latter in the evening; but by an omission a message of invitation did not reach me, and the opportunity was lost.

In the afternoon, 3·20 P.M., the wind failing us, the fires were ordered to be lighted in the engine-room. In 35 minutes steam of 10 lbs. per inch pressure was raised, and the screw being meanwhile lowered and connected, in 65 minutes from the first order the screw was in action. But the wind almost immediately shifting to W. S. W and N. W., and soon beginning to blow strong, we had to disconnect the screw about 8 P.M., and at 9·30 it was hove up. Strong squalls occurring, with heavy rain, we reduced sail, our progress on a S.W. ½ W. course increasing at 10 P.M. to 10½ knots, and at midnight to about 13. Swell heavy from the north-westward.

Monday, Feb. 18.—We made good progress, 13 to 9 knots S. W. ½ W. till daylight. After that time the wind gradually subsided to a light air, and at night to a calm. By Commander Edge's clinometer, which was set in action on the previous evening, we obtained a register betwixt 8·30 P.M. and 8·30 A.M. (12 hours) of 1358 rolls of the ship: maximum angle to port, 22°; to starboard, 10°. Ordinarily the rolling extended from 12° portward to 8° starboard. The heavy north-westerly swell continued, with little alteration, producing much discomfort to the more sensitive passengers.

Tuesday, Feb. 19.—The wind very light all night, and after daylight coming right aft, the ship, under the continued action of the north-westerly swell, rolled rather heavily, making a maximum angle to starboard of 26°, and to port of 17°, and an average rolling, 13° to starboard and 10° to port. Index of clinometer showed, from 10 P.M. to 10 A.M., 2877 movements.

Compasses compared in the forenoon (except standard, not yet being able to be put into use):—

Compass aloft.	Steering.	Companion.
S. W. by W.	S. W. by W.	S. W. by W.
S. W. $\frac{1}{2}$ W.	S. W. $\frac{1}{2}$ W.	S. W. $\frac{1}{4}$ W.

So far as this comparison went, the compass aloft and that at the wheel (which agreed on the first swinging on this course) seemed now to be in their original relation. In rolling, the companion compass usually changed half a point; that is, on rolling to port the needle went half a point westerly, to starboard half a point easterly, from the previous oscillation, or a quarter point each way.

The engine was again put into gear in the afternoon, whilst the wind was fair and light; but the wind soon freshened, and caused the ship to overrun the propelling action of the screw, and produced much vibration and some retardation when the ship's speed had increased to eight or nine knots. At 1·50 A.M. the engine was stopped, and the screw disconnected.

Wednesday, Feb. 20 (N. N. E. to N. by W. and N. W.)—The heavy swell hitherto prevailing from the N. W. (seas rising at times to about 20 feet), with rather a low state of barometer, plainly indicated a hard gale in the middle of the

Atlantic, prevented from reaching our position by conflicting atmospheric currents. The ship, little steadied by her sails, when for the most part the yards were lying nearly square, rolled heavily. Edge's clinometer indicated an extreme inclination of 31° to starboard and 17° to port.

At 6·30 A.M. we came in sight of Cape Finisterre, bearing S. E. (magnetic), about 28 miles distant. The announcement to the passengers of this accomplishment of the passage across and beyond the Bay of Biscay, so far-famed for its cross and tempestuous seas, was an occasion of much interest and congratulation. In our first essay its turbulent characteristic had been abundantly realized, and in the second and successful advance we had had no small discomfort from its turbulent waters; but in neither case, I imagine, was anything specially due to the Bay itself. From the notoriety of the waters of Biscay for turbulency, it is hardly to be questioned that there is something in the configuration of the coasts, or action of the land on the winds, or in modifying the direction of the waves, by which the phenomena is more than ordinarily developed; yet other circumstances, I doubt not, have served to draw attention to the fact, and to give to the occurrence of heavy cross sea in this Bay an estimate far beyond the just proportion.

Of these aggravating circumstances there is one specially to be noted; viz. the effect of the prevailing westerly winds during a considerable part of the year, by which ships proceeding from our southern channels southward are often led to the eastward of their proper course, so as to get hooked as it were within Cape Finisterre, from the natural aversion of the navigators to take a return course to the northward in time to prevent their being embayed. Thus tempted into an easterly position, and encountering perhaps a continuance of westerly gales, much time is spent in knocking about in heavy seas without the possibility of making useful progress; whilst sometimes the ship is driven or urged so near the southern side of the Bay, with a course trending east and west, that when the wind may happen to veer to the northward, a direction which for a ship in the proper track for any country southward would be perfectly fair, the navigator has the mortification of losing the benefit and still enduring, sometimes for days together, an useless or all but useless struggling against adverse elements, heaving turbulent water, and unavailable stormy winds.

The day generally was very fine, with a dry bracing air; the difference of the wet and dry bulb thermometers being 5°. We made good progress, 192 miles by the log in the 24 hours from noon to noon, under all sails, with moderate or fresh breezes from the E. N. E. to N. and N. by W. Sea very blue. Compass comparisons :—

Lat. 42° 20′ N.	Long. 10° 28′ W.
Companion Compass.	Azimuth Standard.
S. W. by W.	S. 36 ½ W.

Compass Experiments.—Up to this time little had been accomplished, or could be, in the way of satisfactory comparison of compasses in relation to the standard; for the pedestal, in the confusion and difficulty of getting at the ballast and restoring the hold, had been inadvertently surrounded by (probably) disturbing influences. Two sheeppens, with vertical iron bars in front and on the sides, had been placed within a few feet aft of the pedestal; and several trusses of hay, bound with iron hooping, occupied a position still nearer to the compass forward. There was every disposition on the part of the Captain and chief officer to make fitting arrangements for satisfactory observations, a disposition which, on my suggestion, was gradually carried into effect, commencing with a substitution of another class of sheeppens, with wooden bars, for those with iron, and ultimately the removal of the iron-bound trusses of hay from their disturbing position.

The altered trim of the ship has proved, as anticipated, most satisfactory. On our former attempt four men were hardly sufficient for the work at the wheel, and now ordinarily, in moderate weather, two were sufficient.

Thursday, Feb. 21 (N. N. W., N. by E., N., N. N. W., variable).— The 21st of February proved to the passengers, as well as others, a very enjoyable day. The morning broke bright and pleasant, and there was no material interruption to the charming character of the weather throughout. All the sufferers by sea-sickness now joined us in the saloon at dinner, and the passengers of all classes appeared on deck. Much pleasant elasticity of feeling prevailed, and the ladies were now, with a smoother sea, at perfect ease, and were seen engaged in much variety of recreation and light occupation—walking the deck, reclining on fold-up chairs, reading, sewing, etc.; whilst below instruments of music, of great diversity, amused their respective performers.

The distance run from noon to noon, with moderate or fresh breezes, was 200 miles on a course (true) of S. 23° W. At mid-day our latitude, by observation, was 39° 14' N., longitude, by chronometer (reduced to noon), 12° 32' W. Lisbon E. by S. true, distant 163 miles. The compass courses steered were S. W. $\frac{1}{2}$ W. to S. W. by W. $\frac{1}{2}$ W.; and after 8 P.M., W. S. W. A bright halo surrounded the moon, of about 23° in diameter. Several attempts were made to determine whether it was in any measure elliptical, but on making the sweep round the moon with the sextant by Lieut. Chimmo, it appeared to be, as no doubt it was, perfectly circular.

Compass comparisons at noon :—

Compass aloft.	Steering.	Companion.	Azimuth.
... ...	... ...	S. W. by W.	S. $42\frac{1}{2}$° W.
... ...	... ...	S. W. by W.	S. $41\frac{1}{2}$° W.
S. W. $\frac{3}{4}$ W.	S. W. by W.	S. W. $\frac{3}{4}$ W.	
S. W. $\frac{3}{4}$ W.	S. W. by W.	S. W. $\frac{3}{4}$ W.	
Sun at noon			S. $8\frac{1}{2}$° W.

The heaviest roll was marked at 13° to port and $7\frac{1}{2}$° to starboard, from noon to noon. The clinometer marked 1300 rolls in the 24 hours.

Friday, Feb. 22 (calm, S. S.W., S.W., W. S.W. to W. N.W.

Lat. 37° 20′, long. 14° 44′ W.)—In the night the wind fell nearly to calm.   At 2·20 A.M. the fires were got up and the screw connected, which increased the rate of progress on a W. S.W. course from two to seven knots.   A southerly breeze sprung up soon after daylight, and towards noon rain began to fall, which continued with little intermission, with a freshening breeze towards night, the rest of the day.   This proved a great and unpleasant contrast to the previous enjoyable weather.

The wind was directly against us for some hours, but the ship breaking off beyond N. W. we wore at 7·30 P.M., and shortly afterwards it veered to W. N. W., and, as indicated by a falling barometer, began to blow very fresh till it reached to a strong gale.

Saturday, Feb. 23 (S. S.W., W. N.W.)—Blowing a gale all night north-westerly, with showers of rain and heavy squalls. After daybreak, set top-gallant sails, and though still blowing a strong gale, the peculiar and splendid qualities of the *Royal Charter* now became conspicuous.   In a gale, which placed two vessels we overtook under double reefed topsails and forecourse, the *Royal Charter* carried her entire double top-sails, single reefed main-sail, mizen and forecourse entire, and main and fore-top-gallant sail !   Yet with all this sail, she yielded ordinarily only in an angle of 10°, whilst her heavily rigged but firmly secured masts never showed the smallest sign of flexure downward from the royal masthead.   Going off the wind a point or two for the sake of speed, she advanced from a rate, at midnight, of eight knots up to a rate of 14 knots, which she very nearly maintained for four hours, and at a rate of 12 or more for about eight hours.   With sea room and a gale of wind the action and qualities of the ship proved magnificent.   With cross seas from south-westward and north-westward in the first commencement of the gale, she behaved admirably.

In this gale we experienced the special advantage of the double topsails enabling us, by the strengthening of two yards, to carry them both throughout the night, and relieving the men from labour and the ship from loss of time and speed, by requiring, in almost the heaviest weather, no reefing. All that is generally needful when, as in their present condition, the foot of the upper topsail is laced to the head of the lower one, is to lower the upper yard down, and leave the topsail to lie in the calm of the other sail, quietly before it. At the same time her motions were quite easy. The maximum angle of rolling from noon to noon, as registered by Edge's clinometer, was 27° to port and 14½° to starboard. The steady heeling, when going wind a-beam 13 to 14 knots, was only from 10° to 11°. The two vessels seen, we passed, as if they had been at anchor, the first indeed, close by the wind, could hardly be making more than three knots.

Compass comparisons :—

Compass aloft.	Steering.	Companion.
S. S. W. ½ W.	S. S. W. ¾ W.	
S. S. W.	S. S. W. ½ W.	
... ...	S. W. by S.	S. S. W. ½ W.
... ...	S. S. W. ½ W.	S. W. by S. ⅙ S.
... ...	S. W. by S.	S. W. by S. ⅓ S.

Note.—Steering Compass about half point more westerly than Compass aloft. Companion Compass about quarter point ditto.

The latitude obtained at noon was 34° 22′, differing but four miles from the reckoning in a run of two days, and longitude, by chronometer (reduced to noon), 15° 41′ W., being about 50 miles of longitude more easterly than the reckoning,* an error, perhaps, partially due to the deviation of the steering compass, which (on south-westerly courses) was nearly

---

* In these two days, the difference of latitude was about 5° or 300 miles; the error of 50′ of longitude was equal to about 40 geographical miles; a deviation of 7° 40′ W. in the steering compass, would be required to produce this error. This change is precisely what was to be expected, for the original deviation of the steering compass on the S. W. by W. course, which the ship was steering, was largely east. This was corrected by magnets at Liverpool. But as the error would diminish by going south faster than the correction, it would be then over-corrected, and give a westerly deviation.—ED.

half a point more westerly than the compass aloft. The position thus obtained, gave the bearing of Porto Santo, S. by W. ½ W. true, distance 86 miles; and of Madeira (centre of the island), about, S. S. W., 122 miles.

For a short time we steered S. S. W. ½ W. in the early afternoon, to give a good berth to Porto Santo on our starboard hand—Captain Boyce preferring with the wind we had, about N. W., to go to the eastward of Madeira, rather than, like other vessels seen, to haul close by the wind, and so make small headway in getting westward of the island. At some seasons, or most seasons perhaps, the westerly route is both desirable and almost necessary for a good passage to Australia or India, but in January, February, and March, Captain Boyce had found an advantage in the easterly route, as generally leading into better winds for some degrees towards the south, and bringing the navigator to the equator in a position, though more easterly than desirable at other seasons, such as 20° to 22° W., yet sufficiently westerly for securing the required benefits of the south-easterly trade winds beyond the line.

After several sharp showers in the afternoon, the sun broke out (about 2·30 P.M.) in resplendent beauty, silvering a large breadth of the somewhat turbulent sea, and exhibiting in the drifting shadows of occasional masses of detached clouds a scene of admirable interest. The waves ordinarily were of considerable elevation, rising frequently to a height of about 20 feet, and every now and then exhibiting an elevation in an unbroken crest of 22 to 24 feet. The colour of the water was characteristically blue. At 8 P.M. our course was altered from S. W. by S. to W. S. W., the northerly direction of the wind enabling us to haul up, without disadvantage to the action of the sails. About 7 P.M. I reckoned we were abreast of Porto Santo, but we did not sight the land, and about 10·30 P.M., we must have been off the eastern point of Madeira, but the sky about the horizon not being clear, we saw no land. The distance run from midnight to midnight, which was near 280 miles, being the greatest run the *Royal Charter* had yet made.

Sunday, Feb. 24 (N.W. to N.N.E.)—We enjoyed a quiet night, with a fair fresh breeze at N. W. by N. to N., making from 11 to 9½ knots betwixt midnight and noon, and a distance by the log of nearly 260 miles within the 24 hours ending at noon. On the whole, our progress, considering that for 24 hours on the 18th and 19th we had made but 60 miles, and that for some hours on the 23rd we had had the wind almost directly against us, was very satisfactory, in having reached the latitude of Madeira in a few hours more than a week.

At noon, we had reached the latitude (by observation) of 30° 45', and were in longitude (by chronometer) 17° 10' W. The day was brilliant and most enjoyable. The air dry and mild. The atmospheric temperature at 9 A.M. was 62°, a true Madeira climate, and an improvement in genial temperature above the probable temperature of the country we had left, of not less than 30°. The first flying fish seen.

The saloon was crowded at morning service, the forward section being occupied by seamen and second and third-class passengers. Unfortunately, it was impossible from the length of the saloon (100 feet), and the interference of the mizen mast, to make myself distinctly heard throughout. But in an address from the first lesson, on the imprisonment of Joseph (Gen. xxxix. 20, 21), I placed myself near the middle, so that the majority of the numerous congregation were brought within hearing. Having announced at the morning service that it was my intention to have a short service in the evening (at 7·30 P.M.) in the third-class mess-room (a large compartment, well furnished with benches, far forward in the ship), I found, on going thither, accompanied by three of the saloon ladies, but little or no preparation, and an appearance of things far from encouraging;—a party of French persons having been just dispossessed whilst playing cards, and others amusing themselves with musical instruments. But the aspect after we seated ourselves at the head of a central table, and obtained a rough appliance of scattered lamps, soon underwent a pleasing change. Numbers of second and third-class passengers, and a good many seamen, crowded in, until we numbered about 100. After reading a selection

from the evening service, with the second lesson (Eph. i.),
I addressed them from the 7th verse as a text: "In whom
we have redemption through His blood, the forgiveness of
sins, according to the riches of His grace."

The attention of the congregation was most striking and
interesting. I had concluded the service with a prayer
adapted to our peculiar condition as adventurers on the wide
ocean, and in anxious solicitude for a blessing on our thus
incidentally assembling, when the possibility of uniting in a
song of praise and thanksgiving was suggested to me. Two
or three serious looking persons immediately adjoining my
position being appealed to, I found it to be quite practicable,
and immediately proposed singing the Evening Hymn. This
was done with a heart and voice so animating that we all
were cheered by it, as we thanked God and took courage. I
then expressed my willingness, if it were acceptable to the
hearers, to meet them again, God willing, on the Wednesday,
for a beginning of a series of week-evening services. On its
being submitted to a show of hands, almost every hand was
promptly lifted up as in hearty concurrence.

Monday, Feb. 25 (N. E., N. E. by E.)—A fine night succeeded by a brilliant and enjoyable day, rendered especially
so by a continuance of fair wind, which, exactly as Captain
Boyce had anticipated, had drawn round to the north-eastward. But the strength of it became reduced, giving a speed
at a minimum of about six knots.

At daylight, Palma, one of the Canary islands, was spied
far off on our port beam.

The change in the weather and climate after so short a
period as nine days from port, was, to those to whom the
voyage was new, a circumstance of marked and extraordinary
interest; with the northerly winds we had for some time
had, indicating by the ocean swell a great extension in that
direction, we were led to infer that, in the country we had so
recently left, a winter of freezing weather was probably prevailing; whilst, with us, we had the genial feeling of an
entrance on summer weather—clear warm sunshine, with a
temperature of 66° at 3 P.M. Passengers' luggage was got up

from below, in order to the bringing out their summer or tropical clothing. In other respects, there was nothing marked in the appearance of the clouds, or sky, or sea; all seemed as usual in the summer of our own country.

About 8 A.M., I got sights for the determination of the local attraction by azimuth with the standard compass, No. 2, supplied me by the Admiralty; the deck around my pedestal not being yet sufficiently cleared (though in progress) for the employment of the best instrument. But the card had considerable oscillation, generally several degrees. A mean of three sets gave the variation $+$ deviation on a S. W. $\frac{1}{2}$ S. course by the compass aloft, of 40° 38′ W., or deducting the the variation of 22° W., a deviation of 17° 38′ W. This did not greatly differ (only about 5°) from the quantity originally due to this direction of the ship's head. At sunset, an amplitude was taken with the same compass, whilst the ship's head, by the companion compass, was S. W. by S. $\frac{1}{2}$ S., the bearing of the sun being W. 2° S. This gave the total compass error 8° 21′, or deducting 22° for variation, the deviation of azimuth compass for the specified course (true magnetic) $= 13° 40′$ W.

At noon, we were in latitude 28° 2′, and longitude (by chronometer) 19° 21′ W. Comparing the compasses furthest aft I had—

Compass aloft.	Steering Compass.	Companion Compass.
S. W. by S.	S. W. $\frac{3}{4}$ S.	S. W. $\frac{3}{4}$ S.
S. W. $\frac{3}{4}$ S.	S. W. $\frac{1}{2}$ S.	S. W. $\frac{2}{3}$ S.

Our favourable progress thus far served well to justify Captain Boyce's anticipations, and to show the advantage of yielding to the north-westerly wind, and taking by a rapid progress the route eastward of Madeira, instead of following the more prevalent usage, as adopted by the vessels we saw, at the expense of two-thirds of their possible speed, of clinging to the wind in order to go the westward of the island.

E

Having in our own case, not only obtained a leading wind
for proceeding westward of the Canary islands but a shifting
of the northerly or north-westerly wind, previously prevailing,
into the N. E. quarter, we gained very strong confidence in
retaining this wind until it should coalesce, probably in an
insensible manner, with the usual N. E. trade wind, into
which at this season of the year, and in the meridian on
which we are advancing, we might expect to enter almost
forthwith.

Universal as the prevalence of trade winds is within certain
belts in the two tropics, and extending frequently beyond
them, the margins of these belts are very irregular, varying
with the longitudes, with the seasons, and with other circumstances depending on yet undetermined laws.   Nor is it
possible in many cases to determine their actual boundaries,
or the place in which a ship first enters within them.   At the
seasons or on occasions when a ship approaches them, through
a region of variable winds and calms, the entrance on them
may be unmistakeable; but on occasions, as with ourselves,
when we advance southward under northerly or north-easterly
winds, we may slip into them without finding any marked or
sensible margin.   There are two indications, however, of the
legitimate trade wind, which serve the experienced navigator
in forming a judgment on the fact; viz. certain peculiarities
in the form and insular detachment of the clouds, and the
prevalent falling of the barometer.   As to the latter token,
we had the marked experience of a previous high barometer
beginning to fall from 30·3 to 30·2, and gradually lower.
This token, noted by Captain Boyce and Lieut. Jansen, was
considered as the best assurance.

The greatest heeling of the ship from 8 P.M. to 8 P.M. of
this day, was 10° to port and 9½° to starboard.

Towards midnight, we had heavy showers of rain, with
light or moderate breezes from the north-eastward.   Our
total progress from midnight to midnight, was about 160
miles, S. W. by W. to S. S. W. ½ W. magnetic.   Variation,
23° to 22° W.

Tuesday, Feb. 26 (N. E. by E. to E.)—A beautiful and

animating day, with clear brilliant evening, the stars becoming more and more lustrous, and the heavens increasingly glorious. About Orion, the studding of bright orbs was so rich and so much advanced in apparent magnitude, that that magnificent constellation was almost overwhelmed amid the surrounding splendour. Ursa Major had become so low, even when eastward of the pole, as to surprise one by its position; whilst Polaris was rapidly descending towards the horizon, and constellations, new to us, were emerging, replacing those so familiar to us, which were about to disappear.

The wind throughout this day, except towards evening when it freshened, was light, seldom giving us, unassisted, a speed beyond 6½ knots. At 4 P.M., steam was got up, and the screw put into action, which gave us eight knots instead of five, and with some freshening of the breeze, 9½ and 10 till midnight. At that time the engine was stopped, and whilst the screw was dragging motionless, our rate fell from 10·2 knots to below 7; and when the screw was raised up, it was 7·3. The slip of screw was about $\frac{1}{13}$th. At 8 P.M. our course (magnetic) was S. S. W. ½ W., and the distance accomplished was 165 miles.

A large ship appeared in sight to the eastward in the afternoon, attracting much observation. She appeared rather light, and was steering some points more westerly than our direction.

At 10·40 A.M., heeling 1° to port—

COURSES.			
Companion Compass.	Azimuth Compass.	Compass aloft.	Steering Compass.
S. 25° 20′ W.	S. 12° 40′ W.	S. 26¾° W.	S. 28½° W.

The clinometer gave for the day, to port 5°, starboard 5°. Barometer, yet falling, 30·3, 30·2. Mean height for this season and latitude, about 30 inches. An amplitude of sun gave the *apparent* bearing by standard compass, S. 90° W., or 270°, whilst the true amplitude for this latitude and date, was 260° 13′; showing a magnetic error, combining variation and deviation, of 9° 47′, or separating the probable variation, 21° 20′ W.; 11° 33′ for the deviation of the azimuth compass on a S. S.W. ½ W. correct magnetic course.

Wednesday, Feb. 27.—Charming weather, chiefly clear, with moderate or fresh breezes; mild but agreeable temperature, ranging from 67° to 69° during the day; the climate of the most enjoyable part of an English summer. Summer clothing was now regularly adopted by the passengers. The saloon deck, fore part, was covered with an awning, which defended the deck from the solar heat, and produced an agreeable and refreshing coolness.

At three o'clock in the morning we crossed the nothern tropic, and consequently entered into the tropical regions.

For the first time since our joining the *Royal Charter*, our provision at dinner-time failed us in the article of roast beef! The admirable service and provision may be worthy of some little notice. The saloon passengers, omitting children, are 54 in number, which, with four officers of the ship, makes an amount of 58 to be provided for. Two tables, separated only by the mizen mast, run down the centre of the saloon, extending together to 50 feet in length. The dinner is served up in silver plate. The bill of fare for February 21st, which is a fair sample of the ordinary provision, ran as follows:—2 joints of roast beef; 2 roast and 1 boiled mutton; 2 roast and 2 boiled chickens; 4 dishes of mutton cutlets; 4 dishes of mutton currie; 1 ham; 2 tongues; 2 roast pork and apple sauce; 2 mutton pies. Vegetables—potatoes, carrots, rice, cabbage, etc. Pastry—4 plum puddings, brandied; 4 rice puddings, 6 fruit tarts, 4 open tarts, 2 sago puddings. Dessert, various.

For breakfast, tea and coffee, with milk from two cows; beefsteaks, mutton cutlets, Irish stew, spiced ham, cold beef or mutton, ham, sardines, rice porridge, stewed mutton, and prevailing articles; with bread, baked on board; and, generally, very capital hot rolls. At tea, on alternate days, the tea and coffee, with toast, plain bread and biscuits, has the addition of marmalade, jam, etc., in ample supply. Luncheon, iu the early part of the voyage, when breakfast was at nine and dinner at four, was a respectable set out, commencing with capital soup.

The latitude, observed at noon, was 22° 7′ N., longitude, by chronometer, 20° 5′ W. The evening was again fine and clear, the stars shining out magnificently.

At 7·30 P.M., as before announced, I proceeded, accompanied by several ladies of the saloon party, to the forward mess-room, where I found a numerous congregation (80 to 100) assembled for Divine service. It was conducted as on Sunday evening, and a portion of the second lesson (Gal. iv.) was taken as a text, on which I addressed a most attentive audience, chiefly of third-class passengers and seamen. Singing had been previously practised, and two hymns from my Seaman's Prayer Book were sung.

The sea was luminous in the wake of the ship, as indeed it had previously been for several evenings; but the phosphorescence, with occasional flashes of light, probably from disturbed medusa, was not greater than or materially different from what we observe on our own coasts.

Sights for azimuth were taken about 8 A.M. Ship's head, by standard compass, was S. 9½° W.; true (mean) altitude of sun = 31°·

Thursday, Feb. 28 (E., N. N. E., N. E.)—A beautiful day throughout. Air temperate, about 70° at noon, with a fine refreshing breeze. Our compass course was S. S. W., except four hours before noon, when it was changed to S. by W. ¼ W. Our speed, under all available sails, varied from 7½ to 10 knots; the distance by log was 208 miles.

The enjoyableness of the trade winds is a fact of sufficient notoriety; but it is only by experience that, as a condition at sea, it is fully to be realized. A fine smooth sea—ship's lateral motion from starboard to port being in our case limited to about 4° to 6° on the 27th, and about 5° on the following day. The *Royal Charter*, indeed, is distinguished for the gentleness of her movements in slight seas, and for the smallness of the angle of her pitching,—neither pitching nor "'send," since our departure from Plymouth, having registered on the clinometer more than $1\frac{1}{2}°$, or, together, $2\frac{1}{2}°$; every one might have appreciated her singular steadiness,—slipping through the water so quietly that, going nine or ten knots, she seemed to be making but little progress. Yet her capabilities as to speed are evidently subdued by the existing weight, much as it has been reduced, of her cargo. For her best capabilities, she is yet evidently too deep by a couple of feet, which being remedied she would probably go from $1\frac{1}{2}$ to two knots faster in light winds, and in her course through the gentler trades.

Compass comparisons were now regularly made twice or thrice a day; but the registry has been transferred to a separate journal, so as to leave only particular results to be here noted. Sights for azimuths, or amplitudes, too, were taken on most days in relation to the standard Admiralty compass, now under convenient arrangements, in a large chest furnished me by Captain Boyce, for readily being placed in position on its pedestal. Unfortunately, however, both the azimuth compasses from the Admiralty proved to be liable to very considerable oscillations whenever the ship was rolling, even within the limited extent of two degrees each way, whilst the steering and companion compasses (fluid or floating) were perfectly steady. In the application of Captain Becher's "Repetition Card," therefore, of which Lieut. Chimmo took the management, there was a decided advantage in testing the two floating compasses. An amplitude obtained by him at sunset, gave the error of this compass, assuming a variation of $20°\ 26'$ W., of $1°\ 20'$ E., on a magnetic course (true) of S. $21\frac{1}{2}°$ W., with which an azimuth obtained by me on the 25th afforded a fairly corresponding result.

Ship under all possible sail; courses steered S. S.W., and from 8 P.M. to midnight S. by W. $\frac{1}{2}$ W.; eight to ten knots.

Friday, Feb. 29 (N. E., N. N. E.)—We had a fresh cool north-easterly breeze until near noon, when it began to fail us. At 1·30 P.M. orders were given (our going being reduced to $4\frac{1}{2}$ to 5) to get up the steam. Though the boilers taking about 20 tons of water were empty, yet the engine was got into operation in about one hour and 25 minutes. Our speed was immediately increased to $7\frac{1}{2}$ and soon to $8\frac{1}{2}$ knots. A large ship which was seen on our port beam steering about S. W. was soon left out of sight. Our course was S. by W. or S. by W. $\frac{1}{2}$ W. The distance run from midnight to midnight was about 180 miles.

At noon our latitude was 16°, and longitude 20° 26', and our position within the Cape de Verde islands about 127 miles east (true) of Bonavista—Cape Verde; Senegal coast bore about S. E. by E., 190 miles distant. In taking this inner passage, not usually pursued by sailing ships at this season, we experienced the light winds and occasional calms to which this region is liable.

The ship having been supplied by the Liverpool Observatory with one of Adie's fine barometers, Lieut. Jansen had it put up in his cabin, and gave admirable and zealous attention to its movements. On our approach to, and entrance within, the N. E. trade winds, its delicate indications, noted for some time every two hours, brought out very beautifully the periodic atmospheric changes—showing a maximum about 10 A.M. and minimum about 2 P.M., a second maximum about 8 P.M. and minimum about 4 A.M. These changes are less in quantity now, the moon being in the quarter, and may be expected to be larger about the full and change.

In the evening, with a brilliant cloudless sky, we had a heavy, and to me extraordinary, dew; the decks at an early hour being as wet as by rain or washing. This, however, did not interfere with the recreation of the first-class passengers in dancing on deck; an exercise which, by perhaps 12 to 20 individuals, was kept up, guided by a diverse succession of available instruments, from dusk until about 9 P.M.

The "zodiacal light" was very distinct. An amplitude of the sun at setting gave the compound of deviation and variation; 9° 40' W. on a course supposed to be true magnetic, of S. 17° W. As the change, if any, of the companion compass (agreeing with that aloft) was as yet very small, we may infer a proximate variation of the needle as follows:—

	°	′
Head by companion compass	198	30
Original deviation on this course	— 1	30
True magnetic direction of ship's head . . .	197	0

	°	′
Add observed amplitude	91	45 W.
	— 6	15 W.
Ship's head by standard compass . ───────	85	30
Supposed true magnetic amplitude	282	30
True amplitude by calculation	262	4
Probable proximate variation . . .	20	26 W.

Though the thermometer was not high we felt the air at night hot and oppressive.

Saturday, March 1 (N., north-easterly, calm).—The trade wind disappointed us greatly, in its extreme lightness and approach to absolute calm. The engine, however, doing us fair service, in relation to its power, we generally made a progress of seven to 8½ knots—of about seven without aid of breeze. Consumption of coals about 14 tons per day. Our course was S. by W. ½ W. to S. S. W. At noon we found ourselves in latitude 12° 56′, longitude 20° 26′ W.

I was much struck with the stillness and strange solitariness of the regions which we now traversed within the northern tropic. Often when I looked scrutinizingly around not a symptom of organic life, beyond the little world comprised on board the *Royal Charter,* was anywhere to be seen! Not a gull was tempted to watch our progress to fish for the oft-dispersed slops of the cooks and stewards. Not a whale, a porpoise, or other inhabitant of the ocean, for long periods together appeared. Nothing could be traced moving in sea or sky, and during much of this day not even a cloud. Sea and wind, in the early day, seemed to sympathise in the characteristic repose. In the high northern latitudes I had been accustomed to a perpetual accompaniment of birds in the air and water, of creatures of various kinds and magnitude, from the great whale down to the animalcules which swam amid and near the Arctic ices.   But here there seemed, for hours and days together, to be an absence of all ; and it was not until towards evening that the solitariness was broken, first by the appearance of a ship in the distance to the east-ward, and a few flying-fish emerging occasionally from the surface water, and after their habitual limited progress again disappearing.

A considerable series of compass comparisons and obser-vations was again made, embracing all the four compasses, at 8·30 A.M., at noon, and at sunset.   They were terminated by an amplitude ; but a swell had risen up from the northward, which causing the ship to roll some three or four degrees each way, greatly embarrassed the action of the Admiralty standard compass.   Some tolerable bearings of the sun were obtained on the approach to sunset ; but at sunset only a tolerable sight could be got, the compass card actually swinging through an extent of 10°.   The floating compass, at the same time, did not oscillate above one-third the like extent.

My attention in these numerous comparisons was directed also to the thermo-magnetic effects of the sun coming alter-nately on the two sides of the ship.   An influence of an observable extent seemed to be indicated.

The evening was quite calm, causing studding sails to be taken in and courses to be hauled up, as impeding the action of the steam power, now doing important service.   It became a matter of some anxiety in respect to our progress and com-fort to ascertain whether our very limited assistance from the N. E. trade winds already experienced had been exhausted for the present.

Reviewing, however, our progress on the completion of two weeks from Plymouth, we find, notwithstanding occasions of light winds, and but small measure of strong and com-manding gales, a result, so far, by no means unsatisfactory.

A phenomenon of astronomical interest was strongly deve-loped in this and the preceding night—the zodiacal light. Commencing below at the horizon, with a considerable breadth and luminousness as of a moderate aurora, it extended upward in the direction of the star Aldebaron, and only disappeared, in progressive alternation, as it approached the Pleiades.

Another astronomical object of much popular interest had likewise come into view two nights earlier than the present, and was seen broadly conspicuous, but far from brilliant, just before I retired for the night—that is, the constellation of the "Southern Cross."

Sunday, March 2 (N. N. E., N. N. W., and N. E.)—A close and oppressive night, chiefly calm, or with a light breeze swelling the sails and yielding an additional knot to the action of the engine, giving us for the 24 hours from noon to noon an average of about eight knots.

The " church" for the devotional services of the day was admirably " rigged" out under awning on the saloon deck. The joiner had, at the Captain's request, prepared a capital desk for books, which was covered with the union jack. All available benches, stools, and folding chairs, were brought up from below, and arranged so as, with the fixed seats, to accommodate the entire congregation, who, from all the departments of the ship, assembled together, to the amount of apparently about 300—a goodly, well-ordered, and most attentive congregation. A few third-class passengers, assisted by some from the saloon, had, under previous arrangement, prepared for singing psalms, and chants for the Venite, Te Deum, etc.; but the latter were, for want of proper arrangement, omitted.

In a discourse, mainly extemporary, I addressed the assemblage from John vii. 17: "If any man will do His will, he shall know of the doctrine." The attention, earnestness, and apparent feeling with many, were most encouraging. There was an impressive feeling that God was present with us, and His presence graciously realized by many a serious enquiring mind.

In the evening, under the Captain's most frank and cordial arrangements, a service similar to that of the preceding Sunday evening was conducted again upon deck. A goodly assemblage once more encouraged our endeavour to sanctify the Sabbath and edify our fellow adventurers. The awnings being allowed to remain, we were defended from the dew and partially shut in within definite bounds. A congregation not greatly inferior in numbers (perhaps equal to two-thirds) of that of the morning gathered themselves beneath the canvass canopy. I could only read a portion of the Liturgy, with one of the lessons (Phil. i.), from the 21st and following verses of which I subsequently discoursed : " For me to live is Christ, and to die is gain," etc. The singing was again pleasantly and devotionally performed, and altogether we had an animating and, as it seemed to be felt, an impressive service.

The day was throughout bright and clear. The awning so shielded the deck from the solar rays that with a gentle breeze blowing in some excess of our speed, or a little on the starboard quarter, the temperature was most enjoyable. Our room on Saturday night had a temperature of 75°; this night, 81°. The highest temperature of the air during the day was 74½°; sea, 77°. Sea blue. No signs of life beyond the ship, except a few flying fish. Where the birds cease to occupy the air the fishes take wing, and assume their province and region !

Monday, March 3 (N. N. by W., N. N. E.)—Variable light winds, rather encumbering the action of the engine than aiding us, though generally fair, so that a great portion of the sails were taken in. A swell prevailed from the N. N. W., and for a time was crossed by a north-easterly swell. The extreme angle of rolling, 7½° to port, 6½° to starboard. The night, to my feeling, was very close and oppressive, the temperature being 81° in the beginning, falling to 79° in the morning.

Two cetaceous animals, of a small species, with a black acutely triangular dorsal fin, were seen about 8 A.M.; they were not near enough for us to judge of their species: their " blowing" was moderate in force, and diffused in its vaporous appearance, very much in the character of the blowing of the mysticetus. The observance of this characteristic result of cetaceous respiration led to a long discussion at the dinner table betwixt myself and passengers near me on the singular

delusion—of such prevalent, almost universal, reception—
that the whale tribe *spent* water instead of a steam-like (air
dense with mucous vapour) exhalation from the lungs; and
it is not a little singular how this erroneous persuasion is
sustained, not only by the individual imagination of thousands
of persons occasionally making sea voyages and seeing whales,
but by writers on natural history of the highest eminence.
For it is difficult to find a work embracing the natural history
of whales in which this error is not propagated, and sought
to be accounted for by the supposition of this ejection of
water being a special provision of nature for the discharge
of water taken in during the process of feeding.   Several
of our passengers, whilst deferring to my knowledge and
experience in respect to this class of animals, affirmed their
unequivocal and decided impression that they had witnessed
the ejection of columns or jets of water on occasion of their
having seen whales in tropical or other regions.   So that one
gentleman said, though he could not resist my knowledge of
the fact, that had he been legally examined on the question
he should have felt no hesitation in making oath to the fact.

The singularly small advantage we have derived from the
N. E. trade winds has been to me and others greatly disappointing.   No doubt our coming southward within the
Cape de Verde islands has caused this; but it is doubtful
whether a circuit westward would, if we take into account
the greater distance, have been any very great gain.   At all
events, our command of steam power rendered the inward
course probably the best, if not specially advantageous to our
progress.   How far we actually enjoyed an advantage from
the N. E. trades is not easily determined, we having apparently slipped into them under brisk northerly breezes.   Our
first north-easterly breeze within the probable range of the
trades commenced about noon of February 24th, when we
were in latitude 30° 45′, longitude 17° 10′ W., which had
certain characteristic signs of a real trade—in the falling of
the barometer and peculiar formations of cloud—on the 25th,
as we were advancing southward of 28° N.

It was only during the four following days that we held

the wind in the strength of a six to a ten-knot breeze. At noon of the 29th the wind began to fail us, and steam was got up early in the afternoon.

Light breezes continued from the north-east and north-ward for a day or two, and then subsided either into unaiding light winds or calms, the effective sailing breeze ceasing, as intimated, on the 29th, when we were yet in 16° of N. latitude, and soon becoming too light even for the purpose of ventilation.

Here, and for a space of some degrees of latitude, up to the effective action of the " S. E. trade winds," our auxiliary steam power, available as we find for a speed of 6½ to 7½ knots in calms or light airs, has its obviously admirable province. But deferring deductions till the fact has been determined in the course of our progress, this general notice of advantage will be sufficient.

Our latitude at noon was 6° 55′ N., longitude 21° 9′ W. The sky now became frequently cloudy ; presently an undefined sheet of thin clouds, an atmosphere filled with vapour ; and in the evening the like state of sky—though it cleared up towards midnight—prevailed; and, whilst it intercepted the otherwise profuse dews, seemed to shut us in within a close unventilated canopy. The coolest and most enjoyable place in the ship was now the " ladies' boudoir," an apartment of small fore and aft dimensions, but as wide as the ship at the stern, and luxuriously cushioned from side to side, and cheered and ventilated by a series of seven cabin windows, overlooking the sea in the rear of the ship. This admirable apartment, which was deserted in cold and heavy weather— the dead-lights being then closed—now comes into luxurious occupancy and use.

Memorandum.—From measurements on a small globe I found the great circle distance from Liverpool to 10° 5′ lat. and 30° W. long., about 67° (latitude measure) ; and from thence 88° + 56° to Port Philip, by the *modified great circle* or " composite route," by the north of the Kerguelen islands, the distance came out about 211°, or 12,660 geographical miles. A full-power steamer would of course shorten the distance by a more easterly course to the Cape of Good Hope, whilst all sailing ships must lengthen the distance by reason of contrary or not directly favourable winds.

There was a feeling of closeness, oppressiveness, as of damp heat, which I do not remember to have before experienced, interfering with the enjoyment of otherwise very fine weather. Our room, small and crowded with the necessary appliances for dressing and changes, etc., was found to be now greatly incommodious—being, indeed, one of the smallest in the whole series of cabins. Though the small glazed port was kept continually open, night and day, and the door, and in part the window, whenever practicable, it was oppressively close, the temperature all night being from 83° to 82°. The effect on the skin was that of perpetual moisture, and not very pleasant clamminess of the touch and sensation.

In other respects our position was agreeable, and, in many particulars, enjoyable. Besides the pleasant association with an intelligent, experienced, amiable, and most unostentatious Captain, we had several associates of much information and agreeable manners. Lieut. Jansen, of the Royal Netherlands Navy, was found quite a master in a large extent of meteorological and hydrographical phenomena; whilst in the ladies' cabin we had several particularly agreeable and accomplished associates. Altogether the saloon society was good and satisfactory, whilst in other departments for passengers I found several very intelligent and pleasant persons, with not a few superior thinking and pious persons.

Our rate of progress, essentially and almost entirely steaming, was from 6½ to 7½ knots. The distance, on a S. W. by S. or S. S. W. compass-course, accomplished from midnight to midnight, was about 170 miles. Sea blue. No external life.

Tuesday, March 4.—Nearly calm, or perfectly so, from midnight to midnight; an oppressive night, and close, damp-feeling, oppressive day. Scarcely a breath of wind refreshed our cabins or the deck. Perpetual unquenchable thirst, and constant clammy moisture of the skin. The sky was generally covered with an undefinable screen of cloud, intercepting all refreshing radiation, or alternating thermometric influences betwixt the sea and the heavens. It screened the direct rays of the sun, but like the curved roof of an oven, seemed but to act as a cumulative screen in preventing the escape of the oppressive condition of the atmosphere. Not that the temperature was actually high, for it scarcely exceeded 81°, but the effect on the feelings was that of languid oppressiveness. In our room at night, notwithstanding an open port and apertures inward for the escape of heat, the thermometer stood at $85\frac{1}{2}$°.

But the men connected with the engine were in a most painful and serious degree the real sufferers. No ventilation could be obtained in the engine-room, wind-sails could not be expanded, and the glazed covering of the hatchway above could only be partially removed. In the coolest part of the engine-room, on the platform just below the hatchway, the temperature was 93°; and near the furnaces, in the position where firemen and stokers had to work, the thermometer rose to 130°, and remained at that point for most of the day and night. Some of the men fainted; but their small corps could not be dispensed with, and they bravely persevered, returning to their trying post as soon as sufficiently recovered. The Captain treated them with much discretion and judgment, as well as kindness. He allowed them altogether fresh provisions, and beer, porter, or weak brandy and water, carefully administered after their period of work.

This trying condition of things fortunately is not irremediable. The arrangement of the sliding windows above the engine-room for being lifted into a vertical position, or entirely removed, would do much; and it would not be difficult, I think, to affix a ventilating trunk and fan—the fan to be worked by the engine—for forcing the air down under such atmospheric stagnation.

The engine, however, small as its comparative power is, did admirable service in pushing the ship forward rapidly through this oppressive belt of calm weather. Our course was S.S.W., and our progress averaged seven knots. The highest temperature of the air was only 83°, of the water 81°.

The evening was like the day, closed in by some slight

showers of rain, cloudy (a thin screen above) with the occasional breaking through of the stars, and a continual stag-
nant air.   Most of the ladies sat upon deck under this
cloudy canopy, lightly covered, and chiefly without bonnets,
which, as the screening of the upper sky prevented the
radiation requisite for the deposition of dew, occasioned
neither inconvenience nor risk.

A ship light was seen on the port beam about 9·30 P.M.
We were then carrying the proper steamer-lights, which
would be visible at a great distance.   She was probably homeward bound.

It was observed by the night watches that the cloudy sky
of the evening had, for several nights in succession, cleared
away before midnight, showing a moderately bright and star-
lit sky.

Wednesday, March 5 (S. E.)—In the morning of the 5th
we found a pleasant change in a steady breeze, recently
sprung up from the S. E. or S. E. by S.   And what was
encouraging, both as to the prospect of progress and comfort, it bore all the characteristics of the "S. E. trade wind,"
though we were yet near 2° northward of the "line."   We
realized in its occurrence the great fact of observation, as to
the moderate prevalent breadth of the belt of calms separating the two trades.   With us, the belt of light winds and
calms, up to this new condition of the trades, might be
reckoned at about 700 miles in breadth.   And, much in proportion to the extent of displacement in particular cases, in
respect to the average position, with the terminal line of the
northern trade, probably will be the general adjustment of
the proximate line of the southern trade.

Our quick transit across the first two belts of light winds
and calms, by the aid of our auxiliary steam-power, is a
result, being one generally to be relied on, which must give
special advantage and popularity in Australian voyages to
ships possessing this appliance, of far more consequence in
passage ships, as a relief from protracted oppressiveness, risk
of sickness, and actual suffering, than even from the gain of
time in the general passage.   But as gain of time and relief

from suffering and risk of sickness must in each case go together, the double advantage cannot but greatly tell upon the success of the plan adopted by the Liverpool and Australian Navigation Company. Passengers on board the *Royal Charter*, in describing the progress on former voyages, cite a variety of cases within their personal experience in which a week to ten days of anxious protracted suffering during calms have been spent in passing one or other of the belts of calms and variables, which we accomplished in a third part of the time or less. In a voyage made in 1834, in an old East India ship, by a gentleman in the saloon, he describes their detention about this parallel during 19 days of light winds and calms. Many emigrants of the working classes were on board; fever of the typh id kind, or ship fever apparently, broke out during that detention, and spread amongst the passengers and others subsequently, so that 22 bodies, including that of a fine young man among the officers (the third officer), were committed to the deep.

During the forenoon and early afternoon we had a steady, moderate, and refreshing breeze from the south-eastward or eastward. Below, indeed, the ship was necessarily warm. Our room, with all possible ventilation, was at $85\frac{1}{2}°$ at midnight, 4th—5th, and in the morning, $81\frac{1}{2}°$. The highest temperature on deck in the shade was $82°$, and that of the sea the same. As on many preceding days, nothing beyond the ship, except occasional flying fish, was to be seen. The sea, as to its visible surface, was like a barren desert. Within its waters, indeed, a variety of minute creatures might be seen, on taking up a portion in a glass; they were generally of the animalcule size and kind, comprising many varieties of radiata. Those brought up by the bucket were very minute; but from the degree of luminosity of the sea at night, and the frequency of the flashes of phosphorescent light, or spots of considerable magnitude, it was obvious that there was a good supply of mollusca of larger size, no doubt many of the well-known kinds of radiata.

The change in the character of the clouds on coming into the verge only of the south-eastern trades was striking and

F

characteristic. Within the equatorial belt of calms (the
"doldrums" or "equatorial doldrums" of the sailors), we
had a prevailing semi-leaden canopy of cloud, of varying
density indeed, and perceptibly patchy, yet not reducible
into any ordinary species. The effect of this, as I have
remarked, is cumulative of heating oppressiveness and lassitude, not however indicated by the thermometer, which,
as well as the barometer, is usually lower here than in the
tracks beyond, especially the particular trade wind region
being then traversed by the sun. The incongruity, indeed,
between the thermometric temperature and that of human
perception, is similar, only converse, to what we find betwixt the thermometer and sensation under a damp chilly
atmosphere in England, in November, or December, or
January, when the chill affects us much more than an actual
depression of temperature of very many degrees, with a
clear elastic state of the air.

The change in the character of the clouds, just referred
to, is as conspicuous as the change in elasticity of feeling
is marked and enjoyable. Whilst clouds of the cirri kind
may be seen in the upper cloud region of the atmosphere,
the cumulus, travelling deliberately in picturesque effects
of shadow on the waters, in separate masses, till forming,
especially at sunset, a dense irregular cloud-bank on the
horizon, becomes the characteristic formation. This gathering of the evening clouds in the horizon has been much felt
in the many disappointments it has occasioned in my attempts
to get evening amplitudes; shutting out the sun, on the
guidance of which I calculated for compass errors, unless I
had prudently taken bearings of that orb when some degrees
high, which, under great meridian altitudes, can easily be
adjusted without calculation, within the limits of error inseparable from ordinary observations.

The engine, which in the intervals of calm or light airs
had given us a prevalent speed of 6½ to 7½ knots, ceased to
be needful as the breeze freshened from S. S. E.-ward, but
was kept on, in aid of our staysails only (all the other sails
being furled) until 4 P.M., when the engine was stopped, and

the screw raised up. As we gradually spread our canvass, we increased our speed (on courses varying from S. by W. to about S. W. by S.) up to 9 or 9½ knots.

Being in a latitude only 85 miles north of the equator, all the passengers and seamen too were on the alert as the evening advanced, and we pretty rapidly came up towards this boundary of the two hemispheres of our globe. A floating light was suddenly called to our notice as it passed the side of the ship, by a shout within board, and it blazed conspicuously on the surface of the sea, where it might be seen for a mile or two astern. This was the usual forerunner or commencement of the ceremonial of crossing the line. But Neptune was restrained by authority, so that even the harmless and more dramatic part of the ceremonial was not allowed to be carried out, because of the difficulty of closing the performance when it might become offensive.

At 9·50 P.M., whilst all passengers were on deck watching for the memorable transit by the ship, we performed the crossing. The pleasurable emotion elicited by the Captain's announcement was increased by the transition of feeling produced by the now breathable and elastic air, and the really enjoyable effects of the free refreshing breeze upon deck. The stars shone brilliantly; the night was light with their influence, and the brightness of the milky way, near the Southern Cross (westward, or rising to the right), was for the first time conspicuous and notable.

March 6 (S. E. to S. S. E., variable).—Though we had entered, as generally believed, the S. E. trade wind, it was yet, as is usual at the commencement, somewhat variable in direction, and still more in strength. At 1 A.M., the wind being very light, the screw was connected, and the engine put in motion. This at once increased our speed from about 3 to 7½ knots. The wind freshening in the evening, and becoming more decided in its prevalence, the engine was again disconnected, and after 9·30 P.M. we proceeded under sail alone. Our course was S.W.; and distance accomplished from midnight to midnight about 184 miles. The latitude at noon was 1° 16′ S., longitude 22° 40′ W. The greatest heeling was 10° to starboard, to port 7°, average heeling 2° to 2½° by clinometer.

Another clear and enjoyable evening, that is, enjoyable on deck, where a fine refreshing breeze was blowing over a numerous assemblage (most of the saloon ladies) on the poop deck. Though there was some dew, no extra dress was adopted, and most of the ladies sat out until 10 P.M. in fold-up chairs, without bonnets or other covering for the head. It was only below where any inconvenience was now suffered from the heat, and though few of the "after cabins" had a temperature above 85° or 86°, yet the effect was so felt by many as to deprive them for several nights of the refreshment of sleep. Yet, comparatively, in relation to other ships, and other departments of the *Royal Charter*, we were well off; nor happily did those in the lower decks, exposed as they were to a much higher temperature, suffer in health.

Friday, March 7.—Generally throughout this day we had a fresh and pleasant south-easterly or S. S. E. breeze, a plain and characteristic trade wind, giving us under all sails that could stand within seven points of the wind, a rate of progress of six to ten knots, on courses (magnetic) varying from S. W. ½ W. to S. W. by S. The weather was generally bright, with passing masses of cumulus cloud, occasionally proving to be reservoirs of slight squalls. The sun was clear till within 2° of the horizon, when it set as usual in a low bank of cloud. Compass comparisons, however, were satisfactorily obtained, as well as an azimuth and sunset amplitude, with the standard compass No. 1, and with Captain Becher's repeating card, in connection with the companion adjusted compass, by Lieut. Chimmo. This gave the error of the companion compass (deviation and variation) 14° 35′ W.; ship's head being S. by W. Sea blue.

Several shoals of flying fish were seen before breakfast, and others subsequently. Their flight was low, and to windward, the whole shoal pursuing a parallel direction on the port side from the ship. Except as to colour, the appearance being of a sparking grey, as reflecting the rays of the eastern sun, the shoal resembled the rising and first flight of the arctic "roaches" (*alca alle*), which from their short wings, fly for a considerable distance near the surface of the water before they can attain a thoroughly aërial position.

Our latitude at noon was 4° 1′ S., and longitude, by chronometer, 24° 21′ W. We were now in near approach to a vertical sun, the apparent altitude obtained by the observers towards the south being about 88½°. Our shadows, therefore, had become reduced to pretty near a transverse section at the shoulders; whilst the shadows of the ladies, having on a popular wide brimmed hat, were represented by a circular disk, except where the amply expanding dresses happened to distort the circle by spreading incidentally beyond it.

The day as usual was spent by most of our party on deck, within the refreshing influence of the fine fresh breeze, and screened from the sun by awnings. In the evening, the ladies, with some of the gentlemen, were gathered in groups, lounging on their low commodious chairs, and amusing themselves and each other by conversation, songs, and stories. In the songs, Miss Holmes, with her beautiful voice, cultivated and admirable style of execution, afforded us a frequent and most enjoyable treat.

Dark threatening patches of cloud occasionally passed over the splendidly star-lit sky, but caused no inconvenience. The warm temperature of the air rendered protection even from the dew, which was but moderate, unnecessary. The sky from the horizon below the "Southern Cross," adorned as it is by the great luminosity of the milky way, and the rich assemblage of stars in the direction of Orion, and beyond it, presented a most glorious spectacle, and an emphatic demonstration of the Eternal Power and Godhead of their Great Creator. The glory of the scene, too, was rendered more attractive by a very minor but interesting phenomenon, now of frequent observation, that of "shooting stars." The luminous projectiles, whilst generally of ordinary magnitude only, were sometimes of higher and more startling impressiveness by reason of their apparently large magnitude.

The night was close, and our rooms somewhat too warm (83° to 82°) for proper refreshment with those unaccustomed to a tropical climate. Many amongst us could sleep but little, and some suffered much inconvenience and deprivation. Nevertheless, we had but little to complain of, and much cause for self-gratulation and thankfulness. The surgeon's report of the state of health of our numerous associates, including but one case of serious illness, and that being one of a female who was ill when she embarked, yielded special cause for praise and thanksgiving to Him who has hitherto so graciously preserved us. This condition of things, under the Divine blessing, is much to be ascribed to the generally good ventilation, and the great height of the 'tween decks (8 feet), and our rapid transit through the region of calms.

Saturday, March 8.—Our room was at $82\frac{1}{2}°$ to 82° during the night. On deck, by day, we had a general temperature of about 80°, never rising in the shade higher than 81°. One solitary gull appeared for a short time in sight; and yesterday, a stormy petrel, also a solitary bird. Flying fish were seen in the early part of the day, flying generally to windward.

The wind throughout this day was prevalently from S. E. by S., enabling us, sailing about seven points from the wind, to make generally but a S. W. course, which, with but little more than a point of westerly variation, yielded no better a direction than S.W. by S. true. Ordinarily, at this season, a far better direction of the trade wind is experienced; so that we are likely to be forced into much nearer proximity to the Brazilian coast than we expected, before we can get the ship's head to point fairly towards the land of our destination. Our rate of sailing was from $7\frac{1}{2}$ to $9\frac{1}{2}$ knots; even at the latter, or higher speed, the small disturbance of the water by the passage of the ship, is a striking circumstance to those, like myself, acccustomed mainly, in long sea voyages, to ships of the old plan of building. In these, as is well known, a rate of nine or ten knots is generally announced by the noise of the disturbed waters as they pass the quarters of the ship, and by the broad belt of chequered or froth-marked surface passing the broadsides from stem to stern. But in ships of the clipper build, a 14 or 15 knot progress is but moderately marked by such signs of disturbance.

To us, inhabitants in previous life of northern latitudes, a new phenomenon broke strikingly on our attention, viz. the appearance of the sun performing its daytime progress to the northward, with the astronomical peculiarity of meridian observations being taken with the face of the observed turned towards the Arctic pole ! Every one knows the fact; but its first realization, nevertheless, is a striking incident.

The evening was as usual, and spent as usual by most of the passengers in our department. With those on deck, as already described; with those below, in playing chess, backgammon, and cards. The sun, as had long been usual with us, went behind a bank of cloud when yet some degrees high. A glimpse for amplitude, however, was caught; clinometer maximum, 13° S.; port, 0. Mean heeling, 7° S., at 8 A.M.

Sunday, March 9 (S. E. by S.)—The wind continued disappointingly shy, allowing us but rarely to make a better compass course than S.W., with a speed of $7\frac{1}{2}$ to nine knots, though a swell considerably more easterly than the wind seemed to indicate a better direction eastward of our position. A large ship seen at daylight, a point or so on the starboard quarter, excited much interest; but she did not come nearer than four or five miles, and steering closer to the wind than we did, she gradually fell astern, and, as night approached, disappeared somewhat on the windward quarter.

The day was fine and pleasant, and favourable for our usual devotional services. The church was again " rigged " under the poop awnings, the deck and windward bulwarks covered with flags. A large congregation again assembled. To our former psalmody was now added, with a view essentially to my personal relief, the chanting of the Venite, Te Deum, Jubilate, and Gloria after the Psalms. The effort of reading, and then addressing the congregation, in this open air position, and under the relaxing influence of a week's previous hot weather, proved very fatiguing.

The subject of the address was taken from the first lesson for the morning (Exod. iii.) In the evening, at 7·30 P.M.,

we again assembled under the principal awning, a numerous
and attentive congregation, comprising a large proportion of
seamen, and second and third class passengers. The second
lesson (Colossians iv. 2, 3) afforded the subject of discourse.
The service was truly (to me, and I trust, to many) satisfactory. The serious attention and solemnity were very impressive on the general sympathies.

Except early shoals of flying fish, I saw no living creature
externally all day. The sea was as usual of late intensely
blue, its phosphoresence in the night only moderate. The
evening was rather cloudy, but stars in intervals of clouds
shone brightly. The black patch below the Southern Cross,
with another near it, were very conspicuous, as well as the
Magellanic cloud, about 20° to the eastward of it (10 P.M.)

Monday, March 10.—Night much as recently, temperature of our room 81½° to 80°. Fine morning. Wind a little
improved in direction, allowing us to lie in our somewhat
broad angle with the wind S.W. by S., approaching to S.S.W.
Another ship appearing, about three or four miles on the lee
quarter, excited great interest among the passengers. Being
nearer us in the afternoon, signals were exchanged: " her
number," giving the name the " *Greenock;*" and subsequent
replies to our enquiries, "from Glasgow, bound for Sydney;"
" 42 days out;" and, courteously and voluntarily, " we wish
you a good voyage."

At noon, we were in latitude 12° 44′ S., and longitude
29° 44′ W., a position approaching a line of much interest
and importance in my magnetic researches, viz. the Magnetic Equator. For we were now but 70 miles or thereabout
from Sir James Ross's determination in 1839, of a portion of
the separating line of the two magnetic hemispheres. A
quotation from his Voyage (8vo. edit.) vol. i., p. 21, will convey in its clear and condensed form, the leading scientific
facts derived from his admirable observations:—

" The geographical position of the magnetic equator where
we crossed it (in December 1839) was lat. 13° 45′ S., long.
30° 41′ W. The regularity as well as the rapidity with which
the alterations of dip occur, is also worthy of notice. At

280 miles north of the magnetic equator the dip was 9° 36′, showing about 2·05 minutes of change for every mile of latitude; at 290 miles to the south, the dip was 9° 51′, or about 2·03 minutes for every mile of latitude. It is to be remembered that this large amount is limited to the region of the magnetic equator; near the poles it requires an approach of about two miles to produce an alteration of a single minute in the dip.''*

At 8 P.M. we were in precisely the same parallel (lat. 13° 45′), and in the meridian of 30° 45½′ W., being only four miles to the westward of Sir James Ross's position. The equatorial line, indeed, has no doubt made some change in its position during the interval of 16 years; but the change cannot amount to any material quantity. General Sabine, the best authority on these subjects, in a tracing of the isodynamic and isoclinal lines, with which he kindly furnished me, from a chart in course of preparation for publication, places the magnetic equator, or line of no dip (on the meridian of 30 W°) in about 14° S.

As forthwith the force of terrestrial induction will begin to be in antagonism with the special distribution or lines of the ship's organic magnetism, so far, at least, as to *tend* to invert the polarity of the vertical force in the iron, we may soon expect to witness a change in the hitherto remarkably regular action of the compasses, especially as to the general agreement of three compasses on and above the poop deck. But such change, it is probable, will not be very conspicuous until the ship comes to sail on a south-easterly course, when her *head* will be nearly at right angles with the magnetic line of our progress from Plymouth to our present position, and in the case of the *Royal Charter*, the vessel will, for the first time, be sailing in a direction opposite to that in which her keel lay whilst building. It is just possible, as in the cases of the *Great Britain*, *Sarah Sands*, and perhaps a few

---

* These results are derived from the formula, *tan dip* = 2 *tan lat.*, which is nearly correct near the magnetic equator, if the latitude be measured from that line, and *tan co-dip* = *tan co-lat.*, which is nearly correct near the magnetic pole, if the co-lat. be measured from that point.—ED.

others, that the *Royal Charter's* compasses may happen to
be in positions of equilibrium amid the ship's changing magnetic forces, and so one or other of the compasses on deck
may possibly continue a useful guide; but the probability is
adverse to such a result. Our regular and careful observations, taken daily, and two or three times a day, however,
cannot fail to inform us instructively what effects are produced by such great changes in the direction and force of
terrestrial magnetism.

In regard to the changes in the magnetic *force,* it may
here be noted that Sir James Ross crossed the line of least
intensity (first) in lat. 19° S., and long. 29° 15′ W., being
200 miles more to the northward than previous observations
had led him to expect (vol. i. p. 22). Again, in lat. 21° S.,
and long. 15° 30′, the same line was recrossed (p. 27). And,
February 9th, 1840, the line of least intensity was a third
time crossed, in lat. 21° S., long. 8° W. General Sabine,
in his admirable discussion of Captain Ross's Magnetical
Observations (Phil. Trans. 1840, pl. v.), intimates that, whilst
the general direction of the line of least intensity, drawn
from observations of Dunlop, Erman, and Sulivan, corresponding nearly to the epoch of 1825, is consistent with that
deduced from the observations of Captain Sir James Ross
in 1840, yet its earlier position is everywhere three or four
degrees south of that which would be inferred from the later
determinations. Its average northerly movement, therefore,
during the 15 years from 1825 to 1840, appears to have
rather exceeded annually 13 miles (p. 30).

Assuming the same ratio of northerly march for the
interval between 1840 and 1856, or 16 years, we should have
the place of the line of least intensity, in the track of the
Royal Charter, in lat. 15° S.; a position which now we were
very nearly approaching, and which we reached about 5 A.M.
of the following day, the 11th.

A difficulty in respect to this deduction appears, however,
in referring again to General Sabine's isodynamic curves, in
which the line of least intensity, on the meridian of 30° W.,
is placed in about 23° S. latitude—reckoning the same

amount of change (4° of lat.) in 16 years, but towards the south instead of the north. It may be that subsequent and more extended observations on the magnetic isodynamics of the globe may have required a recasting of the series of curves, or a revision of the deduction in respect of change quoted from Sir James Ross's Voyage; but here, whilst I write, I can have no means of certification of the actual facts of explanation of the embarrassing difficulty.*

The communication with the *Greenock,* and the fact of her fast sailing and steady advance upon us, produced no

---

* On this passage General Sabine has favoured the Editor with the following note:—

"The line of least force corresponding to the earlier epoch (1825) was first delineated in a map published in the Philosophical Magazine, for February, 1839. From thence it was copied into plate v. of the Philosophical Transactions for 1840, referred to by Dr. Scoresby. Of the observations employed in its delineation, 'the greater part,' as stated in the Philosophical Magazine, 'were made between 1817 and 1836.' Sulivan's observations in the *Opossum* packet, in 1839, were consequently not included in the authorities for that map. This line was again repeated in a faint dotted line in plate iii. of the Philosophical Transactions for 1842, for the purpose of shewing that the line of least force derived from the observations of the *Erebus* and *Terror,* in 1840, was generally 3° or 4° of latitude north of the same line derived for 1825 from the earlier observations; and although an inference was thus drawn that the line of least force appeared to have changed its position intermediately by a movement towards the north, no attempt was made to assign an annual rate of change in its position. Indeed it was expressly said (Philosophical Transactions, 1842, p. 35), that 'many of the observations from which the line of 1825 was drawn were inferior in precision to those of 1840;' and I conceive that it would be quite unsafe to attempt to derive an annual rate of secular change from the comparison of the lines so drawn.

"The tracing of the isodynamic lines with which Dr. Scoresby was furnished by me prior to his departure for Australia was taken from a map then in the press, and since published in Johnstone's Physical Atlas, plate 23. It corresponded to the epoch of 1840 (not to 1855, as seems to have been inadvertently supposed by Dr. Scoresby). The authorities on which I chiefly relied for the position of the line of least force at that epoch, about the meridian of 30° W., were those of Sulivan, in the *Opossum*, in 1839 (Phil. Trans., 1840, art. iv. p. 143), and those of the *Erebus* in 1840 (Phil. Trans. 1842, art. ii). From those observations the position of the line of least force in 1840, about the meridian of 30° W., could scarcely have been placed in a more southerly latitude than 23° S., or in a more northerly latitude than 21° 30' S. Mr. Dunlop's observations in 1831 would indicate a more southerly position at that date by 3° or 4°; but the interval is too short, and the data, especially the earlier ones, not sufficiently precise to justify the conclusion of an annual rate at which the line may have been changing its position."

small excitement among the passengers. Hitherto, though we were perfectly aware that the *Royal Charter's* sailing qualities could not be favourably elicited in light winds, she had never been passed by any of the ships seen. But it was abundantly evident that without a great increase of wind, of which we had no prospect, or the putting into action our auxiliary steam-power, we must soon endure the mortification of being left behind. The ship's rate, however, having gone down to about 6½ knots, the screw was connected, and at 2·30 P.M. the engine was set in motion. We at once started ahead with an addition of two increasing to three knots to our previously attainable speed. We of course shot rapidly ahead of our competitor, so that in an hour she was left about two miles astern, and before sunset, the weather not being very clear, was altogether out of sight.

The evening was again fine and agreeable, and the sea particularly smooth. We might have been in a land-locked sea in moderate weather. The stars were bright; the moon, in the early evening, of a pure silvery resplendent white. I observed, whilst sitting on the port side of the deck, about half-a-dozen shooting stars. One of them was a conspicuous and striking meteor, descending from the zenith in a slightly oblique azimuth towards the east. It was visible through an arc of about 60°, seeming partially to discharge in the progress of its descent, and finally to disappear like the last of a bursting rocket.

Dancing on the poop deck was perseveringly carried on by about a third of the saloon passengers from 7·30 to 10·30 P.M. The warmth of the weather, modified by a light breeze, did not materially interfere with the spirit of the enjoyment.

The sea had been purely blue for some days, indicating no great supply of animalcular existence, an indication which was corroborated by the fact of the slight phosphoresence of the disturbed waters in the wake of the ship, and may explain (or at least is consistent with) the remarkable absence of external animal life.

Tuesday, March 11 (S. E. to E. S. E.)—Our cabin was again at 85°, falling to 80° during the night, and there being little wind, this temperature was found to be productive of much discomfort and lassitude—the lassitude almost of the "doldrum calms,"—occasioning indisposition to mental and physical effort, so that every contemplated voluntary occupation or project falls behind. The ladies, however, seem to overcome the tendency in particular efforts, keeping up the habit of dressing for dinner, and the pleasant characteristics of polite society. A light breeze, becoming better, however, for our southerly progress, still prevailed. "I never saw such light trades," was the substance of exclamations frequently escaping the lips of our more experienced tropical navigators. In truth, neither the N. E. nor the S. E. trade winds had at all fulfilled our expectations. On the contrary, they had thus far greatly disappointed us, both because of their comparative feebleness and the very limited continuance of the N. E. trade.

At noon, still under steam and sails, and now making generally a S. S. W., S. by W., or even more southerly course, we took observations in latitude 16° S., longitude, by chronometer, 30° 30′ W. In 11 days we had made 32° of latitude, with generally moderate or light winds, and passing through the difficulties (as ordinarily experienced) of the intermediate trades calms. This, on the whole, we could not but consider very good work, notwithstanding the westerly meridian into which the scant south-easterly trades had forced us.

The atmospheric temperature at 8 A.M. was 81°. Sea very smooth. Heel of the ship about 5° to starboard; greatest angle for the day about 8½° to starboard, to port 0°. Sea blue. No external life. A beautiful, and under the awning, enjoyable day on deck. Scarcely any phosphoresence at night. Dancing on deck, as on the preceding evening.

Wednesday, March 12 (E. by N., E. N. E.)—Our room was again at 85°, falling to 80° during the night. The breeze freshening in the night, the screw was detached at 1·30 A.M. Our speed ranged from eight to eleven knots, steering under all sail about eight to ten points from the wind, and making a compass course (with 9° westerly variation) of about south. In a squall about 11 P.M. the ship went about 15 knots. The clinometer gave an average heeling to starboard of $7\frac{1}{2}°$, and a maximum of about 9°.   The air and surface of the water equally devoid of visible life as in days past.   At noon, Trinidad, about S. E. $\frac{1}{2}$ S. true, distance 95 miles.   We regretted not being able to sight the island.

Rough magnetic experiments on the polarity of the gunwale-edge of the upper plating, outside the poop, gave (two inches off) forepart poop, opposite the companion—(1), starboard side, slight northern polarity; (2), opposite the wheel, neutral; (3), opposite after capstan, slight northern; (4), aft at corner of curve of stern, moderate southern.   On the port side in the same positions there was generally a vigorous southern polarity.

The day was charming, and very enjoyable.   I made experiments on the dip and magnetic force of the ship, just abaft the pedestal of the standard compass, 4 feet 6 inches above deck.   The direction of the needle had, under the united changing influence of terrestrial dip and the ship's polarity, become very marked and rapid.   On the 8th of March, when in lat. 6° 58′ S., long. 26° 18′, the terrestrial dip, according to General Sabine, being about 14° N., the dipping-needle gave the following results.   Ship's head by standard compass being S. S. W. $\frac{1}{2}$ W. :—

Face *West* ...... dip 35° N. ...... oscillations $3^s.5$ each.
„   *East* ...... „   34° N. ......   „ $3^s.3$ „
Ship heeling $7\frac{1}{2}°$.

On this day, March 12th, latitude 19° 30′ S., longitude 30° 38′ W., dip (according to General Sabine) being about 10° S., I observed near the same place—

Face *West* ...... dip. 6° N. ...... oscillations $3^s.125$ each.
„   *East* ...... „ $7\frac{1}{2}°$ N. ......   „ $3^s.$ „

In the evening, it being impracticable for me, because of the heat, to have a religious service below in the third-class mess-room, arrangements, under the Captain's kind furtherance, were made for Divine service on the poop deck. A large congregation from forward, and second-class, with

most of the first-class ladies, assembled. A portion only of
the Liturgy was read, and, as usual, an address given from a
section of the second lesson (1 Thess. iii. 12, 13).

Thursday, March 13 (E. by N.)—The first reduction of
the temperature below was pleasantly experienced on the
night of the 12th—13th. Our room, commencing with a
temperature of 80°, fell to 78° before morning. The greatest
heeling at night was 13½° to starboard; mean at 8 A.M. 8°,
varying from 7° to 10° during the day.

We had a fine fresh gale throughout from the E. by N.,
by which, under all available sails, with two topmast studding-sails, steering well free, with the wind a little abaft the
beam, we made a varying progress of nine to 12½ knots, but
incidentally in squalls (which occasionally occurred in rain
showers) the ship went probably 15 knots or upwards. Our
course was generally S. to S. ½ E., Captain Boyce preferring
to get rapid southing, rather than clinging by the wind to
make easting, at present, with loss of speed. He was not
anxious to steer more easterly than S. by E. (had that course
been conveniently practicable), whilst northward of 27th or
28th parallel of S. latitude.

At noon, we observed, in lat. 23° 30' S., long. 30° 20' W.,
just crossing, the tropic of Capricorn. Our passage through
the tropics thus gives a most satisfactory progress. The distance of 47° of latitude, and 10½° of westing, has been accomplished in 15⅓ days, a period sometimes spent by sailing
vessels in the passage of the narrow intervening belt of calms
betwixt the two trade winds. An analysis of the passages
through the tropics of a number of sailing vessels of the
ordinary class, chiefly Dutch, proceeding much in our own
track, was made by the talented Dutch naval officer amongst
our passengers, Lieut. Jansen, and published in Utrecht, by
J. Van Googh, in 1855. These passages are arranged according to the season of the year in monthly tables, showing
in the case of each ship registered, the time occupied in passing each 5° of latitude, from 45° N. to 35° S. Taking the
cases for February and March, corresponding to the time of
our passage of the tropics, it appears that on the average of

10 or 12 ships, the accomplishment of this distance occupied
about 26 days. But taking the comparison for the whole
distance from 45° N. to 35° S., the advantage, if it please God
that we continue to be prospered as hitherto, will be vastly
more in our favour.

During the afternoon, I obtained an interesting and satisfactory series of experiments with the dipping-needle, on the
direction and force of the total magnetic force of the earth
and ship, in the several places about the deck where the
instrument was tried. These were at the standard compass
pedestal, on the forecastle, and on either side of the " after"
skylight abaft the mizen mast. The change of the apparent
dip in 18 hours, at the pedestal station, was found to be from
$6\frac{3}{4}°$ N. to $\frac{1}{2}°$ N, or $6\frac{1}{4}°$. Or reckoning from the evening of
Wednesday, 12th, embracing a change of geographical lati-
tude of 8°, the change in the apparent dip, with the ship's
head within a point of the same direction (S. $11\frac{1}{4}°$ W. and S.)
was $15\frac{1}{2}°$.*

A Dutch brig, steering in an opposite direction, passed us
within about a mile to leeward. We gave our number, but
it did not seem to be understood. We could read the name
Thetis on the stern. She was under reefed topsails, without
top-gallant sails, whilst we were carrying royals,—we not
dipping these lofty sails even under the smart squalls which
occasionally occurred.

A halo encompassed the moon, in the evening, of a diameter
such as to pass obliquely between a and γ in Orion, touching
the former with its outer edge and the latter with its inner.

Friday, March 14 (E. N. E. to N. E.)—A somewhat cooler
night. In our cabin, with the port window shut, the thermometer was 82° to 78°; the minimum temperature of the
air at night being 73°. At 8 A.M. the maximum of the
clinometer registered 18°, subsequently going to 20°, and the
mean heeling $7\frac{1}{2}°$ to 9°. The dampness of the air, characteristic of the tropics, still continues. The difference of tem-

---

* This is about the amount which is to be expected, being nearly twice the
amount of change in geographical latitude. See *ante*, p. 73 n.—ED.

perature in the wet and dry bulb thermometers has not, I believe, exceeded 5° throughout.

Having had a brisk gale from noon to noon, described in the ship's log—from the great stability of the ship under all her canvass, including royals—as a " fresh breeze," we made a splendid run, averaging about 12 knots, and actually accomplishing a difference of latitude by observation of two meridian altitudes of 280 miles or 4 degrees 40 miles.

Under this run, with the wind two points abaft the beam, and all available sails, including two topmast studding-sails and two royals, the *Royal Charter* acted bravely. It was a splendid sight to watch this noble ship during a squall, heeling over but 10° or 12° to starboard, with main and fore royals and all other sails in full spread, and performing with powerful effects their propulsive action, until an occasional speed was attained of 14 or 15 knots; yet in accomplishing all this not a mast or yard or boom appeared to complain— not a rope-yarn (to adopt the nautical hyperbole) seemed to be strained—not a spray nor drop of water came on board (the sea being very smooth) from stern to stem—nor were there any of the ordinary signs, so conspicuous in the old class of sailing vessels, in the wide disturbance of the water and in the waves made along the quarters by the ship's progress, marking the force on the one part and resistency on the other in the accomplishment of a maximum speed! These characteristics of the large clipper build of shipping were beautiful and exciting novelties to a sailor of the old school of shipping. In the merchant service of the early part of the present century, and even up to the first quarter of it, a speed of 11 knots was reckoned capital going, and 12 knots I never saw attained; and in the royal navy of the time, 12 to 13 knots constituted the average maximum speed, the latter nearly the special maximum—a speed, too, rarely to be accomplished without straining of gear and spars, if not of the very fabric of the ship's hull. But here, in the *Royal Charter*, we have 11, 12, or 13 knots entirely unmarked by sign of effort, much less of straining; whilst in the more rapid speed in the squalls, the motion was easy and graceful.

G

Birds, excepting a rare example of a solitary one, now
appeared in twos and threes, for the first time (as far as I
observed) since our first entering the tropics. These consisted of the dark Cape hen and an occasional stormy petrel.
Something of the aërial desolation may be due to the season
and the process of incubation with a portion of the feathered
tribe; but the indication is striking of the general paucity of
food in the water in the ocean through which we had passed.
And this illustrates a statement by Lieut. Maury, in his
" Physical Geography of the Sea," in respect to the Greenland whale (the B. mysticetus), that this whale cannot pass
the tropical region of the ocean, which is to it as an utterly
barren sandy desert to the unprovided traveller, or, according to the words of Lieut. Maury, it is "as a sea of fire"
(§ 279); and confirming the impression I had long ago set
forth, that the " right whale" of the Southern ocean is not
the same animal, but a different variety, if not species.

The tropical sea, at least in our track, however it might
afford specimens of animalculæ or minute radiata to interest
the naturalist, was obviously barren, most barren in the sup-
plies needful for the support of the innumerable myriads of
medusæ, cancri, clios, etc., which in their turn constitute the
food, and form the " pasture ground" of the mysticetus; for
this pasture ground in the Greenland seas is conspicuous,
in turbid waters of a deep olive-green colour, to the least
observant fisherman. He never expects " fish," at least he
does not expect to find them in repose, as at home, in blue
water, such as we have almost always had within the tropics;
but in the olive-green water, turbid by reason of its myriads
and crowding of minute forms of life, he has hopes of finding whales, as there at least they might feed, and feed according to their tastes and requirements most sumptuously.

Flying fish, as I have had occasion to notice, constituted
almost the only conspicuous form of life we had seen in any
frequency within the tropics. With these we might, no doubt,
have seen their persecutors—the bonito, the dolphin, and the
shark—had our speed been sufficiently slow; but rarely going
less than seven knots, these larger species could not easily

keep up with us, or be so likely to make their appearance as with ships going at a very slow speed, or occasionally at rest, becalmed. One of the flying-fish, of a larger species, with a double set of flying fins, fell into the fore chains, where it was caught by one of the saloon passengers. It was cooked and served up at dinner, and, as far as could be judged from one morsel, seemed to be delicate eating.

My dipping-needle as well as other compass experiments were continued in the afternoon. The apparatus was tried, 1st, at the pedestal; 2nd, at the central station of the forecastle; and 3rd, on each side of the forecastle, opposite the principal capstan there. At the first station, the dip of yesterday (30′ N.) was found to be changed to 8° 20′ S.; on the forecastle, the central station gave 15¾° S. dip; the starboard side 25¼° S. dip; and the port side, 15½° S. dip.

This, to human feeling and enjoyment, was a perfect day. It was, at least, one of the most perfect days and nights to be realised on the wide ocean of the globe when remote from land. The temperature throughout the day was about 76° or 77½°, falling but little in the evening. From our rising in the morning until retiring at night the weather was fine, sunshiny, or clear, with occasional clouds; a fine fresh breeze animating our graceful, swift-going ship, and refreshing the crowd of passengers under the awning on deck. The moon, in its turn, in the evening sky, shone with brilliant *white* resplendency, producing in the ship's wake, a broad band of reflected splendour, variegated by the surface agitation of the waters under the brisk gale, and extending from our position to the distant horizon.

Nor was the scene simply contemplative. Dancing on the poop was carried on for a couple of hours, with the usual appearance of animation and enjoyment to observers as well as dancers. Our circumstances were a contrast to what might probably be pictured by our sympathising friends in England. This reflection was repeatedly and forcibly suggested. Whilst every occurrence of bad weather since our departure will have awakened deep sympathies and anxieties on our account, we, through the favour of a gracious Pro-

vidence, have thus far experienced neither peril nor storm!
No disturbing anxieties; thank God, no sickness either in
ourselves or those around us; no bad weather of any kind;
scarcely a day of contrary wind to try our patience or inter-
rupt our happy progress.

My own experience never led me to imagine the case of a
passage mainly by sails only, through 80° of latitude (reckoning our progress up to midnight, lat. 30° 85'), in a period of
26⅓ days, or 3° a day continuously, in the several regions of
wintry storm and uncertainty near home, and the variables,
and calms, and trade winds in our subsequent track.   Those
on board, the most experienced in the passage of the tropics,
expressed a doubt whether our present position had ever
been attained before within the same interval of time, by so
slight and rare an appliance of steam-power; or even whether
under any appliances more had ever been accomplished!

Saturday, March 15 (N. N. E.)—Another beautiful and
enjoyable day, with a continual improvement in the temperature and comfort of our sleeping cabins.   Our room was
generally about or below 80°—82° on retiring to rest, and
77½° in the morning.   The clinometer registered a maximum
of 20° to starboard, from a rolling swell from the north-eastward, and a mean heeling of 5° to starboard.   Our course
was chiefly S. by E. ½ E.

Different aërial currents were strikingly marked during the
whole afternoon by three obviously diverse strata of clouds,
the lowest rapidly flying in the direction of the wind towards
the S. S. W.; another stratum of patchy or detached cloud
rapidly flying towards the north from the southward, the
reality of the progress being conspicuously marked by a very
elevated stratum of cirri, which had no perceptible motion.

The weather so fine, and the winds so fresh from the eastward and north-eastward, since our passage of the tropic of
Capricorn to our present position (lat. at noon 32° 9', long.
27° 38' W.) is considered remarkable; variable winds, calms,
and heavy rains being more generally met with in this region
of the South Atlantic ocean.   We were disappointed in both
regions of the trades; the S. E. trades deserting us so soon,

and both trades being so light; but we have been well compensated by what has had the effect of an extra N. E. trade.

My compass comparisons, and determination of their differences (variation and deviation) by azimuths and amplitudes, continue to be diligently and systematically made, the azimuths in primary relation always to the standard compass. The comparison of the companion compass with the results of the amplitudes, taken through the medium of Captain Becher's repeating card, which, for adjusted compasses not provided for azimuths, comes into excellent and effective use, has been undertaken by Lieut. Chimmo, whose observations and results will be elsewhere recorded.

Azimuths, however, in regions where the sun rises so high in the heavens, are by no means satisfactory unless taken at low altitudes. The uncertainty of the verticity of the vanes, augmented by the ship's motion, and the reliance on a reflecting plane, seldom allowed of full confidence, or a reliability of the apparent azimuth within a degree or two. But with amplitudes, where the sun descends so near to the vertical plane, or with azimuths taken at only a few degrees of altitude, excellent observations (except when there was considerable oscillation of the card) were obtained. In high latitudes, on the contrary, especially in winter, the acutely oblique plane in which the sun cuts the horizon renders amplitudes (on account of the uncertainty of the refraction) less stisfactory than azimuths taken at favourable times and under favourable circumstances.

Good dipping-needle experiments were made in five stations, comprising two new positions, the starboard and port sides of the poop deck, 7 feet 8 inches abaft the "fore" rail. The needle, indeed, is very sluggish in the pivots, not coming to rest in a uniform position within an angle varying from 1° to 2°. But by drawing it aside by a weak magnet in alternate directions, a tolerably satisfactory mean was found to be determinable; the more satisfactory indeed, from the rapidity of the changes and largeness of the differences (sometimes 5°, 6°, or even 8°) of the observed dip from day to day.

The day, after a little threatening of rain which passed off, concluded with a fine evening, leading to a repetition of the dancing, with an improved volunteer band. The dew was heavy. Some of our party sat up beyond their usual hour to observe the fine constellation of the " Scorpion," as it now came fully and conspicuously above the haze near the horizon.

Sea still richly blue, and atmosphere very barren of life, an occasional Cape hen or stormy petrel alone giving indication of improvement from the almost utter deadness of the tropics. Still a remarkably calm sea, and for weeks we might have been passing the characteristically quiet waters of the famed Pacific.

Sunday, March 16 (N. N. E. to Southerly, calm).—Our room throughout the night was about 78°. The clinometer had registered 10° to starboard and 1° to port; the mean at 8 A.M. was about 3° to starboard. Sea smooth, as indeed it has usually been for weeks past. The wind blowing a fine fresh and steady breeze we made 230 miles on course (magnetic), deflecting from S. by E. ½ E. to S. E. by S. Our latitude at noon was 35° 3' S., longitude 24° 57' W.

During the day, our long enjoyed favourable wind continued to further us; but in the evening it failed us, and a new wind (a light breeze from S. to S. S. E.) sprung up, and cast us for some hours upon an easterly course. The steam was got up, and about 8 P.M. the engine, after a long period of repose, was again put into action. There was a great change, too, in the state of the weather, fog or rain beginning to prevail.

Our usual devotional services, however, under the shelter of the awnings, were held still upon deck, notwithstanding the dampness of the evening. At the morning service, I took for my subject two incidents recorded in the second lesson and gospel for the day (Matt. xxvi. 75, and xxvii. 3—5), the repentance of Peter and of Judas. The analogies of their position and sin, and the contrast of their repentance and its results, constituted the leading subjects of consideration.

Monday, March 17.—Our room getting gradually cooler, 77½°—78° at night. Clinometer, maximum to port 10°, to starboard 10°. The morning was calm and foggy or hazy, with a long rolling swell from the south-westward. Ship under steam only until 8 A.M., with all sails furled or clewed up, making six knots on E. S. E. courses. But a fresh breeze, increasing to a brisk gale, springing up about 8 A.M. from the N. E., we were enabled to make sail, first in staysails, and subsequently in courses and topsails. Our speed gradually increased to nine knots, and at 1 A.M. of Tuesday, the screw was detached and taken up.

The first albatross appeared this morning, associated with many Cape hens. Many of the former were about us throughout the day, presenting a new scene in aërial life after the deadness of almost the whole of our track from entering the tropics. Graceful and majestic on their expansive and elegantly shaped wings, the albatross floats in the air, and follows without effort the fastest sailing ship.

It was Saint Patrick's day. Several of our saloon associates being Irish, arrangements were made for a champagne supper and ball, and formal cards of invitation issued. The arrangements for shelter, under the direction of Lieut. Chimmo, who was all life and energy, were admirable. A space of some 70 or 80 feet, protected above by awnings, on the sides by curtains of canvass, on the forepart by a vertical screen of canvass, and on the afterpart by flags, was arranged for the promenade and dance, whilst an augmented and now more practised band from among the passengers played a variety of familiar airs. The scene as one of innocent and animating enjoyment to the participators, and scarcely less enjoyment to numerous spectators of all classes, admitted within the foremost barrier, was singularly curious and striking. Such a scene on the deck of our splendid ship whilst forcing her way with a speed of ten knots through the Southern Atlantic waters, a half gale of wind blowing around, and occasional light rain drifting against our screens, and the parties assembled yielding themselves to unreserved enjoyment and all but unconsciousness of the sea and wind without, would, we thought, be scarcely realised in imagination by friends at home, whose anxious sympathies were, perhaps, anticipating equinoxial gales, rather than picturing the lively throng on the deck of the *Royal Charter!* The supper table was ranged across the poop deck from side to side near to the stern, and its taste, variety, and adornment, were alike creditable to the cooks and stewards. Toasts were drunk in champagne, some short speeches made, and a general cheerful pleasantry prevailed amongst the select party, numbering, with Captain Boyce, and some of the officers, and the ladies who had been engaged in the dance, etc. about 25 or 30 persons. At midnight orders were given to dismantle the special arrangements, and in little more than quarter of an hour the whole of the enclosings and apparatus of refreshments had disappeared.

A considerable series of dipping-needle experiments were made, and were generally satisfactory, though from the want of free action of the needle, not so complete and accurate as I could have wished. It was mortifying, indeed, to have to bestow the same attention and time, or, indeed, considerably more, in making experiments yielding only proximate results, especially in reference to the magnetic force in the direction of the dip, when with the instrument in a more perfect state, data for the best possible determinations of the nature and phenomena of the magnetism in iron ships might have been obtained. The great changes of the apparent dip, however, from day to day, and the striking differences in degree, yet in most particulars the consistency of the results in different stations about the deck, lead me to hope that the observations and experiments will yet turn to considerable practical account.

Tuesday, March 18 (E. N. E. to N. E.)—The temperature of the air had now undergone so much change as to feel cold on the deck, though agreeable below. Our room was 70° to 72° throughout the night. The air in the morning and most of the day was about 65°, the sea 62°. A strong short sea met us on the port bow, causing, with a sea from the north-eastward, some rolling and more pitching than we have hitherto had since leaving Plymouth. The clinometer had registered for the night a maximum of heeling to starboard of 18°, to port 10°, average heeling at 8 A.M. 10°.

The wind being nearly east for many hours after noon of

yesterday, we made less easting than, in our present increasing southern latitude, we could have wished. We made
good, however, a course (true) of about S. E. ½ S. from noon
to noon, with a distance of about 210 miles. The compass
courses steered varied from S. by E. to S. E. ½ S. No
higher course being attempted, even when the wind came to
N. E., Captain Boyce wisely preferring to make rapid headway in a direction proximate to his proper course, to hauling
up higher at a sacrifice, perhaps, of one-third of our speed,
generally from 10 to 11 knots, and not unfrequently 12.

The *Royal Charter* again called forth our high admiration
of her performance this day against a strong awkward head
sea. So little did she pitch, in comparison with the class of
vessels I had been accustomed to, that we were able to carry
royals and whole courses, topsails and staysails, notwithstanding this formidable hinderance, under a half gale of wind—
a gale that would have put the common class of ships under
double reefed topsails. But the *Royal Charter* cut her way
through the seas opposing us most beautifully, carrying her
large spread of canvass with no visible complaining of spars,
and often going 12 knots or more, scarcely throwing a spray
upon the poop, and but rarely wetting the forecastle, where
for a period of near half an hour I watched her (to my
experience) wonderful performance.

The birds flying in our wake were now increased in numbers and variety, comprising many albatrosses, Cape hens,
and petrels. A cetaceous animal (which, however, I did not
see) was observed at some distance to leeward.

The day was by no means so pleasant on deck as usual,
and from the great increase of the ship's motion, signs of
sea-sickness or headache appeared amongst several of the
passengers. On deck it was not unfrequently damp, and the
air felt chill. Much of our very light clothing had disappeared, and warmer habiliments were being gradually
brought into requisition.

The compass comparisons excite, from the absence on our
present courses of any marked change, not a little surprise.
The impression is suggested that, not improbably, the *Royal*

Charter's compasses may have happened to be fixed under these naturally compensating influences of the ship's magnetisms on either side, and forward and aft, by which the compassses of the *Great Britain,* the *Sarah Sands,* and a few other other vessels, have, even in the different hemispheres, performed so well. Sailing, however, as we now were, with the ship's head very near to the original direction of the points of change in the compasses (our course nearly S. E., and one of the points of change being S. E. by E. $\frac{1}{2}$ E), it cannot be ascertained until we happen to take a very different course,—whether or not, this first considerable mechanical action of the sea encountered since leaving Plymouth may have changed the ship's original magnetic condition and magnetic lines. That her magnetic condition has changed, but possibly so equally on the two sides as to occasion no compass disturbance, is to be inferred from the alteration of polarity in the upper plating running out vertically outside of the gunwale of the poop deck.

Wednesday, March 19 (N. N. E.)—The temperature of the air, even with a northerly wind, falling rapidly. In the morning the air was 65°, water 62°; our room 68°. In the evening it fell to 59°. Maximum heeling to starboard about 18°, mean heeling 9° to 10° to starboard.

Again the ship performed beautifully, going with the wind two points abaft the beam, all square sails, and foretopmast studding-sail, she averaged from noon to noon $12\frac{1}{4}$ knots very nearly. The total distance made, though the wind sometimes fell to a moderate breeze, was 293 miles; compass course S. E. $\frac{1}{2}$ E. Under the strongest breeze, such as with ordinary merchant ships would have been designated a " double-reefed topsail breeze," or " fresh gale," we carried the three royals gallantly, and when running 12 to 13 knots, there was no sign of unquietness or strain. She slipped through the water with singular ease and gracefulness, not giving the impression of rapid, but of ordinary speed. Scarcely a drop of water in the form of spray reached the poop deck or deckhouse, except once or twice in the evening.

Dipping-needle experiments were pursued during two

hours in the afternoon. But the needle did not oscillate freely when near the exact line of the dip, sometimes coming to rest in positions differing from it by 2° or 3°. This was very mortifying. But still, by taking many averages, and observing the fair agreement of the results with the face of the instrument E. and W., the observations to a considerable extent promised to be useful. At the standard compass position, the change of dip was about 10° in 24 hours sailing, and the same quantity for the previous day.

Our favourable breeze continued throughout the day, giving us an average speed on a S. E. ½ E. (magnetic) course of full 12 knots. From the drift of the clouds, strongly setting *from* the N.W. or W.N.W., on the first appearance of the moon after sunset, I suspected we were about to have a change; and this took place suddenly about ten hours afterwards.

Thursday, March 20 (S. S. E. to S. etc.)—A cruel sport, much indulged in on the previous day, was happily interrupted this morning by rain. The shooting at albatrosses, which in great numbers follow the ship, appears to be a prevalent usage in many ships of our class voyaging to Australia and other considerable southern latitudes. From half-past six until half-past eight A.M. the afterpart of the poop of the *Royal Charter* was yesterday occupied with "sportsmen" and lookers on, who with rifles or other guns were every now and then firing at the unconscious elegant birds gracefully hovering about our rear. I fancy 50 to 100 shots were fired, happily with rare instances of their taking effect; but in one case I saw, on being induced to look astern by a general shout, a poor stricken bird struggling on the surface of the water apparently mortally wounded. This useless infliction of injury and suffering on these noble looking birds, where there was no chance of obtaining them as specimens for the museum, nor for any other use, was to my feelings, and, I believe, the feelings of many others, particularly painful. The inducement was "sport," the object, practise in shooting. With some it was, one would earnestly hope, indulged in from want of consideration, and from

example.   But in attempting remonstrance, it was curious to
hear the excuses for the cruelty on these confiding or unfear-
ing creatures;—"they were savage birds;" "they attacked
their unfortunate associate, as soon as it was sufficiently clear
of the ship, to prey upon it;" "a boy who had once fallen
overboard was attacked by an albatross, descending upon
him with immense force, and striking him with its bill on the
head, so that when a boat reached him, his face was found
covered with blood, and the skull fractured;—therefore," it
was inferred, "it was proper to shoot them." I quietly applied
the argument to the case of dogs.   I had known a dog attack
a person and injure him dangerously; it was therefore proper
to amuse ourselves as we had opportunity by shooting dogs!
But the case was not admitted to be in point.

Falling nearly calm about 10 A.M., a line with a baited
hook was put over the stern by one the passengers, which
before long, whilst the engine and screw were getting into
action, was seized by an albatross, and the fine majestic bird
hauled almost unscathed upon deck.   It showed a disposition
to defend itself, whilst in vain attempting to rise from the
deck, and frequently drove the crowd of lookers on before it.
Captain Boyce had seen a trial of power betwixt an active
terrier dog, and an unfortunate albatross that had been
hooked and drawn upon deck; but the bird, though but
little adapted for rapid movement, soon gave the dog an
impressive lesson by seizing him by the nose and biting him
so severely that he ran off howling most piteously!   The
power in the bill, a strong gull-shaped weapon of eight inches
in length, with a sharp bent point, is singularly great.   It
required caution, and justified the retreat of our little crowd
of spectators in avoiding the huge bird in his attempts to
make way through their ranks with a view to escape.   Its
dimensions were as follows:—Spread of wings across the
back, from tip to tip, exactly 10 feet.   Length from the tip
of the beak to end of the tail, 3 feet 7 inches.   Circumference
round the body and compressed wings, 2 feet 9 inches.
Length of the bill (as bare of feathers), 8 inches.   Web
foot 7 inches long and 8 inches broad.   These webs are often

used for the construction of purses, by being separated betwixt the skin or web, above and below, and leaving the claws
as ornaments. They are flexible, have a fine yellow surface
when dry, and are considered at once curious and ornamental.

Shortly before daylight, our northerly favourable wind
suddenly deserted us, and rapidly went round to the southward, so as to bring the ship on the starboard tack, the first
time for a long period in which the wind had been on that
side, or decidedly unfavourable. Sometimes, indeed, the
course, aided by the engine which was put into operation
about noon, and enabled the ship to lie closer to the wind,
was E. S. E. and S. E. by E., but generally more towards the
east. Hence we began to lose some of the advantage we had
gained in southern latitude.

The sea was singularly smooth, and aided by a freshening
breeze, we went nine to ten knots, sometimes more; but at
the latter speed, the engine was of no advantage, whilst the
screw produced a very disagreeable vibration and noise.
The temperature of the air fell in the afternoon to 54°, in the
evening lower. Winter clothing began again to creep into
use, the temperature already feeling sharp and cooler than
was quite agreeable. The barometer, on the other hand, rose
gradually, and for this region, where the average is not so
high as in the northern regions, to an unwonted elevation,
attaining in the evening 30·5, and still rising.

Magnetical experiments and compass comparisons continued to be made regularly. The polarity of the upper
plating of the ship—as to the edge rising to the top of the
gunwale outside—indicated more decided change both on the
poop and on the forecastle, where the upper edge came into
sight and was easily accessible for trial with a pocket compass. On the port side the polarity of the iron-top edge was
northerly, both forward and aft, and partially on the starboard
side; but this polarity did not seem to descend far below the
very edge, and, as yet, I had not arranged for letting a larger
gimbal compass down the side for ascertaining the actual
extent of the change. Iron standards, *forward*, set in the
gunwale for supporting an iron chain running in place of a

railing on the sides of the forecastle, such standards being detached from the side plating, all gave northern polarity at the top, being the converse of what would occur in European latitudes, or considerable northern dip.

But the most curious and interesting experiments were those made with the dipping-needle. The standard (central) station now gave a southerly dip of $40\frac{1}{2}°$, the terrestrial dip, according to General Sabine's dip chart, being 46°S.; showing a gradual approximate of the ship's dip to the true. At the magnetic equator (March 12), the ship's dip at this midship and central station was about 20° N., the difference (behind) about 20°. Taking a series of observations made March 13, 14, 15, 17, 19, 20, I find the following gradually diminishing series of differences, roughly taken of 19 (repeating that of March 11), $16\frac{3}{4}°$, $15\frac{1}{2}°$, $14\frac{3}{4}°$, $15\frac{1}{4}°$, $12\frac{3}{4}°$, 11°, $5\frac{1}{4}°$. The ship's head at the time of observation varying from S. 11° W. to S. 53° E., and the ship's heeling from 10° to starboard to 3° to port. The changes of the dip on the *two* sides of the poop deck, and their differences, were found more perplexing. The differences appeared to be much affected by the amount and direction of the ship's heeling; but the apparent results will be deduced more safely by waiting for a more extensive series of experiments. As to the compass comparisons, there is yet a remarkable agreement, though not so close as hitherto. The companion compass, nearer to that aloft, was now a quarter point more easterly (on an E. S. E. course) than the wheel compass, a difference which seemed to be on the increase.

March 21, Good Friday (S. S. E., S. E., variable, calm.)— Our cabin was now sufficiently cool, the temperature at night passing from 62° to 59°, that of the air fell to 49°. It was a fine bracing, but cold feeling air. Sea singularly smooth. Maximum heeling 5° to port; average (8 A.M.) 4°.

The wind freshening about 4 A.M. to a rate of ten knots or more, the engine was stopped and the screw raised, the screw for some hours whilst the ship was going above nine knots producing a most disagreeable noise and vibration, amounting sometimes to a violent shaking of the ship and the machinery of and about the engine. Above the speed of seven or $7\frac{1}{2}$

knots, indeed, the engine acts disadvantageously, and often perhaps to disservice. Below that speed it works smoothly, and always when propelling the ship and taking the burden alone, its action is smooth and beautiful, the motion of the screw and machinery being hardly felt.

Though our day's run from midnight to midnight produced a distance of about 210 miles, it was on a course making above a degree of northing, which was considered disadvantageous, notwithstanding the rapid reduction of our westerly longitude. The strength of the wind varied from a gentle breeze to one of $9\frac{1}{2}$ or 10 knots close hauled.

Being Good Friday, we had Divine service in the morning in the saloon, the weather being too cold for the open air. There was a comparatively small number of worshippers, perhaps 60 to 80, there being many amongst our associates of diverse theological views. The service, however, was interesting, and the attention fixed and solemn whilst I addressed them all but extemporaneously from our Lord's touching exclamation whilst He hung on the "cursed tree," (Matt. xx. 46): "My God! my God! why hast thou forsaken me?"

The wind fell to nearly calm in the afternoon, and the engine was again put into operation. With the yards sharp braced, and sails, except staysails, withdrawn, the ship made seven knots, steering E. S. E.; but hauling two points *into* the wind, without sails, against a light breeze, we then made $6\frac{1}{2}$ knots, on a preference course of S. E.

The steering and companion compass now differed half a point or more, and the steering compass about three-quarters of a point to a point from that aloft.

Saturday, March 22 (S. W., W., W. N. W.)—We had but a moderate breeze the whole of this day, that is, for the *Royal Charter*, our speed on a S. E. to S. E. $\frac{1}{3}$ E. steering-compass course, being from $6\frac{1}{2}$ to $9\frac{1}{2}$ knots. The engine was stopped at 1 P.M., after which we proceeded under sail alone. Our course was much more easterly, for the sake of a *full* sail, than was desired, Captain Boyce being anxious to make his course northward of the Kerguelen islands; but for two or three days we had not been able to make any southing.

The moderate and scant wind in this region, with such a high barometer (30·7 to 30·5), was considered as unusual as it was disappointing to those who were anxiously and sanguinely looking for a very short passage, if not the shortest yet accomplished.

Magnetical experiments were again pursued; but the dipping-needle for this day and Thursday did not act so well as on former occasions. To come at satisfactory results was very troublesome, required many repetitions, and still without the desirable success. As, however, the changes of the dip continued to be considerable, they were not uninteresting, nor, it was hoped, unuseful.

The temperature continuing to fall, we felt our room and the saloon unpleasantly chill. A swell coming on from the south-westward, the ship began to roll more than usual. The maximum for the night of Friday–Saturday, was $16\frac{1}{4}°$ to port, $12\frac{1}{2}°$ to starboard. Many birds now followed in our wake, or were seen on the wing around us, particularly albatrosses, Cape hens, Cape pigeons, the stormy petrel, and a gull, said to be the fulmar or mallimuk, which it certainly much resembled.

There was a dance below, the lower saloon being cleared, and a champagne feast, given by one of the passengers on occasion of (to him) some memorable anniversary. So long as the ladies remained below, all (as far as we could judge) went on with fitting propriety, but subsequently, the indulgence became uproarious, and terminated in a very disgraceful scene.

March 23, *Easter Day.*—The disturbance of the preceding evening was not unfelt in the morning of this eventful commemoration. Some who had been participators did not appear at breakfast.

The wind having freshened during the night, though becoming more scant about 8 A.M., we made rapid way, steering S. E. by E., to E. S. E. and S. E. Our speed was from 8 to 13 knots, reckoning from midnight to midnight, yielding a distance by log of about 276 miles, but giving us no southing. A strong south-westerly swell gave the ship a maximum roll to port of 21°, to starboard 7°; mean heel, at 8 A.M., 9° to port. The thermometer now fell to 46°; the sea being 42°.

The evening was fine; sea smoother; wind less, and speed reduced to 10½, and ultimately at midnight to 8 knots.

Divine service was performed in the morning in the saloon. The address was from 1 Cor. xv. The attendance was not so large as usual, but might be 60 to 80.

In the afternoon, as previously announced, I administered the ordinance of the Lord's Supper in the ladies' boudoir. No gentleman offered or attended; we assembled, ladies and servants, eight or ten altogether.

In the evening I held a third service in the "third-class mess-room." The attendance was not large, but it was an interesting one. Deep attention and feeling were manifested by many of the men there, especially under the address, which was taken from the second lesson (Acts ii.) Many, I trust,—if I might judge from the manifested feeling and solemnity, or my own sympathies,—might have said, "It is good for us to be here."

A sperm whale was seen by the Captain very near to the ship, just before the conclusion of the morning service.

Monday, March 24 (W. N. W. to N. W.)—A tolerably quiet night—a moderate breeze and but little swell. Our room 60° to 56° or 55°. The morning was chill, and wet and unpleasant on deck, and cold and comfortless below. The lack of stoves or fires in a voyage of this kind, said "not to need them," is certainly a great mistake, and an omission which ought to be corrected. Coming out, we had it sufficiently uncomfortable for a week; but after passing the tropics, in a few days to pass into a temperature of 45°—46°, expected to go down to near the freezing point, and this to be endured for days and weeks together, is a change which ought to be remedied. From the transitions endured on the outward voyage, and the severity of the passage round Cape Horn, I hardly know a voyage more demanding this ordinary comfort.

The maximum heeling registered by the clinometer for the

H

night was 16° to port, and 7° to starboard. Mean at 8 A.M.
about 2° to port. In the evening, however, the wind having
increased to a strong gale, with a rather high sea from the
W. N.W., the ship rolled occasionally to the maximum of
24° to port, and 22° to starboard; but this was only occasional, for ordinarily the rolling was very small.

Our courses for the day (by companion compass, which
was more correct by three-fourths of a point than the steering compass,) were S. E. ½ S., S. E. by E., S. E., S. E. ½ S.,
and S. S. E. ½ E., with a speed varying (from noon to noon)
from 7¾ to 12½ knots. The true course was reckoned at
S. 72° E., and distance run 242 miles. From midnight to
midnight the distance was pretty nearly the same. Cape of
Good Hope, at noon N. E. ½ N. (true), distant 630 miles.

The steering compass was now found to differ just a point
from that aloft, being a point more southerly, and from the
companion compass (also adjusted) three-fourths of a point.

In the evening the wind freshened to a gale, with an increasing sea, sometimes in waves of 20 feet from the N.W.
or W. N.W. The ship, under a crowd of sail, and scudding
with the wind about a point on the starboard quarter, again
excited our highest admiration. Carrying two royals, studding sails below and aloft up to those of the top-gallant sails,
and running, at 9 to 10 P.M., 13 and 14 knots, she steered with
the greatest precision, and was as easy and ordinarily as upright
and steady as if she had been running in a moderate breeze.
Occasionally, as above noticed, she took a series of three or
four rolls of 10° or 15°, rarely of 20°, inclination; but, in
the intervals, would often go from two or three to five minutes
not rolling more than 2° or 3°, so quiet, so free from creakings
of the joinings of the wood and metal below, that one would
have supposed it to be calm weather, or light winds with a
smooth sea! Her rolling too was with the smoothness of a
swing, and it was evident there was no straining that could
sensibly affect the perfectness of the general fabric of the ship.

The only circumstance of rational anxiety in this rapid
and beautiful progress was the haziness of the weather, and
some hours of darkness before the rising of the moon. For

now we were traversing a region where icebergs are sometimes met with, and the falling in with any, whilst scudding
under such a crowd of sail in the night, could not be contemplated without feelings of some anxiety.

Tuesday, March 25 (N. W. to W., N. W., N. by W.)—
Scarcely a spray had come on board the ship, except incidentally in the waist or on the main deck, during the preceding
day and night. But about 2 A.M. a high crossing of seas struck
the ship under the starboard main chains with such force as
to make the whole hull and masts stagger. Some of the
passengers were awoke by it; one thought (the first impression) that the ship had struck against a piece of ice!
But beyond the shock and the coming in of a small body of
water upon the main deck, the result was perfectly indifferent,
and the effect was regarded by the lookers on only as an
amusing incident of sea life to the poor drenched sailors who
happened to receive it in the shape of a shower bath. I
should not, indeed, have thought the incident worth recording
had it not been for its rareness with the *Royal Charter*, and
for the fact of the notable shock and quivering staggering
effects upon the hull and rigging of the ship.

The sea being considerable, occasional waves 18 to 20 feet,
with a width (estimated) of about 400 feet, we had a result
in rolling (at 8 A.M.), registered by the clinometer for the
night, of $18\frac{1}{2}°$ to port, and about 18° to starboard; at 10 P.M.
a maximum of 15° to port, and 10° to starboard.

At 6 A.M. it was discovered that the truss of the mainyard
(patent) was, at one of the clasps round the yard, detached
by the loss of a screw-nut. The lower topsail and mainsail,
etc., had to be hauled up for the securing of the yard. This
reduced the ship's speed above a knot. But at 10 A.M., the
defect having been made good and the yard secured, the sails
were again set, but the breeze being less, we went at the
diminishing rate of ten, nine, eight, down to four knots, from
10 A.M. till evening. The day's work from noon to noon, on
a course S. E. $\frac{1}{2}$ S. by the elevated compass, the steering
compass being half a point different from the companion,
and a point more southerly than that aloft, was 240 miles.

The Captain and passengers were much disappointed in
the failure of our free wind; strong fair winds being usually
expected at this season; it, however, continued quite fair,
varying only from W. N.W. or W. to N.W., and ultimately
to N. by W. Our speed varied from 13½ knots to 4 knots
during the (civil) day. For the nautical day (noon to noon)
we made, by log (for we got no observation of the sun
either this day or yesterday) 274 miles, on a true course
S. 73° E. by reckoning, allowing as above, a point westerly
deviation, and 27° W. variation. This gave the latitude (by
log) 44° 36′, longitude 15° 43′; but by the Captain's larger
allowances, the latitude was 42° 52′ S., and longitude 15° 37′ E.;
and even this position I am inclined to think will be found
more southerly than the true, on account of the increasing
deviations of the compass on deck.

The day was gloomy and often wet; the evening rainy,
and the early part of the night, till the moon arose, dark.
Albatrosses, Cape hens, gulls, petrels, etc., still accompanied
us in considerable numbers.

Wednesday, March 26—A fresh or brisk gale prevailed
during the night, the wind with drizzling rain, at first from
the north and suddenly shifting to N.W., and after veering
to W. N.W., so remained all the day. Our course was S. E.
by S. till noon, by companion compass, S. S. E. ½ E. by
wheel compass, and S. E. ¼ E. by compass aloft; the differences twice tried being 7¼° and 7° betwixt the steering compass and that aloft. Our rate of going varied (betwixt midnight and midnight) from 9½ to about 14 knots. Clinometer
at 8 A.M , port 14½°, starboard 12½°.

At noon and 9 A.M. we fortunately got sight of the sun,
which gave the latitude 44° 51′, and longitude (reduced to
noon) 20° 22′ E. Distance run from noon to noon 211 miles.
The position of the ship thus obtained verified the supposition
yesterday made as to the probable error of our previous
reckoning from increase and excess of deviation.

We were now in the Indian ocean, and in a region and
parallel subject at this season to stormy weather; of this, as
the barometer indicated by falling to 29·3, we had soon an

example, a gale commencing and gradually increasing to a hard gale, with heavy squalls in the frequent showers, in the afternoon. As it was the first in our voyage after leaving Plymouth, and as it brought out additional proofs of the splendid qualities of the *Royal Charter* as a " sea-boat," it may not be without use to attempt a description of its commencement and progress.

OUR FIRST GALE ! We had been till now without a gale, and without having occasion to take in the top-gallant sails, with wind a-beam, except I believe on one occasion soon after starting, and that only for a very few hours and without requiring to take in royals, on account of wind, when going more free from the wind, more than twice, and for only short periods !

The gale might be considered as commencing about two in the afternoon, when scudding under royals and studding sails, some of these lighter sails were taken in, and our speed (with the wind two or three points on the starboard quarter) had increased to 14 knots.

The appearance of the sky at the commencement was of a character which I had been accustomed to observe in the North Sea towards the conclusion of a gale. Large masses of detached clouds floated in the air, passing majestically across the sky, amid which, and in the intervals of the showery squalls, the sun broke out in cheering clearness, giving to the surface of the fast rising waves, those picturesque contrasts of sun and shadow which constitute so beautiful a feature in the lessening and ceasing of the storm. The storm waves in their rise and progress, not being able to keep pace with the rapidly increasing and boisterous gale, were rebuked for their sluggishness by their tops being driven forward in white sprays and waves resembling the prevalent speckled surface in a " white squall," thus overlaying the general blue surface with a snow-white embossing of streaks in the direction of the wind, and with speckled patches of distributed foam. In the wake of the ship—where for some hundreds of yards astern the water had been disturbed, thrown into myriads of whirls, and mixed with

bubbles of air, entangled by the occasional foam on the bows,
and buried by the overriding of the huge bulk of the ship,
and Maelstrom action of the disturbed waters,—these sub-
merged optical resistances to the free passage of the light,
reflected the silver rays from beneath, and painted the whole
breadth of our track, in transparent richness, with emerald
green, whilst the real and universal colour of the water was
a rich marine blue! The scene, at this early period, combined as it was with the easy and graceful, yet swift and
(looking on her as a thing of life, one might say) unconscious
progress of the ship—a progress without appearance of risk
or damage—was one of singular beauty and enjoyment.
Almost every lady of our saloon party, as well as the gentlemen, turned out upon the poop, where in rapid and not
difficult walking they were enjoying the novelty and splendour of the scene.

But the interest was further enhanced when a little before
six o'clock we were called up to witness the sunset—a rich
and glorious exhibition. The greater part of the heavens
towards the N., N.W., and W., was covered by a dense
body of uniform leaden-coloured cloud—a sort of *squall
cloud*, not sufficiently charged with moisture for the usual
discharge of rain—which had left, on the rising of its north-
western edge, a small elongated space of clear sky, horizontally spread betwixt the overhanging curtain and a low
horizontal band of cumulus-like cloud, pushing its irregular
margin just above the distant line of the horizon, in which
aperture the sun showed itself before its retirement for the
day, overspreading the contiguous clouds with a rich and
golden blaze of singular beauty and glory. The richness of
the colouring, the overlaying, as it might seem, the cloud
masses with a bright haze of glorious radiance, subduing the
ordinary opacity of the cloud, and resolving it into a species
of luminous incandescence, presented characteristics in which
the pencil of a Turner would have delighted to revel, and
which that great painter would have boldly thrown into effect
by the contrast of colours, which in themselves would have
seemed extravagant, but which, under his bold conceptions

and management, would have justified his work to the admiring artist!

From the increase of wind and the rapid rise of the sea, the scene, which at first was of the character of the beautiful and picturesque, soon passed into that of the grand and sublime. The highest waves I had estimated at 2 P.M. were from 16 to 18 feet above the depression, but at 4 P.M. waves of 20 feet and upwards were observed and carefully estimated; whilst the flight of the ship before the storm had become a grand spectacle, as under a cloud of sail—whole courses and double topsails up to the elevated royals and wide spreading studding-sails—she cut her almost noiseless path at a speed which had occasionally increased, according to the register of the log, up to 16 knots; for 15½ knots had been fairly run off the reel in intervals of the squalls, when her rate was decidedly by no means the fastest.

A gale at sea, and the foam and elevation and character of the stimulated waters, are matters of familiar knowledge with all who have observantly made a voyage across or along the main ocean; but the way of a large and splendid clipper ship like the *Royal Charter* is not one of familiarity, even to greatly experienced seamen. It is a case by itself. My own experience of one and twenty years in merchant ships, of what used to be regarded as the best build, and of several voyages in packet ships and liners of the modern class engaged in the transit of the Atlantic, afforded no comparison with the character of the action of the *Royal Charter;* nor did the considerable experience of Lieut. Jansen, of the Netherlands Navy, or of Lieut. Chimmo, of the British Navy, in war ships even of the modern day, appear to have at all prepared them for the performances of this ship in a storm!

Thursday, March 27 (W. N. W.)—The first gale therefore encountered by the *Royal Charter*, since her improvement in loading and trim, may deserve something more of particular description; and in attempting such description, I will suppose myself writing for those who have not witnessed a storm at sea.

Of the commencement of the gale I have already written—
taking the external aspect of sea and sky from 2 P.M. to
sunset of the 26th. The wind continued to blow from the
same quarter (N. W. to W. N. W.) all night, increasing to a
heavy gale, with severe squalls in the showers. Notwithstanding this, full courses with topsails and two top-gallant
sails were carried all night—the studding sails, with the
mizentop-gallant sail, being alone taken in, and the remaining top-gallant sails having their halyards relaxed, and the
sails partially lowered down during the continuance of the
squalls. The showers, as the thermometer fell to 39° during
the night, consisted of the different varieties of rain, hail,
and snow. In the morning we had a heavy sea, and in the
intervals of the showers in the forenoon most splendid effects
on the scene by brilliant sunshine on the mighty disturbed
waters. The waves were not regular, but thrown into occasional peaks and wedges of water, at the intersection of the
cross seas, and raised to majestic heights. For estimating
the heights I placed myself successively in a variety of
positions about the poop, and at various elevations from the
deck. The highest position I took, from whence I watched
the rolling surges for about half an hour, was standing on the
spanker boom, close by the mizen mast. The height of the
boom was 6 feet from the deck and of my eyes 5½ feet,
giving together 11½ feet. The poop deck from the ordinary
sea level, at the place of my look out, was about 18 feet high,
so that the point or level from which I looked abroad and
compared the wave summits with the visible horizon, was
about 29–30 feet high.

Taking my observations when the ship was pretty fairly in
the hollow betwixt the waves—the distance of the waves
being apparently about 500 feet—the intercepting of the
horizon by the advancing summits (the *depression* from the
height of the eye being of small consideration) gave a fair
scale of measurement of the height. At this elevation a
considerable number of waves and unbroken water—broad
solid masses of water, spreading 50 to 100 yards or more in
width of irregular but mainly equal height—were observed

to eclipse the horizon, and in a few instances to rise, obviously some feet, I might almost say several feet, above the horizon line of the sea. These waves, therefore, were not less than 32–33 feet from the hollow to the crest; and in some cases might be reckoned as reaching to about 35 feet. But the highest, coming within my scale of measurement, were of course comparatively few. Sometimes a series of two or three, once four or five, in succession, rose to the full height of the horizon, and some above it; at other times, the higher wave came singly and at intervals of some minutes. But the contingencies attendant on subjecting an irregular and rapidly moving body of irregular height to the appointed scale of measurement, so as to catch the higher crests in the proper position of the ship, were so numerous, that the fewness of the loftier waves which came within range could afford but a very imperfect estimate of their actual proportion.

These storm waves, it may be observed, have rarely the parallelism and breadth which might seem to be indicated by the regularity and extent of the swell rolling in and breaking upon a wide and sandy beach. The storm waves are generally of limited extent of ridge, though I have estimated the distinct lines of summits of a width of 200 or 300 yards, or perhaps a quarter of a mile; but these summits, cast into an irregular waving line of various elevations, generally rise towards the middle into the loftier wave summit, sometimes of a breadth, as I have said, of 50 to 100, or even 150 yards or more, of solid water; sometimes, especially in the case of crossing seas, cast into lofty irregular peaks; and these peaks not unfrequently overrunning, by the dynamics of their extraordinary elevation, the velocity of the general wave, are thrown over as snow-white masses of roaring breakers. The spaces betwixt the great run of waves are, as every one has observed, filled in by minor undulations of every variety of figure of the wave formation.

We did not always escape the assaults of the more threatening waves, however, with the impressions of unmixed enjoyment to personal feelings; for sometimes the rushing

wave, ready to fall over in a breaker, or taking the side or
quarter of the ship in a sort of knot of water, would strike
the already yielding side, and, as if in spite at the avoidance
of its main onslaught, throw a considerable quantity of water
upon deck. Thus seas, not in dangerous, only in wetting
quantity, were shipped occasionally over the greater part of
the extent of the deck, from the middle or afterpart of the
poop forward; and though these seas might strike smart
blows, sometimes with a staggering force on the whole frame
of the ship, yet the effects of the sea that came on board
were always so remote from danger or damage, that a general
shout of merriment would be indulged in by the observers
of the incident at the expense of the well-drenched recipients
of the shower, in which the suffering parties themselves,
after shaking off the surplus wet, would not unfrequently
join!

During the gale the *Royal Charter* performed her duty,
as a first-rate " clipper," bravely ; and whilst being thrown
necessarily about in " sending" and rolling by the heavy
cross sea, she made admirable progress, the log for the day
(from noon to noon) indicating a minimum speed of $13\frac{1}{2}$ knots
and a maximum of 16 knots. The total distance registered
for the 24 hours was no less than 352 geographical miles, or
an average of $14\frac{2}{3}$ knots—a run, as comprised within a single
nautical day, rarely if ever exceeded by the process of sailing !
The same average speed was, indeed, maintained during a
continuous period, from noon of the 27th, of very nearly
30 hours.

This, as I have said, was our first gale, and a most prosperous one in the furtherance of our progress during a period
of 40 days, and comprising a run of the aggregate " days'
work" of 8115 miles; and what is also worthy of note, this
distance, averaging 200 miles a day, was effected without
even encountering a " head sea," and so without any actual
technical " pitching" of the ship being experienced !

Our progress in the gale, however, was not without its
special anxieties, which were principally as to the failure of
our wheel ropes and the falling in with ice during the night

—the first part of which was dark from the absence of the moon. The *Royal Charter's* steering wheel, for the advantage of a superior action on the binnacle compass, is placed *before* the mizen mast, and therefore at a considerable distance from the stern. The advantage sought has been well realized— the compass action under so great a change in magnetic latitude being yet (so far as could be tested by the courses steered) limited to an error of about 21°, or a point and three quarters; but the attachment of a long rope of *hide* to the chain working in the tiller purchase has proved insecure, having already given way more than once. Though a rope of this kind, more substantial than the first, has been put into connection with the wheel, yet I could by no means feel confidence in its security. Nor did the precaution of " relieving tackles," as they were applied to the tiller, give very much confidence, as, in case of the breaking of the wheel ropes, the ship would probably " broach to" into the wind with all her cloud of sail, only possible to be carried whilst scudding before the storm, before the hands stationed on the main deck could reach the position on the poop for the applying and bringing to bear on the helm this precautionary provision. It was matter, therefore, of much thankfulness that not only did the wheel ropes stand firm under this perpetual and violent straining, but that no apparent peril arose from the falling in with icebergs or fragments of ice—the occurrence of which in the region we were traversing, and at this season, is by no means uncommon.

We were grateful to a gracious Providence in being preserved from both these perils.

Towards night of the 27th the gale gradually subsided, till our speed was reduced at midnight to ten knots, or somewhat below it. At 7 P.M. made all sail, setting royals, etc.

Commander Edge's clinometer gave the maximum rolling of the ship from the night of the 26th, 10 P.M., to the night of the 27th at the same hour, 28° to port, and 19° to starboard. From 10·30 P.M. of the 26th to 9 A.M. of the 27th, the register for the " rollings " had advanced from 6500 to 8000, or 1500 for 10½ hours, or 25ˢ to each roll.

At 9 P.M. of the 27th, the number was 9784; and at 8 A.M.
of the 28th, 10,904,—giving for the 11 hours, 1120 rolls of
the ship; 35° to each roll. The barometer at the commencement of the gale fell to 29·482; it was now at the close of the
27th (29·931). At "tea-time," the wind having fallen, a
common enquiry by the passengers was "What is she going?"
The reply was "eleven knots;" this called forth the exclamation "only eleven knots!"

The action of the compasses during the gale deserves some
notice. That aloft, being unfortunately fitted with an opaque
card, though lighted from above, could only be seen from the
deck by the help of a superior opera-glass and good eyes;
and I did not happen to send a person aloft to observe it.
It was obviously as steady, if not steadier, however, than
the floating adjusted compass which had usually acted so well.
These were observed to swing considerably, sometimes greatly.
A range of two and a half points was observed by Lieut.
Jansen in one of them, at a time late in the gale, when the
compass aloft swung only a point. Another observer noted
a range of five points in one of the deck compasses, the
card swinging at the time two and a half points each way.

The compass aloft, however, when I subsequently got an
officer to observe it for comparison with the others, acted
well, even at a time when the ship was occasionally rolling
to an extent of $18\frac{1}{2}°$ and 20°, and when I had found the sea
still so heavy that very many waves rose to the height of 23
to 24 feet, whilst those of 22 feet were quite frequent. A
swing of a point at the time of observation was rarely if ever
exceeded, and frequently the card was quite steady. The
most decided evidence of the "observableness" of this compass, or its efficiency for direction in heavy weather, was
derived from the sets, of about three observations each, for
comparison with the other compasses. Thus, on repeated
occasions, on the 26th to the 28th, when the ship was steering nearly the same course, viz., S. 48° E., 38°, 39°, 40°,
by the compass aloft, and its position was taken by signals
from me whilst observing a compass on deck; the differences
or *apparent* errors of the steering compass, on the several

occasions, came out thus, nearly in accordance,—13¾°, 14°, 14°, 12½°. And this agreement was by no means unusual; whilst it established my confidence in the compass comparisons, made at all our obvious disadvantages—yielding the conviction that, by taking the mean of three or four observations made consecutively, the guidance of the compass aloft (so far as errors of observation and occasional unsteadiness were concerned) could be relied on for reference, within a degree or two, at most, of the truth even in boisterous weather and heavy seas! This was a determination, sustained by the whole extent of our experience from leaving Liverpool, of vast importance, and to me of deep satisfaction, in the question of safe compass guidance by means of a reference compass aloft.

Friday, March 28 (N. to N.W., and N. N. E. or N. E.)— Our courses for this day (by companion compass) were S. E. ⅓ S., S. E. by S., and S. E. by S. ½ S. Reckoning from noon to noon we made 267 miles of distance by log, estimated on a mean true course of S. 80° E. Our rate of going for the nautical day varied from 14⅓ to 9½ knots.

Good observations for latitude and longitude, important to us on approaching the situation of Prince Edward's islands, were happily obtained, which gave the latitude at noon 45° 37', and longitude, by chronometer, reduced to noon 35° 52' E.

Most of the day the wind was pretty nearly right aft, blowing fresh, but allowing studding sails on both sides forward for some hours, and at all times royals.

The sea was yet considerable. From the height of the poop deck, a good proportion of waves rose high enough to intercept the horizon, indicating a height exceeding 22 feet; but on taking a position two feet higher only, I found scarcely one *near* wave reaching 24 feet during a quarter of an hour's observation.

I took much pains during the afternoon in endeavouring to get good compass comparisons, the results of which I have anticipated in remarks on a preceding page. But an important set of observations consisted of azimuths taken, with great difficulty on account of the ship's rolling and the com-

pass swing, with the Admiralty compass No. 1. By continuous attention for near half an hour, with the help of the
second officer looking out for the altitude, I obtained two
good and reliable azimuths, with the ship's head, by the
standard compass, at S. 48½° E. and S. 54½° E., the mean
being very nearly that of the course given, as reduced to the
pointing of the steering compass. These azimuths worked
separately gave the error of the standard compass for deviation and variation 32° 36′ W. and 33° 36′ W., mean 33° 2′ W.
This close agreement, verified moreover by an amplitude (or
azimuth near the setting of the sun), caught when the sun
was in the exact line of the ship's keel, under guidance of
my midship position and that of the mizen mast, was very
satisfactory and conclusive.

The application of the true azimuth, thus obtained to the
various compasses, gave the declination of the compass aloft
from the true meridian 35⅓° W.; of the steering compass
49⅓° W.; of the companion compass 42⅓° W. Applying the
variation as given by the most recent Admiralty South Atlantic
Charts, viz. 33⅓° W., and corroborated by the best Indian
Ocean observations, the error due to deviation came out,
compass aloft 2° W.; steering compass 16° W.; companion
compass 9° W.; and standard compass ½° E. These results,
which I consider to be reliable, to within a very close and
insignificant approximation, shew, as I had long been of
opinion, that we were sailing, and for many days had been
sailing, on or very near to the point of change of the standard
compass. Hence it is obvious that, with our present very
limited experience in the facts of the circuit of the compass,
we can form no reasonable judgment, or even probable guess,
as to what the errors of the adjusted compass will be found
to be, whenever the opportunity, please God, of swinging the
ship in a high southern dip of the needle may be afforded
us. In the meantime, and whilst the wind continues fair, we
must be content with making the best observations which our
determinate course and particular circumstances may admit.

Saturday, March 29 (N. N. E. to N.W.)—Happily, during
the period of night and darkness before the rising of the

moon, which still for a part of the time did us good service, we fell in with no ice. The wind fell at midnight to a seven-knot breeze, but shifting to N. N. E., it soon freshened again, and we made increasing progress up to $11\frac{1}{4}$ knots near midday and still faster in the afternoon. From noon to noon we accomplished a distance of 240 miles by the log. About 4 A.M. our position was reckoned to be abreast of Prince Edward's islands, and within a distance, northward of them, of perhaps 40 miles. These islands being low, with a prevalency of hazy or bad weather about them at this season, are dangerous to the incautious navigator. They have been at different times made sadly memorable as the scene of calamitous shipwrecks. One of these, the loss of the brig *Richard Dart*, occurred on the 19th of June, 1849. She was from London, and bound to Auckland, New Zealand, having on board Lieut. Liddell, R.N., with a party of sappers and miners and several other passengers, including several females and children. Having had no observations for five or six days, the brig's position was erroneously reckoned more than a degree to the northward of Prince Edward's islands. Among the various tracks to Australia from the region of the termination of the S. E. trade winds, which may be prudently pursued from the neighbourhood of Trinidad, the position from whence the navigator is usually able to commence the plan of sailing he may have decided on, the Captain of the *Dart* adopted the northerly "composite" course of 45° to 46° S., which led him into this calamity, the loss of his vessel and about 50 individuals on board, including the gallant officer Lieut. Liddell, who lost his life in his generous efforts to save a lady (Mrs. Felton). The saving of distance over Mercator's sailing should have been (to Adelaide) about 400 miles, besides the gaining of stronger and better winds. (*Nautical Magazine*, 1850, pp. 228, 273).

The clinometer gave a maximum heeling to port of $10\frac{1}{2}°$, and to starboard of 14°, with a movement in the register for rolling of 490 in the 10 hours, or 73ˢ to each roll. The rolling had increased at 10·30 P.M. to 14° port, and 14° starboard.

The day was damp, frequently hazy, and sometimes rainy

—sufficiently though not excessively uncomfortable. Our direction (mast-head compass) was about S. E. ½ E. or S. E. by E., the result being, as I reckoned it, about an E. course true for the day.

Sunday, March 30.—We ran onward S. E. by S. about E. ¼ S. true, with the wind shifting from quarter to quarter or aft, and a speed (the wind sometimes being very moderate) of 7½ to 14½ knots. Observations for latitude and longitude were happily obtained satisfactorily. They confirmed rigidly the conclusions arrived at on the 28th about the compass errors, and would have rendered our progress at once sure and satisfactory, had not a set of compass comparisons (unfortunately not reckoned up until too late in the evening to be repeated) indicated an increase in the error of the steering compass, on a S. by E. ½ E. course by it, from 16° to 20¼° W. deviation. As no such discrepancy had ever before occurred since our entrance into southern latitudes, I was doubtful whether it was due to the error of the quartermaster sent to observe the compass aloft, or whether (one course by the binnacle compass being from one to two points more southerly than usual) there might not be a larger error on that course.

The clinometer gave maximum rolling for the night of 15° to port, and 12½° starboard; mean, 0. The register was at 4247 at 8·30 P.M., and at 10·30 P.M. 4653, with a maximum angle to port of 9°, and starboard 17°.

The morning was pleasant and bracing, the air drier than usual; but about 2 P.M. hazy showers commenced, which soon became more and more wetting, up to regular rain. As we were now (in the evening) in the parallel of the Crozet islands, and were likely, at our increasing speed, to near them soon after daylight, care was taken in steering to allow for the possible change in the deviation, and to get a fair berth to the southward of this second group of dangerous rocks and islands.

Divine service in the morning was conducted in the saloon. Among the passengers of the first class a considerable number preferred to walk the deck or keep in their cabins to assembling with the congregation. It was painful to observe

so many, not only averse to join in Divine worship with the
"Church" on board, but too obviously utterly indifferent
to spiritual things. Nevertheless, we had a goodly assemblage—many plainly decent or earnest participators; and the
service, the chanting and Psalms being very well performed,
was subduing and interesting. I was guided, as usual, in the
subject of address by the Church's selections of Scripture, dis-
coursing from 1 John v. 11, 12, out of the Epistle for the day.

In the evening I did duty in the third-class mess-room—
a large space, quite crowded; probably 100 to 150 were present. The singing was good and cheering. I had left the
selection of one of the hymns to a little party who led the
singing, and was pleased by the spiritual indication in the
selection :—

> " Give me the wings of faith to rise
>   Within the vail, and see
> The saints above, how great their joys,
>   How bright their glories be !" etc.

The second lesson was Heb. i., from which I selected verses
1, 2, 3,—showing the glory of the nature of our Saviour
Christ, and His acquisitions on our behalf and for us, and the
inconceivable blessings attained in His being made *heir* of
all things, and in our joint participation with Him (Rom. viii.
16, 17);—also, " all things are yours," etc. Christ was
viewed as the *Word*—the Word that was God; as the
brightness of the Divine Glory also (John i. 14) ; as the perfection of man, and the image and Godhead of the Messiah;
as the Author of faith and grace, and Preserver of His
people; as now holding the treasures of His life and travail,
His righteousness and death *for us*—for joint participation
with believers—for their felicity, glory, honour, immortality,
eternal life. The confidence (as illustrated) given us by a great
leader : a great captain—Wellington, or admiral—Nelson.
How superlative this comfort and assurance in the " Captain
of our Salvation," God's dear Son ; Emmanuel, God with us !

The attention, earnestness, and feeling, were at once deeply
interesting and touching. The Lord, I doubt not, was in
the midst !

I

Monday, March 31.—A night of good speed—13 to 15
knots—was accompanied by rain and thick weather, the most
ordinary condition of this wet and boisterous region. Through
the Divine mercy and guidance we went on safely—though
for a few moments somewhat in jeopardy from a jerk of the
sea on the rudder having thrown one or two of the helmsmen
over the wheel and forced the wheel adrift; but, happily,
before the ship had time to spring into the wind, as she inclined, the weather helm was got effectually to bear on the
ship. She was scudding at the time with the wind about
three or four points on the port quarter, under a large spread
of canvass including studding sails and royals, blowing a
gale of wind, with a high cross sea—seas from the N. W. and
W. N. W. (true) being still prevalent. But whilst so scudding
at the rate of about 15 knots, I was led to remark, in the
middle of the night when in bed, the astonishing quietness
of the ship's action, repeating the admirable performance
described under date of March 24; she often went for minutes
together, at this great speed, with hardly observable motion
or rolling—though a sea with 20 to 22 feet waves was agitat-
ing the waters. Sometimes the quietness had something in
it quite impressive. During 10 or 15 minutes (about 2 A.M.),
when I was giving special attention to the fact and counting
in intervals of quietness the seconds of time, I noted the
special intervals of 7, 8, 10, 12, 20, and 24 seconds, in which
not a roll could be perceived, nor the creaking of a single
joint or fastening could be heard. There was absolute stillness within board as in a ship in dock, or in the attainment,
in the Greenland seas, of perfect shelter from ice. Fifteen
knots were being run with the only token on these occasions
of a *slight* vibratory tremor in the ship—which from experience I had learnt was the sure sign of a rapid speed—when
everything else seemed to indicate entire rest! The brisk
gale continued to urge us on rapidly.

Previously to this, Captain Boyce had remarked on the
extraordinary fineness of the weather and moderateness of
the winds in this region, usually characterised by storms and
bad weather. Summer and winter, and at all seasons, he had

found the weather alike; on the present occasion it was the finest he had ever experienced.

The region or parallel, indeed, comprising a broad and extended (longitudinally) belt, containing many degrees of latitude north and south of our present or more southerly tending track, is selected by the enterprising and zealous navigator on account of the prevalence therein of strong fair winds, ordinarily blowing hard from the W., S. W., N. W., or other westerly directions. The northern limit of it varies with the season of the year like the trade winds. Hence, for gaining the advantage in respect of probable progress from strong favourable gales, a course of the "composite" kind may be adopted, extending southward in a considerable variety of parallels according to the season; so that, according to Captain Boyce's experience, when the sun is in far south declination, or in the season of summer, a route south of the Kerguelen islands rising to 52°, or further south, may be the best for a short passage, whilst in winter, or in the months of June, July, and August, a height of 40° S. might suffice for the attainment of sufficiently strong and advantageous winds; intermediate tracks, say from 45° to 48°, having the same advantage in the seasons of spring and autumn of these regions. The inducements in these cases to go further south-ward than the winds may require being the shortening the distance by an approximation to the "great circle" course. Against the greater probable advantages from fair strong winds and the shortening of the distance, by taking a high southern latitude, as the tangent of the southward curvature on the Mercator's chart, are to be set the boisterous and often bad weather, the greater risk from icebergs, the risk, in certain cases where observations of the sun may fail, from islands, such as Prince Edward's, the Crozet, and Kerguelen groups, which may lie in or near the route selected, with the increase of discomfort from cold and stormy weather pro-verbially prevailing in the vicinity of these islands.

At noon of this day, we reckoned ourselves (the sun not appearing) in lat. 47° 30′ S., and long. 52° 43′ E., near the meridian of Possession island, the largest of the Crozet group,

and about 60 miles to the northward of it. Soon after daybreak we considered ourselves to be within about 40 miles of
Penguin island, the first of the group we had to pass.

The wind at N. W. (compass) or W. (true) increased in
the afternoon to a hard gale, and ultimately in squalls with
showers of rain and sleet it blew very fiercely. An object
of some alarm and much excitement to be met with at night
in a heavy sea and gale and showery weather was announced
at 7·30 P.M. as being seen. Two icebergs, indeed, showing
ominously in the showery sky, were then on the port beam ;
and soon afterwards a third appeared on the port bow. But
for convenience of description I defer further notice of them
to the record of the 1st of April,—merely adding in this
place that, through the protecting guiding care of a gracious
Providence, we passed safely through the anxious night.
The mixture of feeling and contrast of conduct among the
passengers was very striking and surprising. Some naturally
timid, thoughtful, reflective, were in much serious anxiety at
the realising of a peril hoped to have been escaped ; whilst
others, after a crowding of the decks in the excitement of
the moment, seemed to see nothing in the formidable un-
mapped islands but objects of curiosity, and returned in brief
space to their cards and play and recreations,—indulging in
as much unsubdued levity as if they were enjoying the best
circumstances of a sea voyage, or the perfect security of a
social party on shore ! It might have appeared severe to
have searched for a text expressive of my own feeling, with
the knowledge and experience of ice navigation, of this
exceeding inconsideration and reckless levity of some whose
disregard of everything serious, or aversion even to the
ordinances of religion, had painfully shown itself under much
variety of conversation and incidental intercourse ; but the
text in Eccl. vii. 6* was forced emphatically and repeatedly on
my recollection. Without indulgence in verbal profaneness
or. intemperance, unless two or three incidentally ; without
any distinct violation of the proprieties of conversation or

* " For as the crackling of thorns under a pot so is the laughter of the fool;
this also is vanity."

the volunteering of infidel or ungodly principles or views,—there was such a thorough negative in their manner or conversation as to religion or spirituality—such a marked avoidance, whilst not professing to object conscientiously to the Liturgy of the Church, to the assembling themselves at our or any Sabbath-day ordinances, that no reasonable exercise of christian charity could evade the painful conviction of the state of men of whom it is written, "God was not in all their thoughts."

The birds, which had for many days been numerous, diminished greatly after Sunday evening, especially the albatrosses, a solitary bird of this majestic class being but seldom seen. The prevalent birds were the Cape hen, a small petrel, and a few gulls.

It may be here noted, that on the commencement of severe weather, I was somewhat surprised to see the sailors, men, and officers, in regular costume of our Greenland crews. Caps with protection for the ears with the helmsmen; mittens, strong water and wind proof dresses from head to foot, and, still more characteristically, large draw-up fishermen's boots of the strongest leather and ample dimensions. It was striking to see the youngsters, among the officers, thus practising in heavy seas and sleety decks, the difficult task of managing these heavy feet and leg protectors in walking and pursuing their duty about the decks!

CHAPTER III.

MEMORANDUM.—The preceding pages give winds and courses by the steering compass, uncorrected for deviation or variation, except where otherwise expressed. In the following pages, the winds, courses, sea, etc. are reduced to the true meridian.

April 1, 1856.—I commence this third chapter of my journal, whilst blowing a hard gale at west, with heavy occasional squalls and showers of soft hail or sleet, the *Royal Charter* scudding directly before the wind, or making an E. ½ N. course, under double main and foretop sails, foresail, maintop-gallant sail, fore staysail and mizen topsail; a heavy sea, at the same time rolling after and beyond us with waves reaching in very many cases to 40 feet in height of broad solid-water crests, besides peaks and broken crests of many feet greater elevation!

The notable topics calling for special description under this day's records, were the falling in with icebergs, our second and fiercest gale, and the astonishing performance and behaviour of the *Royal Charter* under this severe trial of her qualities as a sea-boat.

For convenience of description I have deferred the account of the icebergs and the storm, as just noticed under the previous day's remarks, to the present period, when an iceberg clearly within view and a sea at its most majestic attainment of grandeur, afforded more precise means of attempting the description satifactorily; that is, if any description suggested by personal contemplations could accurately convey a picture of one of the sublimest operations of "the stormy wind and tempest," with the exciting and solemnising incident of the dangers of our progress in the period of darkness amid the mighty icebergs of Antarctic formation! These topics lead us backward, a few hours in time, to the evening of 31st March.

Though the line of track up to or beyond the latitude of

46 degrees south had led us within a well-known range of icebergs, yet, when we had advanced so far to the eastward as the 59th degree of longitude—beyond which it appears icebergs are but rarely seen in our parallel,—we were led to hope, and *earnestly* to hope, as the evening closed in with a heavy gale, hard squalls and perplexing showers of sleet or hail, and a greatly rising sea, that the risk of falling in with icebergs had happily been past. But we were disappointed. The "look out," stimulated by promise of reward from the Captain, reported at 7·30 P.M. an iceberg on the port bow. All was now excitement—a mixture of curiosity and apprehension—and the forepart of the poop became speedily crowded with the passengers from below. The object of interest and peril could only be recognised, however, as it gradually came nearer on the port beam, by that characteristic luminosity by which this peril to the navigator is happily and, I may add, providentially indicated. But the disappointing indistinctness in this case, or as some might have thought, the doubtfulness of the reality, were very soon superseded by another announcement from the chief officer, that he thought we were coming to another iceberg ahead of the ship or a little on the port bow. The ominous luminosity now left no doubt of our nearer approach towards another of these formidable incumbrances and perils belonging to this track of navigation. The ship's course being altered somewhat to starboard, brought us sufficiently far clear of it; but, as it approached the beam of the ship in direction, it became sufficiently conspicuous, shining out of a dense black shower in that quarter, to enable us to judge both of its general form and magnitude. Its form was in a waving outline of considerable breath, it might be 200 to 300 yards, with a height possibly of 100 feet or more. It was a white luminosity on a dark or almost black background of a rain or snow shower.

Altogether, three icebergs were discerned through the darkness and haze; but whilst Captain Boyce pursued his course amongst them, some reduction of the sails was judiciously made; and for a time the maintop-gallant sail, mizentop sails, and forecourse, were taken in   At 10 P.M., however,

the weather being clearer, the sky, indeed, by no means so
dark as usual in the absence of the moon, with a slight
Aurora Australis yielding us obvious advantage, the foresail
was again set, and throughout the night our course, E. $\frac{1}{2}$ N.
(true), was pursued mercifully without interruption either
from icebergs or the extremely heavy sea, which had been
greatly augmented under the continuance and violence of the
storm.

A fair estimate of our danger from the icebergs could not
be easily made. In a gale so heavy, with the occurrence of
ice in uncertain and undeterminable quantity and distribution,
no considerate person could be free from solemn thought or
apprehension. But the regulation of these feelings under
such circumstances is a matter of much importance to personal comfort, and with the Christian man matter for peculiar
submission to and repose on the Divine providential will and
guidance. Rationally considered, the special dangers on the
one side are meeting with a succession of icebergs after
shifting the course or partially hauling to, so as to render
the *weathering* of a second or third iceberg, or the getting
a large ship of the clipper class sufficiently off the wind in
time to wear clear of the danger, impracticable,—besides the
possibility of another of these formidable enemies to security
appearing in the line selected for escape. Over and above
this must be noted the special and even more formidable
danger of falling in with detached fragments of ice; lumps
from fractured and wasting bergs lying low in the water, and
without the provision of luminosity for making themselves
visible, yet in masses sufficiently great to stave in the bows
or utterly destroy the strongest existing ship. Such dangers
encountered in darkness, aggravated by showers, or haze or
fog, are not to be rationally contemplated without some
serious or anxious thought.

But on the other hand, the question of danger ought to
embrace the probable extent, number, magnitude, and frequency of icebergs on any given track, in comparison with
the magnitude or width of the spaces amongst them. To the
proportion in favour of safety must be added the effects of

watchfulness on the part of the navigator, forming some pre-vious estimate of his position of danger; the chances of the icebergs being passed in daylight, moonlight, or in moderate weather, or revealed when in the way by their luminosity; and, finally, the results of experience in the comparative fewness of the accidents demonstrated by the small additional premium, and the moderateness of the premium generally demanded by Underwriters for the insurance of ships voy-aging to Australia or New Zealand. So estimated, the result comes out there is risk, and considerable risk, to ships voy-aging through seas liable to the danger of icebergs; yet the risk, taken on an average, and in a merely intellectual or commercial point of view, is but small. In our particular case, the risk from separated fragments of the bergs we passed was lessened by the course we pursued and the steady direction of the gale; as fragments in a gale and heavy sea will always be found nearly astream of the iceberg—that is, in the line of the wind from the berg—to windward or to leeward; but we, in keeping far on one side, did not intersect or come near this usual line of icy debris and fragments. Hence, as a cautionary rule for sailing among icebergs at night, I would recommend the passing them, if possible, well on one side, in respect to the direction of the wind, and not to intersect the stream-line of the ice.

Not seeing more of icebergs during the night, it was hoped that we had passed beyond their region into an entirely free sea. But about 7·30 A.M. I was called by the Captain with the announcement that there was another large iceberg on the port bow. So far as the sight only was concerned, this was an announcement of considerable interest, as I naturally felt it in the opportunity of comparing these mighty *flotations* of the Antarctic with the familiar ices of the Arctic regions. Hastily equipping myself for the exposure to the gale,—for the thermometer was at 39°, and the decks sprinkled with hail, and the wind blowing violently,—I arrived on the poop in good time to see the berg in its best approach and nearest position. It was already nearly on the port beam; and although perhaps a mile to a mile and a half off was abundantly conspicuous, and, as viewed with an opera-glass, discernible in its particular features.

Of the mass, when covered as sometimes it nearly was with the broken dashing waters of a heavy sea, it was difficult to judge. Probably it would be hardly less than 300 yards to quarter of a mile in its greatest width, and possibly 150 to 200 feet high in its loftiest peaks; for it had two pointed peaks rising up like spires out of the solid mass, with one or more smaller elevations of a like kind betwixt or annexed to them. Pictured as it happened to us on the dark storm-like face of a snow shower, its native whiteness shone out with characteristic conspicuousness, though no longer displaying the luminosity of those seen in the darkness of night. The spectacle of this floating ice-island in the assault and burst of the white waters of an incidental highest wave—a wave probably of 40 feet altitude, and much breadth and massiveness of water—striking its steep or vertical face towards the west, and flying upward, throwing a splendid white canopy over the greatest part of the windward elevations,—was magnificent.

We now turn to the performance of the *Royal Charter* under the irregular and very heavy sea, which the continuance of the gale and the hardness of the squalls had stirred up. The prevalent waves were about one-third higher than those of the 25th of March—our first gale. It may be of some interest to describe the modes adopted for obtaining scales of measurement for the height, distance of the undulations, and velocity of the waves. They were, in the main, similar to those I had adopted in inquiries on the waves of the Atlantic—the results of which were communicated to the British Association, at its meeting in Edinburgh, in 1850. As the ship was scudding directly before the gale, and was astonishingly steady and upright for considerable intervals of time; and as, notwithstanding her great length, she found an ample extent for being evenly cradled, as to her general bearings of flotation, betwixt wave and wave, it was easy to find a scale, as I did, by ascending the mizen rigging, by which the height of the ordinary and highest waves, as they might be found to correspond with or to intercept the sea

horizon, could be satisfactorily measured. In this way, ascending by steps, after a first position taken at a guess, I found
that with my feet on the seventh rattling (the eye at $12\frac{1}{2}$),
I had the line of elevation of the prevalent wall-formed
waves. But I attained the interval betwixt the 16th and 17th
rattling before I obtained the measure of elevation of the
waves of the maximum class. Of these, several rose still
higher, but I kept my position till one or more of great
breadth and solidity of mass was seen rising boldly above
the horizon along an unbroken extent of some hundreds of
feet. The height of the eye in this position above the eye on
the saloon deck below was $15\frac{1}{2}$ feet, and the height of view
on the deck, as otherwise estimated, being about 23 feet—
the sum of $38\frac{1}{2}$ feet was obtained for the maximum class of
the prevailing waves. But, as the wave specially noted was
at least two feet higher than the horizontal level, its actual
height could not have been less than 41 or 42 feet. But
from the station of observation, several incidental peaks or
points of waves, often thrown up by the intersections of cross
seas, were observed to exceed considerably the ordinary
maximum, whilst the *tops* of breaking crests might be seen
rising still higher—indicating an ultimate maximum which
could hardly be less than 50 feet.

The next points of inquiry were the distance of the waves
from crest to crest and their general velocity. These inquiries,
however, from the want of regularity in the sea—a peculiarity,
I believe, of the waves of the Southern ocean—could not be
so conclusively determined as the height; but, proximately,
the results were very satisfactory as compared with similar
investigations in the North Atlantic.

The time occupied by the waves in successively passing
the ship, was found, when the ship's rate of going was *about*
12 knots, to be generally (in the case of distinct series of
heavy rollers) about 18 seconds.

With this fact for guidance I ascended the main rigging to
a convenient height, where, being in the centre of the ship,
I could observe when the crest of a wave went beneath the
stern and when it emerged from beneath the bow. This

observation frequently repeated gave me a pretty regular result, as shown by the counting of seconds with a watch at my ear, of nine seconds for the interval occupied in the passage of the ship's length, say 320 feet, and of nine seconds more for the arrival at the stern of the following wave. (The length of the *Royal Charter's* keel is 308½ feet; length over all 336 feet. But I estimated, as I believe, a medium of the rake and overhang, and therefore take about a mean measure or estimate of 320 feet for the ship's length.) This indicated plainly a distance of just about twice the ship's length, or $320 \times 2 = 640$ feet (the length of the keel, or extent from the places of measurement attempted to be taken), for the prevalent space, transversely to their axes, from crest to crest of the waves.

A tolerably correct estimate of the velocity of waves became fairly deducible from these data. A space was seen to be passed over by any particular wave, with reference to the ship's position, of about 640 feet in 18 seconds; but this was of course too small by the proportion of the ship's retreat from the wave, or advance in the same direction, during the same interval. Taking the geographical mile at 6075 feet, we find the quantity of 1 69 feet per knot to be due to each second of time ; or 20 28 feet per second as the ship's progress when going 12 knots. This multiplied by 9, the number of seconds occupied by the sea in running the length of the ship, gives a product of 182·52 feet for the advance of the ship in the same line and within the same interval. To the velocity of the waves, therefore, reckoned on the ground of the ship being stationary—say 320 feet in nine seconds, or 35·56 feet per second, we add the advance of the ship or 20·28 feet per second in the same line of direction, and thus obtain the result, as to the actual velocity of the wave, of 55·84 feet per second, or 33·9 geographical miles per hour—a result the more satisfactory seeing that in making a like experiment on very high waves (30 feet), measured in the North Atlantic, in 1847, I obtained for these waves a distance from crest to crest of 559 feet, and a velocity of 32·67 miles per hour.

Now, as to the action and performance of the *Royal Charter* under this hard gale and mighty disturbance of the waters, the experience we again derived was truly astonishing, and, compared with all my previous experience, what I should have deemed impossible; for by far the greatest portion of the time, I should say four minutes out of five, we had no observable motion, the ship being steady, quiet, and often apparently absolutely still. A minute or two would often pass whilst these heavy waves were rolling harmlessly forward, and but just raising in a slight degree the stem and alternately depressing it, when we might have seemed to be sailing in a sea of extreme calmness in the finest weather. In these intervals of dead quiet, no wood-work, joint, or junction of iron and timber, emitted an audible sound—no creaking was heard,—and at night there was sometimes a quiet most striking in its stillness. Of cases of this perfect quiet in time of heavy sea, squalls, and storm, I frequently noted intervals of seven and eight seconds, of 10 to 12, sometimes of 20 up to 24 seconds, where there was not motion sufficient to break a silence of repose like that of dock or harbour. Hence, notwithstanding the lurches or rolling, extending sometimes to 15° or 20° on one side, and perhaps once in several hours to 30°—the maximum never exceeded up to this time,—a rolling inseparable from a progress directly before the wind, in difficult steering and with squared yards, —yet most occupations below, with ladies as well as others, went on as usual; and, when the state of the decks as to dryness would admit, exercise on deck likewise. Thus when the waves were at the highest—when elevations of 40 feet and upwards were rolling around and beneath the ship— Mrs. Scoresby accompanied me on deck for exercise, and to view, in an instant of bright sunshine, the sublime scenes around, and found no difficulty in walking the poop deck, which was unencumbered and dry. She accompanied me, too, along the gangways extending from the poop to the deck-house, and from thence to the broad and spacious forecastle up to the very bitts, within a few feet of the stem,—and even to this extent, and along a range of 320 feet of deck

and platform, the progress was perfectly easy, and at the
time the whole extent was clean (unusually so, almost to
whiteness), and dry from end to end. Again I may remark
that our meals were always served up to the minute, in the
handsome services, covers, and appendages, before noticed.
Everything cooked with the same effectiveness and complete-
ness in storm as in calm—fresh provision, roast and boiled,
in fowls, mutton, pork, etc., unfailing and abundant,—pastry,
puddings, and the variety of niceties for each particular
course, always ample and good of their kind; so that in
speaking of the servants and cooks as part of the ship, and
of the ship as a thing or creature of life, I may say that the
Royal Charter had no consciousness of bad weather, and
made no signs of complaining in storms or heavy seas. During a heavy squall, for instance, at dinner-time on this day—
a fierce snow storm for a period, the wind blowing tremendously—no effect whatever was produced on the comfort of
those who sat at table; and a wine-glass I had emptied stood
for many minutes entirely unsupported betwixt the protecting bars of the table, and it was only liable to be disturbed
by some particular lurch which might happen to occur.

Again, in regard to pitching and " sending," the action
of the ship was equally remarkable, both for the easiness of
the motion and the smallness of the inclination of the keel
from the horizontal level. A 40-feet wave, on its entrance
below the stern or counter of the ship, whilst the bow was
exactly in the lowest or most depressed portion betwixt crest
and crest, should raise the stern, as from the simplest view of
the case it might seem, to at least its own elevation, or give
an angle of inclination to the keel of about 7°; but no such
measure of pitching or " sending" motion was ever observed
—probably not above half as much.   For, in no instance in
scudding, did I ever observe the bow of the ship plunge nor
the stern rise to anything like the position apparently due to
the elevation of the passing waves.   The action, indeed, was
obviously of this nature: from the admirable adjustment of
the ship's *lines* of construction, forward and aft, the loftiest
wave, on its reaching the stern-post below, exerts its lifting

tendency, not abruptly or suddenly, as where the *quarters* are heavy and the *run* thick, but very gradually, so that the disturbing force, passing beyond the place of greatest influence before its due action is realised, becomes modified and reduced.

These principles are no doubt in operation in every tolerable model of marine architecture, but not to the degree of perfection in which the tendency to assume horizontality of position, and to receive the least possible disturbing effects from the most formidable disturbing causes in the action of rough, irregular, or heavy seas, has been attained in the modelling and building of the *Royal Charter;* and whilst similar results in kind will be found to have been obtained in very many or most of the scientifically constructed and splendid clipper and other first-class ships of this important age, I should much doubt whether in any single instance the approach to perfectness of the model of the *Royal Charter* has been exceeded, or even—in all the elements of the perfect "sea-boat," as adapted for these southern regions, proverbial for turbulent seas and boisterous weather—been equalled.

The view from the poop and forecastle which my wife and some others of our ladies witnessed for considerable periods together, even in the height of the gale, was one, especially during the favourable occasions of bright sunshine, of sublime magnificence; whilst the general view of the tumultuous waters as we looked astern, as the ship was scudding before the storm, and as we marked the waves rolling perpetually onward, and overtaking in succession the swift-sailing ship, presented a picture of striking grandeur. The more threatening storm seas, as every now and then they rose high above our position, and intercepted (astern and on the quarters of the ship) every other portion of the mighty waters, could hardly be contemplated,—I ought to say, could not *rationally* be contemplated, without awe ! Nor was the action of the ship under the mighty disturbance the least impressive or least striking feature in the general picture. As if endued with life and instinct, the ship seems to anticipate

the approach of the threatening mountain wave, rising gradually abaft before it reaches her, yielding as if in respect
her threatened quarter, and gracefully depressing her opposite
bow as if doing courteous obeisance to the sovereignty of
the waters, so that the infuriate-looking wave passes harmlessly beneath her stern or starboard quarter, and bursting
out from below on her port beam or bow, rushes furiously
away and falls over, in its haste, into a foaming breaker,—
as (to use reverently the sacred emphatic figure) a lion from
the swelling of Jordan!

In the general glory of the scene we saw repeated, with
additional majesty and grandeur, the effect of the sun's light,
illuminating portions of the waters *within* their mass, showing out the transparent colouring of emerald green, and
throwing a broad blaze of glory from the surface of the
leeward waters—spreading, by reason of the vast variety of
reflecting surfaces and infinite modifications of the obliquity
of their reflecting planes, a silver-light expansion embracing
within radii from our position, as a centre, to the horizon of
more than three points of the compass in breadth.

Our picture, or attempt to sketch the sublime scene, has
been contemplated as without; in ourselves we were a lone
vessel in the midst of an apparently immeasurable expanse
of waters. Yet we were not altogether solitary. Our companions, the sea birds, were not only numerous, but their
habits, freedom of action, and powers of motion, superior
even to the speed of the extraordinary tempest, was an interesting feature in the picture. Among these the albatross
always excited admiration for his ease of motion and gracefulness on the wing.

Our day's progress from noon to noon (nautically) comprised a distance by log of 299 miles, on a course which
should have been just *east* (true), but in consequence of the
frequent "yawing" of the ship to starboard, and the general
prevalence of her steering that way, we made *southing* this
and the preceding day (no observation being had yesterday) apparently to the extent of half a point or more. Good
observations for the day gave our meridian latitude 48° 8,

being considerably further south than, on our projected track by Prince Edward's island, had been desired or intended. Our longitude was 59° 28′ E. Though the gale was so strong we made by no means the same progress which would have been accomplished in finer weather and easier seas.

The ship had indeed gone 14 knots or 15 within the civil day (from midnight to midnight); but for the greater part of the day, her speed, by reason of the reduced sail (no studding sails, on account of the heavy and frequent squalls, being fitting or safe) and the wind being generally *right aft*, and the sea so heavy, did not exceed 11½ knots, and was rarely maintained at 12.

The clinometer, at 8·30 A.M., had marked the maximum rolling for the night of March 31—April 1 at 32° to port and 24° to starboard, with a register for rolling of 8607; and advance in 24 hours of (8607 — 5500) 3107.

The barometer, which had been as low as 29·145, had risen by 10 P.M. gradually to 29·483. The temperature in the early morning was 34°; at 9 A.M. 39°. The soft hail and snow were so abundant in the showers that snow-balls were made, and amusement extracted out of the (to us) ungenial curiosity.

Few birds were now with us; the great proportion, with almost all the albatrosses, had left our company on Monday.

The night again, happily for our anxieties on account of icebergs, although moonless till past midnight, was peculiarly light. This seems to be a peculiarity of the southern hemisphere. The stars in intervals of the showers blazed out brilliantly. No compass comparisons could be made on account of the swinging of the cards—the steering and companion compass being rarely steady, and often swinging through an extent of three or four points.

As to personal feelings, we had a very unquiet night and comfortless day. Whilst the high seas from the west were yet unsubdued, the wind shifted to the S. W., and though the barometer had risen to a rather high position (30·010), the gale continued with little abatement and the squalls almost as fierce as ever, and even more frequent, soon raising

K

up a second and cross sea from the S. W., which before noon
exhibited fresh waves of 36 to 40 feet! Not always to be
marked distinctively, yet every now and then a series of
waves from the west forced themselves into notice by the
turbulence and height of the peaks of intersection with those
from the S. W. The ship, though still acting wonderfully in
comparison of most other ships, became *uneasy*—that is, for
the *Royal Charter*. The clinometer marked a maximum (at
8·30 A.M.) of 35° to starboard and 24° to port, with an advance
of the numerical register to 10,401. And as to discomfort, we
had to endure an external temperature of 37°—40° for most
of the day, with little amelioration below; whilst from the
wetness of the decks from frequent showers of snow and
sleet, and occasionally spray flying on board, the taking of
exercise, with any degree of ease or security, was very difficult, often impracticable. Wrapping up in our winter coats,
when below, did not suffice for the retention of warmth or
the smallest enjoyment of comfort. The heating of the saloon
may be a difficulty, but surely not an insurmountable one.
I never before sailed in a ship without stove or stoves; yet
we are told these Australian vessels, changing the temperature from 84° to near freezing in 20 or 30 days, and having
to encounter severe frost, storm, and snow, in returning by
Cape Horn, do not need stoves! In our case—as far as I
can understand what might be practicable and safe—I cannot
see any difficulty in heating the saloon by a stove or other
heating apparatus in the engine-room (if stoves *in* the saloon
are objectionable), where the distance to be conveyed, of hot
water or air, would only be a few yards to the forepart of
the saloon.

All our compasses, under the heavy cross sea of this day,
went wild as to steadiness! Both the adjusted compasses
on the floating principle oscillated continually, sometimes to
the extent of four, or as I was told even six points, or more.
The compass aloft was equally unsteady. I proposed to
weight the cards of these below, and in trying the effect of
two copper pennies on one of them the oscillations became
greatly reduced. The excessive unsteadiness, indeed, might

have been reasonably ascribed to the changes of the induced magnetism on the two sides of the ship, so heavily and perpetually rolling whilst now under a large southern dip and much increase of the magnetic force, and was probably not wholly uninfluenced by this cause ; but the corresponding unsteadiness of the compass aloft—which could not be materially affected by such transient and distant magnetic changes —seemed to indicate that a great part of the disturbing action lay in the swing of the ship and its synchronism with the oscillations of the cards.

Our course during the day was intended to make due east (S. E. by binnacle compass, with 30° W. variation and about 16° W. deviation), but the bad action of the compasses and the tendency by defect of steering to swing off, under the high sea and heavy squalls—the gale continuing, with only partial abatement, until the afternoon or evening,—produced an actual course of about a point of northing. This gave our latitude at noon 47·23, long. 66·42 E., as obtained by good sights of the sun. Subsequently (P.M.) guarding against this error, we made our intended course with considerable accuracy, which brought us to the meridian of Bligh's Cap, a northern and western islet associated with the Kerguelen islands, about seven in the evening, and within a distance, by estimation, of 60 miles. Our speed from noon to noon was 11½ to 12½ (once for a short time 15) knots, so as to accomplish a distance for the nautical day of 301 miles. But the heavy cross seas and hard squalls much hindered our progress, keeping us back as to a full spread of canvass as well as reducing our speed by the turbulence of the surface over and through which we must pass. Damage was done to the maintop-gallant sail in one of the squalls, and a lower studding sail forward ripped away from its sustaining yards in another, but the damages were not difficult of reparation.

The evening was milder, and though the night was darker than usual, we passed, under moderated wind and sea, a tolerably quiet night.

Thursday, April 3 (W. to N. W.)—During the night we completed the meridian of the eastern limit of the Kerguelen

K 2

island—having passed the nearest part, Christmas harbour, at the distance of about 80 miles—a distance which probably tended to our comfort if not advantage, as storms of the greatest severity are proverbially associated with its vicinity. Sir James Ross describes the gales he experienced there, even in summer, as extraordinary in their fierceness; for no season seems to reduce or subdue its tempestuous character. Our near approach to this island brought back our feathered associates: albatrosses, with Cape pigeons and small petrels, appearing in much greater numbers than at any former period of our voyage. Their numbers and boldness tempted out our sportsmen again to pursue the inconsiderate and cruel sport of firing at them, and I believe several hundred shots were fired at these confiding creatures.

The wind had become somewhat steady in the N. W. quarter during the day, veering, however, more aft at night. Our speed, taking the whole day, varied from 8½ to 13½ knots; for the nautical day we made 249 miles of distance on a true course of E. 2° S., giving a latitude at noon of 47° 31', and longitude of 72° 28', just about half the *eastern* longitude to be accomplished, please God, in our contemplated route to Melbourne.

The rapid rising of the seas, in this extraordinary region, the height and irregularity of the waves, are circumstances of singular curiosity, and, in a hydrodynamic view, of considerable interest,—the first-mentioned peculiarity having been well indicated in the descriptive remark of an intelligent friend, that "the Indian ocean and seas are always ready for any wind." The cross seas are almost constant; and in cases of sudden or rapid cessation of a gale, the waters seem to rise in heaps from the innumerable peaks thrown up by the crossing of diverse waves. Thus, before long after the shifting of the wind to the N.W., we had a new sea from its quarter, a residue of the old great waves from W. and W. S.W., with additional heavy rollers coming in conspicuously at intervals from the S. S.W. In foggy or hazy weather, or in the midst of a snow shower, the turbulence thus elicited seems almost marvellous—occasioning the popular notion that whilst (truly) the sea is subdued by heavy and continuous rain, it gets up in a fog; a fallacy no doubt suggested by the limitation of the view, and to the definition of the contour of waters so circumscribed within the limits of vision, showing, perhaps, a single wave only, and this apparently magnified from the misjudgment of distances.

It is not therefore that fog magnifies objects, as the popular opinion erroneously goes, but that, producing a deception as to distance and intensity of colour, the judgment by the eye conveys an erroneous impression. Thus, in the Arctic regions, the erroneous judgments formed of objects obscurely seen in a fog by inexperienced or inconsiderate persons are sometimes quite ridiculous. Small lumps of hummocky ice on first coming into view, especially from a boat, have been mistaken for huge masses or pieces of icebergs. I have known a seal or little narwal, on rising quietly to the surface of water, to be mistaken by some of the crew of a " boat on watch " for a whale, and the order given to pull away whilst the officer of the boat was getting his harpoon in hand. Nay, I have known in cases where I was personally present in the boat, a bird—a fulmar gull—on quietly floating indistinctly within vision in a dense fog—to be mistaken for a narwal (which might be 15 or 16 feet in length), or even for an animal of larger magnitude.

Hence, there can be no doubt, I think, that the notion of an elevating action of fog on the waves is a complete delusion; the result of a false impression of increase of magnitude by reason of misjudgment of distance, and colour or quantity of light.

Occasionally, for a few hours, there was some diminution in the sea; but it seldom was allowed to decline much—never, that we have yet seen in these regions, to repose. So that within this day, at least at 8·30 A.M., the clinometer registered a maximum angle of 38° to port and 14° to starboard; with a register of 4018. My impression, however, is that the ship did not roll to anything like the extent of 38°, as on occasion of decidedly heavier rolling, maxima have been indicated of at least one-third less. The pendulum receiving perhaps a very sudden impulse, or being resisted at first by the friction in starting the wheel-work, oscillates beyond the true direction of terrestrial gravitation, and so registers a maximum beyond the truth.

The morning though chill and cheerless had been partially dry, but rain set in about 1·30 P.M., and hardly ceased afterwards for the rest of the day. The waves had fallen to from 16 to 22 feet; the temperature which had been 37° yesterday now rose to 41°; we became comparatively warmer, and had a tolerably quiet night.

For the first time, in an interval of six days, did the state of the weather, either from storm or rain, allow conveniently the undertaking of dipping-needle and other magnetical experiments; even compass comparisons during three days had been suspended because of the wild swinging of the compass cards of all three fixed compasses, rendering any useful observations impracticable. The weather, however, being more favourable about noon, I began a series of dipping-needle experiments, which proved so interesting, in respect to changes, that I was led to pursue them after the setting in of rain. The dip at the standard station (near the middle of the ship) which on the 28th of March had (on the same course, S. 41° or 42° E.) been 62°, was now found to be 70½°; and the magnetic force in the ship, as indicated roughly by the time of oscillation of the dipping-needle, had changed in the ratio, inversely, of $2·25^2$, $2·08^2$, since March 17th; the change in the terrestrial force, according to General Sabine, being from about 0·89 to 1·51. But an important fact, already referred to, of the gradual reduction of the *differences* betwixt the direction of the dip, as modified by the ship's disturbing influence, and of the true terrestrial dip, after a considerable continuance of the ship in a new magnetic hemisphere, became in these last experiments most striking and conspicuous. Thus, comparing our *observed* with the (supposed) *true* inclinations of the needle for the period, only when the ship's head was always in a direction betwixt S. and S. E., I find approximately the following series of differences:—

Terrestrial Dip South	10°	15°	23°	28½°	36°	44°	47°	50½°	57¾°	62¼°	70¼°
Observed Dip .	6¾ N.	½ N.	8⅛ S.	13¼	23¼	33	40½	43	52	62	70½
Difference .	16¾	15½	14¾	15¼	12¾	11	6½	7½	5¾	½	—¼
Angle of heeling .	7 st.	9 st.	9 st.	4 st.	5 st.	10 st.	3 pt.	3½ pt.	3 pt.	0	6 st.
Days of interval .	...	¾	1	1	2	2	1	2	3	3	6

Whilst the series of gradually diminishing differences or of the gradual reduction of the ship's disturbing influence on the dipping-needle is sufficiently conclusive, there are discrepancies, within a limited extent, which it is not easy to explain. The defects of the needle made use of might account for *small* discrepancies, to an extent possibly of a degree or even a degree and a half, but, I think, not much larger. Yet two other elements of disturbing or changing magnetic action remain, which are the peculiar and modified effects of induced magnetism on the ship by varying heeling positions, and the effects of more or less mechanical action impressed on the ship's fabric, and tending to facilitate the adjustment of the ship's *central* dip to that of the terrestrial dip.

The experiments made on the two sides of the poop deck, and at two separate stations, gave results perhaps more curious, which when duly considered and discussed may be more important. At present I shall only risk two observations: that not only does the dip like the deviating action differ essentially on the two sides of the ship, but that the differences seem to be subject to progressive alteration, and that the *sign* of the differences appears also to change.

Friday, April 4 (W. S. W., N. W., N. N. W., S. W.)— A quiet wet night—wind and sea more moderate, but an uncomfortable day because of almost constant rain. Everything below cold and damp; scarcely any possibility of taking exercise; no means of drying wet clothing, nor any mode, whilst out of bed, of gaining heat. Yet this deprivation of fire or stove throughout ships from England to Australia seems to be the rule and system—if not a *bad* one, certainly an uncomfortable and wretched practice.

The range of the clinometer at 9 A.M. for the 24 hours previous was much less than usual of late, 14° to starboard, and 15° to port.

We made an east course for the whole nautical day—from noon to noon,—making 259 miles of distance, at a speed ranging from 8½ to 13½ knots. At noon we had reached the longitude of 78° 48′ E.

The wind, which had been moderate in the day, shifting from the quarter to aft and the converse more than once—began to increase towards evening, and as indicated by the declining barometer, which fell to 29·368, soon increased to a heavy gale. At sunset, and for some time afterwards, it was at W., right aft. The night altogether was dark and more severely stormy than we had yet had it. The squalls were tremendous,—I might say terrific. Once, after the the shifting of the wind to the W. S.W. to S.W, the ship, by the incaution of the helmsmen, in a sudden and severe squall, came up into the wind when scudding under double topsails and maintop-gallant sail, and occasioned for several minutes, whilst the top-gallant sail was being got down, and the head-yards braced forward, no small anxiety and alarm,—as she lurched heavily to leeward. But, happily, she soon "payed off," without suffering loss of canvass or damage, an escape from danger which we could hardly have hoped for, so completely had the sea risen to the formidable high and cross dangerous character to which it attained during the night. It was wonderful and merciful in Providence to enable any ship, under such a quantity of canvass in a furious gale, so to escape unscathed!

Saturday, April 5 (W. S.W., W.)—It blew hard all night with a frequent repetition of the now prevalent fierce squalls. The sea, by reason of a shift of wind to the S.W., and heavy rolling waves coming nearly on the ship's beam whilst there was no abatement of those coming upon us abaft or astern, became higher or at least more strangely turbulent than we had yet seen it. It hove the ship about in rolling and lurching violently for such a ship and sea-boat; and not at night only, but during the whole day, even the *Royal Charter* seemed to have set aside her good behaviour, and was, as any other ship under like circumstances must have been, very uneasy. The manner in which the high pyramids or peaks of water, thrown up at the intersections of the two classes of seas, assailed the ship occasionally in concave or bent masses just in the act of falling over into breakers, rendered it impossible for any mechanical structure in the world altogether to evade their assaults. Several seas were thus shipped, the heaviest of which, or what appeared to us to be the heaviest, struck the ship a heavy staggering blow on the quarter, just as the breakfast-table was being spread, and threw on board a large quantity of water, some of which (considerable in amount) penetrated the various apertures of the sliding windows at the sides of one of the saloon skylights, and rendered it necessary to renew the table cover and other appliances which had just been arranged. I notice this small incident for the purpose of mentioning that for the first time since our sailing from Plymouth was the breakfast delayed for five minutes beyond the specified time (now half-past eight), and then it was found as complete, both in manner, sufficiency and quality—comprising hot rolls, eggs, and ham, porridge, and various other dishes—as in the calmest weather.

In respect to the peculiar turbulency of sea, it may be noted, that not only were the more regular waves seen rising to the height doubtless of 40 feet or more, but the irregular crossing peaks rose vastly higher. The cold and storm prevented me, unfortunately, attempting accurate measurements early in the day, but I could not be but satisfied, and it was the conviction of Captain Boyce and others, that at no time had we had to encounter such vast disturbances of the " mighty deep " The quantity and magnitude of the white crests—the " combings " of the waves — toppling and breaking over in every direction obviously exceeded by far anything we had previously witnessed. Ordinarily, I think, these white combing crests do not extend very wide in continuity—perhaps 50 to 80 feet? But here, in this case, I remarked crests of white

foaming water extending in continuous breadth to 200 or 300
feet,—for I could not estimate some of the broadest, which
were sufficiently near for the comparison, at less than the
extent from stem to stern of the *Royal Charter*, or above 300
feet! As some guide in the question of what the height
of the waves in the early part of the day and height of the
gale must have been, and of the magnitude and breadth of
the breaking crests or summits, I may mention that near
the period of sunset, after the gale had long subsided, and
eight or ten hours after royals and studding sails had been
set, I still found, on careful measurement, the majority of
the regularly formed waves now coming from W. S. W.,
reaching a height of 24 to 26 feet, and the larger waves
rising considerably above the horizon, when, in the mizen
rigging, my level of vision was elevated to above 33 feet,—
so that the higher waves could not be reckoned at less than
35 perhaps 38 feet in unbroken crests; and at this time I
noticed the breadth of some of the breaking or combing
crests as yet extending to from 200 to 300 feet!

With all this roughness of the element through which the
labouring ship had to force an undulating pathway, we yet
made, as certified by observations, a good day's work. From
noon to noon we had made good a distance on an east course
of 318 miles, and had run down a meridional breadth of
nearly 8° of longitude. This brought us—through the furtherance of a gracious Providence—to within 58° of longitude,
and within a distance of about 2600 geographical miles of our
contemplated port, during a voyage from Plymouth of only
49 days—a progress thus far never, I believe, equalled by
sailing ship or even full-power steamer.—D. L.

A celestial phenomenon which occurred on this day deserves at least a passing notice—the solar eclipse; which,
had we been ten days sail in advance of our present position,
we might perhaps have observed as a total eclipse. The day,
unfortunately, though not so black and cloudy as many to
which we have been lately accustomed, was generally cloudy
—admitting only at considerable intervals a clear view of the
sun. This deprived us of the chance of noticing the com-

mencement, or observing the largest measure of the obscuration by the body of the moon. I got, however, a capital view
of the phenomenon, by means of one sight of an opera-glass
screened by a coloured glass.

Sunday, April 6 (N. N.W., variable, W., W. S.W.—A
storm of a new order—as far as my personal experience went,
as well as that of many of the oldest sailors on board—fell
upon us. The weather had been more moderate in the previous evening, reducing our speed for seven hours to 9½ down
to 8½ knots. But the state of the sky—covered generally with
a dark grey stratum of cloud, varying in density of patches
and of no very definite denomination, and frequently darkened
in a further degree by the drawing over of squall-showers—
was not indicative of a return to fine weather; whilst the
great and rapid fall of the barometer threatened a renewal of
the storm. But the result exceeded even for this boisterous
region all our expectations. The progress of the barometer
and thermometer I here note:—

| Civil Time. | Barometer. | Attached Therm. | Temperature. | |
			Air.	Water.
4 P.M. . April 5th	29·609	51°	43°	50°
6 P.M.	29·561	52		
8 P.M.	29·533	54	42	44
Midnight, April 6th	29·320	57	41½	44
4 A.M.	28·895	54	48	45
5 A.M.	28·869	54		
6 A.M.	28·825	54		
7 A.M.	28·816	54		
8 A.M.	28·803	53	50	47
9 A.M.	28·725	55		
10 A.M.	28·725	55		
11 A.M.	28·769	55		
12 Noon.	28·798	55	46	44
2 P.M.	29·020	56		
4 P.M.	29·083	54	41½	44
6 P.M.	29·220	54		
8 P.M.	29·228	54	40	43
10 P.M.	29·285	54		

The storm, probably an extension beyond the usual limits

of the cyclone, not unfrequent in the Indian ocean at this the " hurricane season " of the year, may be said to have commenced about 2 A.M. in a fierce black squall, that is, a squall of marked severity with rain of such an obscuring thickness as to involve sea and sky in a general darkness. Had this occurred whilst we were in the region of icebergs, our situation would have been a fearful one. It came on at N. N. W., and after blowing fiercely for a couple hours— occasioning the maintop-gallant sail and mainsail to be taken in—it lulled and became variable, whilst thick rain continued to fall. But at 5 A.M., veering to the W., it came on blowing very heavily, with fiercest squalls reducing the sails to *lower* topsails, foresail, and foretopmast staysail, occasioning the lowering down for the first time since leaving Plymouth of the upper main and foretop sails,—by which the effect or result of close-reefed topsails is produced without the ne- cessity of a man going aloft. Under this sail we scudded through the height of a storm, amounting in violence to that of a hurricane, and mercifully were carried on our way safely, though not without peril, during six or seven hours of tremendous violence, with a continuance of little abated intensity until about 4 P.M., when it sensibly moderated.

In attempting to describe gale after gale, and tempest after storm, I have not contemplated these grand operations of the elements, or their effects at sea, as anything remarkable in *their kind*, or as exhibiting any phenomena unfamiliar to the sailor or to the practised voyager across the great oceans of the globe,—but I have given the description as part of a particular journal, written not for sailors or experienced sea adventurers, but to put on record for any who may hereafter read them the impressions made on my own mind, and the results of some habit of observing and experience in noting the more remarkable features and phenomena of storms at sea. Hence the descriptions heretofore given, however particular in the details and accompaniments of the tempest-stirred waters, exhibit no features, except as regards magnitude of waves, but what belong to what may be called the every- day experience of sailors, and imply no particular danger to a good well-found ship, properly navigated; but the storm now described, was one of those more terrific disturbances of the elements, which, though by no means unfrequent in certain tropical regions of the globe, necessarily involve, particularly when a ship is attempting to avail itself of such mighty forces for the pursuance of a voyage by scudding before it, risks to ship and passengers, whatever the class or perfection of the ship may be, of no ordinary kind.

The more obvious and characteristic risk in scudding before a cyclone or hurricane, or other tremendous disturbances of atmosphere and ocean, is the " broaching to" of the ship on coming suddenly up with broadside or bow to the wind by failure of the helm or steering. This may possibly happen by neglect or ignorance, or trepidation, under a stroke of the sea on the rudder or afterpart of the deck, of the helmsmen; by failure of any of the gear, especially the wheel ropes, connected with the steering arrangements,—and the effects might be most disastrous. In our case, when scudding before this hurricane of wind and tremendously high and cross seas, there was one special peril—the wheel ropes having more than once failed during the voyage. " Relieving tackles " indeed, were in readiness, and men kept " aft " to manage them; but as the tackles from the encumbrances about the tiller could not be kept in full connection, we could not but feel great misgivings as to whether, in case of necessity, they could be hooked on in time to prevent disaster. This, I say, (from circumstances, however, which on arrival at port may be in a great degree remedied) was our particular source of anxiety and danger; for whilst the wind howled with terrific violence, the sea was lashed into the most tremendous forms of disturbance. Under the best application of watchful and intelligent seamanship on the part of the Captain, supported by an active mate or two, and a selection of the most able and effective seamen on board, we passed—under the blessing of a good and gracious Providence—the ordeal of many hours of perilous scudding, subjected to many hard blows from heavy breaking cross seas—in safety,—and without damage (except to one of the quarter-boats) to ship, sail, or spar.

Suffering from cold and sore throat, I could not venture to brave the storm on deck as I could have wished, or do more than observe the effects of the furious storm from the " companion " and poop deck, and after some abatement of the wind from the mizen rigging.

At 10 A.M. the scene was awfully sublime, and shortly afterwards, about 11 or 11·30 A.M., it was in its highest condition of terrible magnificence. The continuance of the wind for several hours steadily at W. (the direction of a previously existing swell) produced waves of the most formidable magnitude; whilst the sea, from its commencement at north, and a former sea from the south-westward, threw up perplexed waters into the most strangely tumultuous peaks and crests and other forms of waves. The sea was to me a new phenomenon. Even in the terrific and devastating hurricanes of which I had so often read descriptions the sea has rarely time to gain the enormous height it now had with us—a height frequently of 40 feet—regular waves rolling in the direction of the wind, and incomparably higher peaks and crests produced by crossing waves. Here, too, every feature of the tempest was set forth in grandest and most awful magnitude and sublimity. The fearful force and grandeur of the waves—the fierce howling of the storm—the novel and majestic magnitude of the crests and peaks and broken summits—the peril to ship and life in the event of an accident to the helm in scudding—the glorious action, as I may call it, of the ship under these tremendous disturbances—and the drift sprays, confounding sight as an atmospheric haze,—gave the deepest interest to this memorable scene! These features of grandeur were made more impressive by not unfrequent gleams of bright sunshine, penetrating amid the broad cumulus-like masses of cloud which drifted across the upper sky, and throwing beams far from cheerful into the midst of the exciting scene— an incongruous glare of heavenly light which threw the rest of the picture into more striking contrast—and which, on the coming over us of the rain or snow shower of the fiercer squalls, painted the dark threatening astern with more ominous blackness. Of these mighty tempests in the Indian ocean, and hurricanes and tornadoes elsewhere, Lieut. Maury, U. S. Navy, in his usual graphic and forcible style, says,—" The winds " (at this period of the year, and under the specified circumstances of " the Mauritius hurricanes, or the cyclones of the Indian ocean,") " breaking loose from their controlling forces, seem to rage with a fury that would break up the very fountains of the deep."

As to the condition of the mighty disturbed water, with reference to Him by whose power it is so raised, and by whose Providence it is so restrained, the expressions in the " Forms of Prayer to be used at Sea " in our admirable Liturgy, are striking, where we say, " O most powerful Lord God, at whose command the winds blow and lift up the waves of the sea, ... now we see how terrible Thou art, and all Thy works of wonder; the great God to be feared above all." And then, in the thanksgivings for preservation under such threatening circumstances, we are instructed to say, " O most mighty and gracious God, Thy mercy is over all Thy works, ... Thou hast showed us how both winds and seas obey Thy command," etc.

But let me attempt to describe more of the scenes aboard as they appeared to me from my posts of observation, looking first at 10 A.M. from the main companion, and, subsequently, in the height of the hurricane from the forepart of the poop deck. The sea, as I have remarked, was to me a new unseen phenomenon, or aggregate of phenomena. Two or three circumstances were of this peculiar and characteristic order. I had seen and observed the action of a heavy gale coming on suddenly many times in the Arctic seas, but it was generally in a smooth or but moderately undulating sea, where, before waves of the first class had time to form, the secondary and missive seas, first obeying the influence of the wind, would be perpetually crested with white water, and this lighter portion of the liquid element carried forward and spread abroad in a low stratum at the surface of drift spray. But here the phenomenon, whilst corresponding in its source and nature, was of an extent, in quantity, density, and height of distribution, to which I had seen nothing approaching, or giving an im-

pression of the strange effects I now contemplated. The
drift spray was the produce of every wave and variety of
wave. No wave could keep pace with the legitimate demands, in hydrodynamic law, of the wind's terrible vehe-
mence. Waves of 40 feet height, which satisfy the greatest
demands perhaps of any of our North sea or high northern
Atlantic storms, bore no adequate relation to the impetuosity
of this hurricane tempest. A sort of surface impetus seemed
to be given, forcing the crests of the loftier waves into a
velocity so much beyond the motion of the regular undulations, as not only to cast almost every peak and summit into
the form of a breaker, but in some cases to give such a degree
of magnitude and breadth to the breaking summit—as one
mass of white water labouring forward after another and
retarded by the dimimished velocity of that before it—that
the main surface *behind* some of the mightiest waves would
present but one unsubdued and wide spread breast of foam—
a phenomenon I had never seen but in waves breaking over
broad masses of ice or over an insulated shelving rock! Then,
as if impatient still of results proportionate to its mighty
forces—or as if indignant at the tardy and imperfect obedience—the vehement storm not only blew off the lighter
summits of the foaming crests, but actually seized upon great
masses of the roaring peaks of crossing seas, cut them off as
it were from their legitimate support, and drifted them away
with the measure of its own vast speed in that form of storm
spray to which I have referred. Thus the quantity of drift
from this source and that of the minor waves constituted a
stratum of haze so thick and dense that, whilst the sun's rays
obtained free ingress, vision from the poop of the *Royal
Charter*, where the eye was about 22 feet above the level of
the sea, was limited in all directions, and for hours of continuance, to about the third part of a mile !

Another noticable particular arising from the phenomenon
just described, and illustrating the peculiarity and severity
of the storm, was the extent and manner in which the
whole surface water was embossed by the residual foam of
the curling and breaking waves. In ordinary gales this em-

bossing of the surface is limited mainly to broken parallel streaks, taking the direction of the wind—for which it forms a convenient guide—with occasional patches not so cast into shape; but here, besides the prevalent wind streaks of residual foam, the sea generally was so thoroughly marked by it, that on careful observation I could scarcely detect a single square yard of surface that was free from it. Such a circumstance I had never before remarked, at least in the midst of a gale, where regular waves of the highest measured magnitude had had time to form.

The primary waves on this occasion were in many cases of great extent in one undulation and crest. Many were estimated as spreading out to an extent of a sixth, a quarter, if not of half a mile. The extreme height in the fiercest of the gale I could not, being so unwell, attempt to measure. But in the course of the afternoon, after two or three hours of perceptible diminution of the storm, I found the prevailing seas of some 30 feet height, and numerous higher waves of broad extensive solid-water crests or summits, ranging to 40 to 42 feet or more, with knots and broken crests some yards higher.

On board the ship, we notice the sails spread to this tempest, comprising lower (corresponding with close-reefed) main and fore topsail and *whole* foresail, which was carried without difficulty; for the sail, being far too broad at the foot for full extension, became self-reducing as the heavier squalls forced the body of it forward into a position almost horizontal. The topsails, too, presented an admirable instance of the great advantage of Captain Howes' (U. S.) plan of double yards and sails on the topmasts, avoiding the necessity of reefing by the mere lowering down of the upper sails; and so, whilst saving the labour of a large number of men, enabling the navigator to carry sail fearlessly, even in contemplation of bad weather, from the facility afforded him of reducing, almost by self-acting arrangements, his topsails into their smallest dimensions.

But there was within board a picture full of life and interest, which may well claim a place for special description.

L

This was exhibited on the platform of the poop in the steering, and was fully displayed in broad and striking aspect at my station in the companion. We had no anxiety about the security of the sails and spars of this admirably rigged ship; no one had a fear, after our much and gratifying experience, but that the *Royal Charter*, if her head could be kept the right way before the hurricane, would, as a seaboat, do her duty nobly. But no one experienced in scudding before a fierce gale and heavy sea, and knowing how the safety of the ship depended on the steering, could contemplate the possible failure in apparatus, gear, or management of the helm of the *Royal Charter*, when so scudding under the violence of this terrible cyclone and tremendous sea, without much anxiety. Every ordinary precaution had indeed been taken to guard against the breaking of the wheel ropes, which in a former part of the voyage had all but happened —men being stationed at hand on the poop, and relieving tackles being placed in the most advantageous position for being attached to the tiller in case of necessity,—but still the uncertainty of putting into effect this appliance in time to prevent the ship broaching to, left a risk open, which no reflecting person could overlook or fail to be impressed with.

This well apprehended risk it was which gave character and sharpness to the picture of energetic life about the helm. There you saw four men—the best class of seamen—supported by others on either side of the deck, superintended and sometimes vigorously assisted by the hand of a principal officer— keeping the wheel in active play, as they endeavoured to counteract any sideway tendency of the ship's head, or to anticipate the probable swing from previous movements of the wheel or impulses of heavy seas. Every man there was a picture of energetic manly life. You saw in his face an expression, to be read of every one, that he felt that in the management of the wheel he held the destinies, under Providence, of ship and human life in his hands,—that the ship broaching to by his failure in seaman-like tact and management would be destruction; hence the picturesque adaptation of his strength and figure showed in Nature's true developments of expression and beautifulness of manly excitement, that the importance of his performance and duty was fully realized. Then there was the Captain, standing a few yards forward of the helm, in front of the skylight—his figure and features characteristic in expression of an intelligent percep-tion of his responsibilities and his reliance on his experience of direction in difficulties and perils—his position partly side-ways, so as to be able with equal facility to watch every movement of the ship's head and every turn of a spoke of the wheel—and thus giving impulse and guidance to the helmsmen in usual emphatic words, " Starboard! starboard! —steady so—port a little—meet her again—mind your star-board helm," and so on,—the tendency of the ship indeed from the wind being slightly on the starboard quarter, was to spring to in that direction,- requiring the greatest possible watchfulness on the one part and anticipation by the helm of such movement, to keep the ship's fiercely-urged progress, especially on occasion of not unfrequent hard blows of the sea on the starboard quarter, on a safe and pretty steady course.

Were I a painter, there is no scene which, since my aban-donment of Arctic adventure, has come under my personal observation, that I should more earnestly attempt to place on canvass than the poop deck of the *Royal Charter*, with the immediate elements for a picture without, during the height of the hurricane. First, in the afterpart of the ship, looking upward, we should have the mizen mast of the ship denuded of all sail, with the cordage swelling out forward under the force of the wind—then the ship herself cast into an oblique heel towards the port side, the stem raised high by a moun-tain-like wave—then the living pictures at the helm—the attending officer and the directing Captain standing sideways, in the foreground of all; then externally the assailing moun-tain-like wave, following close on the starboard quarter, and giving the direction and angle to the ship's inclined position, yet threatening, as many such waves do, to overwhelm the ship in mightiness of waters; then the atmospheric part of the picture, the mistiness of the storm-drift—the sun throw-

ing a lurid glare through an aperture in the dense masses of
cloud flying above—eliciting in the sea-spray of some immediate breaking crest a striking and brilliant segment of a
prismatic arch ; and finally, beyond this, astern, or on the
left hand of the picture above, an approaching squall shower,
thrown by the contrast of the penetrating sunbeams into the
aspect of consummate threatening and blackness !

The action of the ship in this fearful tempest may be
expressed in two words, she "behaved splendidly !" Her
buoyancy, her liveliness, her ease of motion, her all but intelligent efforts and yieldings to avoid the assaults of the sea,
were as astonishing as they were admirable. Seas indeed, in
repeated instances, were shipped over the quarter, on the
poop, or into the waist. But it could not have been otherwise. Knots of seas of terrible threatening, the intersecting
product of two high prevailing seas—one with the wind, as
we have noticed, from the westward; the other from the
south westward, raised above the height of the highest waves
into lofty peaks, curving inwards, or with a just-breaking
concave in advance, came careering towards us, and striking
perhaps the quarter of the ship with a quivering blow, threw
up a mass of water over the poop which, finding unfortunately
long and numerous chinks and apertures in the extensive
skylights of the saloon, poured down a no small body of
water, to the vast discomfort of the passengers and the satu-
rating of matting, carpetting and coverings within its reach,
with penetrating and enduring wetness ! So again, seas
shipped over the waist sometimes, on the lifting of the ship's
head, or on her rolling from side to side, entered the " forward " doors of the saloon, opening on the main-deck, and
rushing along the corridors communicating with the upper
saloon-cabins, obtained entrance into many—rising to the extent of some inches depth of water, to the no small damage
and distress of their occupants. These mischiefs and dis-
comforts, being all remediable, have fallen to our lot as the
first experimenters of a new ship; others in future will reap
the profit of our experience. But none of these things de-
tracted in the smallest degree from the admirable qualities of

the *Royal Charter* herself, for as to the ship and her high qualities and splendid capabilities her character remained untouched.

Some minor and very different circumstances came in for a share of observation. I allude to them as of some curiosity. One might have expected that with the main deck frequently bearing a foot depth of water, and the saloon,—its tables, floors, and swinging frames of glasses and table auxiliaries,—that some simple form of provision would have sufficed for the satisfying of Nature's requirements, and that little formality would be looked for in serving it up. But neither of these probable results were realized. The breakfast-table had indeed been drenched by a sea after the service was laid, sugar and milk spoilt and cloth drenched; but the activity of the stewards overcame the accident, and breakfast was still served at the proper time, much in the usual way. About 10 A.M. there was quite an inundation from the skylights; but the stewards again on the alert had soon baled out the body of water, in buckets, and removed and replaced the hundreds of suspended glasses and other apparatus above the table, in dry, clean, and proper order. No sooner had this been accomplished and everything got right, than another sea (it being the height of the hurricane) came over the poop-deck, and again sent its large contributions to the discomfort below. Again were the men at their posts—industrious and active, as we have seen the colony of ants, when their hill structure has been damaged by the foot of a passer by,—and the same round of renovation and replacement carried into effect as before! Notwithstanding these seas and the water coming in by the forward doors or main-companion of the saloon, the arrangements of the day were little interrupted, and at 3·30 P.M., just half an hour after the usual time, we heard the dinner-bell sending forth its wonted summons. To our surprise the tables were covered as in favourable times: a dinner of considerable variety,—in roast and boiled mutton, pork, cutlets, stews, currys, vegetables,—with plum-pudding surrounded by brandy flame, rice puddings, tarts in variety, which in due course came on,—and all served up in the ordinary handsome order. Excepting soup and an accident to the fowls, the hurricane and peril occasioned no other lack in the provision for the table.

Neither Captain Boyce, with all his extent of experience in navigating these boisterous seas, nor other officers,—including naval officers having had knowledge too of the Indian ocean,—had ever witnessed, taking it altogether, such a hurricane and sea. Lieut. Jansen, of the Netherlands Royal Navy, who had spent many years in surveying and scientific investigation in and about the Indian ocean, designated it at once as a cyclone, having its axis southward of us, and travelling towards the S. E.

The performance of the ship in the way of progress towards her designed port—though it could not be so rapid as in a smoother sea and more manageable strength of wind might have been effected—was yet satisfactory. The distance accomplished from noon to noon (nautical day) in a speed extending from 7 to 15 knots, was 265 miles, bringing us to longitude 73° 2′ E , but for the civil day, from midnight to midnight, making 306 miles. The abatement of storm in severity was sufficiently decided at 5·30 P.M. to cause the upper main topsail to be hoisted up, and, as the evening advanced, the upper foretopsail and lower mizen topsail. Towards midnight we had strong gales with more rare or passing squalls. The general log-book of the *Royal Charter*, which has run in a peculiar nomenclature,—designating winds, not as ordinarily described, but as to manner of their action or impression on this magnificent ship,—launched forth into unusual phraseology in reference to the gale of this day. Ordinarily, winds propelling the ship 11 or 12 knots are designated as "moderate breezes;" winds that would put ordinary ships under double-reefed topsails, whilst running "free," are styled "fresh breezes;" such gales as would put ordinary ships under close-reefed topsails and reefed foresail (by the wind) are put down as strong breezes, or fresh breezes with strong squalls; occasionally we have "fresh gales," squally or hard squalls, in such gales as we had on the 1st of April; but in reference to the period from 10 to 11 of this morning, we find the notification, "blowing a hurricane;" and at noon, "hurricane continued, with tremendous sea, no horizon."

On an occasion above all other Sabbaths since the commencement of our voyage, in which those of pious or even sober minds might have regretted the lack of opportunity of assembling together for Divine worship, we had to regret the deprivation. I had indeed announced my personal inability, from sickness, to conduct the public service, when the pouring in of water into the saloon, on one assault of the sea after another, showed that at any rate a public service would have been impracticable. In our private cabin, however, where we assembled two or three together, within the limits of the promise, we had a short morning service, gathered mainly from the "Forms of Prayer to be used at Sea," and in the evening, commencing with selections from the "Order for Evening Prayer," and adding portions of the former prayers and thanksgivings, a similar small assemblage had the privilege of again uniting in supplications to "the God and Father of all our mercies," through and in reliance on Him who is "the Way, the Truth, and the Life." In conclusion, an excellent and comforting discourse was read by my wife, from Robinson's "Scripture Characters,"—being the Exposition in the Life or History of our Saviour Christ, of chapters xiv. and xv. It appeared (though taken incidentally in our progress in former reading) to be peculiarly adapted for consolation to those who duly felt the condition and risk of the tempest through which we were passing.

Monday, April 7.—It blew still a gale the whole day ; but only in such measure of force as gave little or no anxiety to any one on board the *Royal Charter*. It was attended by strong squalls, with showers of hail, snow, or rain. The wind, W. by N. to S. W., was for several hours so comparatively moderate as to allow the carrying of topgallant sails, whole maincourse, and two studding sails forward on the starboard side, but at night it again blew very hard. Courses steered E. by S. and E. Rate of going from 12 to 15 knots, for the civil day, and distance made good from midnight to midnight (viz. 23½ hours), about 310 miles.

The sea, still very irregular, from W., S. W., and S., at night, continued to run high. The ordinary waves being estimated at 25 to 30 feet, with others much loftier. A fortunate lull favoured the shifting of the wheel ropes.

The clinometer, at 2 P.M., noted a maximum heeling, for the night and morning, of 28° to port, and 19° to starboard, with a numerical register of 10,000.

Being very unwell, I could not look out but for a few moments, to observe the weather and waves and the sail of the ship.

Tuesday, April 8 (W. S. W.)—The gale continued throughout this day also, occasionally blowing in heavy squalls, but on the whole, with a fair wind, being to our perceptions fine weather. Indeed the clouds broke away in the early day, yielding gleams and periods of sunshine, unmarred on this occasion by those unpleasant adjuncts which, on a previous day, had deprived this heavenly radiance of its cheering influences. We continued to make good and indeed rapid progress on our desired course of 12 to 15 knots, and completing a distance, from noon to noon, of 322 miles—our greatest for the day, except once.

Our progress and advance hitherto had, to use the expression of one of the officers, been "glorious." I am not aware that it had ever been equalled in the world by any ship, or by any route, or by any appliance, even of fullest steam power. At midnight of this day, reckoning by Greenwich time, and the day of 24 hours, we had, in 52 days out from Plymouth, accomplished a distance (and that in a course very nearly approaching the shortest practicable route for sailing ships, as modified by the lines of Continental coasts, and the essential deviation constrained by the trade winds) of about 11,637 miles! In this period, taking a southerly and westerly direction, to the latitude of 20° S., and longitude 30° W., and then sweeping round, as the wind usually allows to the southward and eastward, we had now reached by the "composite" track, ascending to 48° S., the longitude of 111° E., or within 1500 miles of Melbourne!

Reckoning up to two days back, when we had just completed our 50th day out, I find the distance made good in the usual and best track of progress to have exceeded 11,000 miles, or more than one-half the circumference of the globe!*

Reckoning again, for the last 12 days, ending at this midnight, I find we have made 3,522 miles by the log, or 293 miles per day; on taking strictly the extent in *longitude*, accomplished within that time, the distance, in nautical or geographical miles, comes out at 3430, or 286 miles a day.

I was yet too unwell to go upon deck, as the air was cold —about 45°, with showers of hail, and blowing hard; I could do nothing therefore personally in compass comparisons, nor did the continued turbulence of a cross sea render the comparison very easy or satisfactory. The sea during the day, however, had caused us but little inconvenience; but at 9 P.M., a knot from crossing waves, struck the ship a hard and quivering blow on the starboard quarter, sending such a quantity of water over the deck, that some ladies and others in the saloon got completely drenched, as by a broad profuse shower-bath, with the portion of the sea that made its way through the crevices of the skylights. For a while the prevalent games and other pursuits were necessarily suspended.

Wednesday, April 9 (S.W., S., calm, S.E.)—The barometer having fallen again very low, below 29 inches (28·811), the Captain expecting a renewal of heavy weather, made prudent preparations for it; but no gale occurred. On the contrary, the wind began to subside with the advance of the night, and by daylight of the 9th it had fallen almost to a calm—now quite a novelty with us for weeks past, and thousands of miles of progress. At noon the engine was put into action, which advanced our rate of going from 3 to 7¾ knots. But a breeze springing up and freshening towards evening, the steam-power was withdrawn, and at 7 P.M. the screw raised up.

The ship having in this day's progress broken off in her course to E. N. E., or more northerly, I was anxious, whilst

---

* *i. e.* geographical miles, of 60 to the degree, and 10,800 to the semi-circumference. In English miles the semi-circumference is between 12,406 and 12,450. —ED.

confined to my room, to ascertain how the compasses had been affected. Lieut. Chimmo undertook to obtain a set of comparisons, of the three fixed compasses, when the following results (3 sets) were obtained :—

Compass aloft.	Steering Compass.	Companion Compass.
E. 17° N.	E. 21° $\frac{1}{2}$ N.	E. 25$\frac{1}{4}$ N.

These results were interesting and important, as showing what I had always expected, that whereas the increase in the westerly deviations, on a south-easterly course, had been moderate (about 16° in the steering compass), that it was probable that other differences, possibly much greater changes, would be found in other directions of the ship's head. This single experiment showed at least a marked change in the deviations of the adjusted compasses, for though the quantity (compared with the compass aloft) was not great, the *sign* of the difference was changed. The difference betwixt the steering compass and that aloft, on an E. by N. $\frac{1}{2}$ N. course, appeared to be 4$\frac{1}{2}$° E., and of the companion compass (now the greatest), 8$\frac{1}{4}$° E. ; a change as indicated by this one set of observations, of 4$\frac{1}{2}$°+14°=18$\frac{1}{2}$° in the steering compass, and of 8$\frac{1}{4}$° +7°=15$\frac{1}{4}$° in the companion compass. But the real alterations will, with a prospering Providence, be satisfactorily determined at Melbourne.

Our distance accomplished, from noon to noon, was 243 miles, on a course of N. 83° E. The longitude at noon 112° 57' by chronometer, and latitude by observation 45° 40'. In the evening we had heavy rain with strong squalls.

Thursday, April 10 (S. E., S. S.W.)—Though the weather appeared fine at midnight, and the barometer had risen very considerably from its very low and threatening position on the night of the 8th, yet the great disturbance of the atmospheric equilibrium which had taken place, did not allow us altogether to escape. For, at 3 A.M., the wind having previously chopped round (as the Captain had predicted as pro-

bable) from southwesterly to S. E., a gale suddenly broke
in upon us, and laid the ship down at a greater angle of steady
heeling (the clinometer marked 25° to port) than we had
ever before experienced.   Many ladies got greatly alarmed.
But the ship being relieved, after considerable labour, of her
large and powerful mainsail, and of the upper topsails, soon
assumed her usual position of satisfactory stability. Here, in
a special manner, the admirable efficiency of the arrangement
of double topsails was realised in a most important manner.
The halyards being let go, the yards (contrary to the ordinary result with full single topsails) ran down of themselves,
and depositing, by the swell of the canvass, these upper sails
under the lee and shelter of the lower topsails, virtually took
themselves in without a man going aloft, or the engagement
of the hands upon deck—the reef tackles alone requiring to
be hauled out, and these not until the other sails occasioning
difficulty or threatening, were safely disposed of.   One of
these, the main topmast (or main) staysail, fortunately broke
a bolt sustaining the halyards, and then most amiably ran
down the stay, and placed itself quite snug, without other
damage !

About 8 A.M. the gale, now about S. E. by S., was found
decreasing, and at 11 A.M. the upper topsails were hoisted up
and the inner jib set. Towards evening the wind had greatly
moderated, and we had a fine night, not unfrequently clear.

The distance accomplished this day, from noon to noon,
only 162 miles, was the shortest since March 22nd; yet it
brought us (at noon) within the meridian of the western part
of Australia—King George Sound bearing N. by E., 580
miles off, and Melbourne being only about 1,250 miles
distant.

The weather being cold and I not sufficiently recovered to
risk the exposure, I did not venture upon deck. Soon after
sunset, happily, the wind veered to the S. and then S.W., so
that we resumed our required course with partially squared
yards, and a progress of about eight knots. A quiet night and
moderate fine day—quite a novelty with us who have for so
many weeks been navigating the region of tumultuous seas

and hard gales! After luncheon, the day being drier than
we had had for a long time (5° betwixt wet and dry bulb)
and the sun shining, I ventured upon deck, and found it
cheering and beneficial in its effect on my recent severe cold
—thank God, now passing, I hope, satisfactorily off. The
clinometer, the first instrument noted, showed a maximum
angle of heeling of 15° to port and 7° to starboard. The
wind was again fair, about W. by S., the ship making generally good and sometimes rapid progress, extending from
8½ to 13 knots. This renewal of our good speed put all on
board in excellent spirits. The course steered by binnacle
compass was E. by S., which, under the greatly reduced
variation, now only about 7° W., with a reduction too of
westerly deviation in that compass to about 3° or 4°, gave us,
pretty nearly, a due E. course for the *civil* day. At noon
we had made a distance of 212 miles from the previous noon,
on a course N. 86° E., true, bringing us to longitude 121° 4′ E.,
in latitude 44° 24′ S.

Compass comparisons and dipping-needle experiments, in
three stations, were made in the course of the afternoon. The
ship being on a more easterly course than formerly, and not
far from what originally was one of the points of change—
our true magnetic course being about E. ¾ S., and the point
of change, when at Liverpool, being E. S. E. ½ S.—no satisfactory conclusions could be drawn from them. The dipping-
needle experiments, however, continued to be interesting.
The more so as we were now pretty nearly in our *highest*
dip—about 73¾° S., with an intensity of the magnetic force
represented by 1·86, or an increase from about 0·8, what we
had it on the 13th of March, where it was the least. The
results of my observations very well corresponded with the
values so ably deduced by General Sabine. The observed
dip at the "standard station" was now 75°, having advanced,
probably by virtue of the ship's newly induced magnetism,
to being in excess of the terrestrial dip. From my first observations in 10° N. inclination (March 8), where the apparent
dip at this station seemed to be about 24½° behind the terrestrial dip (34½° N.), the difference gradually decreased up to

March 28, when the terrestrial dip being $62\frac{1}{2}°$, the observed dip was found to be very nearly the same, viz. 62°. In two subsequent trials, however, in dip $70\frac{1}{4}°$ and dip $73\frac{3}{4}°$, the differences apparently have changed their sign. Further observations, indeed, will be necessary for the complete establishment of this apparent result.

The evening was alternately brilliant with the bright rays of a fine crescent moon, and stars, and shrouded with passing showers of hail or rain, having their usual accompaniment of squalls of wind. At one time, even our small moon (seven days old) yielded a lunar-rainbow by the brightness of its radiance.

After the setting of the moon, a considerable number of beautifully brilliant medusæ were seen floating apparently on the very surface of the water. They seemed to me (not having witnessed the phenomenon before) as the gigantic "glow-worms" of the waste of waters through which we were passing. They appeared to be of the size of a large orange—highly or intensely phosphorescent, and shining not by their being agitated or stimulated by the ship's passing them, for many of them were at a considerable distance, extending to some hundreds of yards. At one time the chief officer, speaking of them, remarked, " we passed through quite a fleet of them—18 or 20 in a group!"

Saturday, April 12 (W. S.W., W., N.W.)—A quiet night —the ship making capital speed, 11 to 13 knots—with a fresh breeze at W. by S., with little change, and the ship steering east, slightly northerly, in pursuance of the Great Circle route. The agreeable night was followed by a pleasant enjoyable day, with a dry bracing air. After luncheon I went on deck, and pursued for a couple of hours my dipping-needle and magnetic experiments. Of the former I repeated, under considerable motion, the experiments at the standard station, with similar results to those obtained yesterday—the mean dip being $75\frac{1}{2}°$, or thereabout. Experiments were also made at the three stations on the forecastle and two on the forepart of the poop-deck. The experiments on the port side were embarrassing from the greatness of the dip, the dip

being about 85½° forward and 87½° aft,—so that it was difficult, with so much motion, to determine even proximately the
compass direction of the ship's head and the proper plane in
which to place the needle. But this was pretty well effected
by first finding the azimuth in which the needle assumed
a vertical direction, and then taking the azimuths of 90°
E. and W.

On many occasions I had made trial of the magnetic condition of the top edge of the ship's iron plating, especially
on the poop and forecastle. The *indications* of the results
have already been noticed. But now I found them perfectly
decisive. Thus a small pocket-compass, being taken almost
round the forecastle on both sides and forward to near the
stem, indicated strong *northern* polarity in the upper plates
wherever tried—both outside and inside—the compass being
held an inch or two from the iron. The like results were
obtained on the poop in all the places tried, and on both
sides of the ship, and inside as well as outside—all the plating
attracted powerfully the south end of the needle, indicating
a reversal of the polarities, as far as the upper edges of the
plates could testify. All standards, an anchor-stock standing
upright on the starboard bow, the entire tops or heads of the
iron capstans forward, had all changed their original magnetism—the tops now having northern polarity instead of
southern. The extent of this change in the general fabric
of the ship's hull outside, I hope, please God, to be able to
determine more satisfactorily at Port Philip. So far as the
results now obtained could be relied on as indicating a general
change in the ship's magnetism—to me they were especially
interesting, as corresponding with my theoretic views so often
and particularly published, and so much resisted by some
men eminent in science. I had never contended indeed for
more than what the present experiments might serve completely to vindicate, viz., that in a voyage of this description,
the reversal of the dip must *tend*, under the mechanical straining and violence to which a ship is always more or less subjected, to a reversal of the original organic polarities, and
that such tendency must produce a sensible change; but to

what extent in respect to quantity, I did not venture to predict.

The distance for the nautical day, made at noon, was 268 miles, on a course N. 85° E. The latitude by observation being 44° 10′ S., long. 127° 14′ E. For the rest of the day till midnight we made 8 to 11½ knots—the breeze which subsided at noon having freshened again in the evening. The night was fine and brilliant—the stars shining out in glorious resplendency.

An interesting experiment was made a day or two ago, in pumping the ship by the engine, and with the most beautiful smooth working, making 110 strokes per minute. This plan, it is obvious, with a little appropriate arrangement, may be rendered of vast service, not only for pumping from a general reservoir of the leakage, but by means of a connecting pipe and separate tap for each water-tight compartment, enabling the action of the engine to be brought to bear upon any one of the compartments separately. Such an arrangement might be of infinite value in case of damage within the range of any particular compartment. The engines, it should be noted, consume a singularly small quantity of fuel—ordinarily 13 to 14 tons per day, whilst 18 or 19 is an excessive quantity. During the whole voyage hitherto the consumption of coal has only been about 180 tons! The only drawback has been the lowness of screw, when raised, and the consequent danger to the stern, and retardation of speed and embarrassment of steering.

Sunday, April 13 (N. W. to W. S. W.)—In the subdued language of the *Royal Charter's* log-book, there were "strong breezes" during the night, yielding, on an E. ½ N. course, with the wind on the port quarter, a speed of 13 to 13½ knots, until obliged to take in studding-sails, and otherwise to reduce sail, on account of the increase of wind to a strong or hard gale. This became so heavy in a long continuous squall, about 2 P.M., that the upper topsails had to be lowered down, and two lower topsails, with reefed foresail and foretop staysail, alone displayed. But the gale, as expected, from the medium position of the barometer, did not last. At 5 P.M.

additional sail was set, and by 7 P.M. we were under lofty sail, including studding-sails—our course now being shaped more northerly, so as to make, as was intended, N. E. by E., true; but a change of the deviation, and increase of it easterly, in the binnacle compass, considered probable in my remarks on the 9th, threw the ship considerably to leeward of what was intended—the compass aloft, in consequence of the cross heavy sea and the general unsteadiness of the other compasses, unfortunately not having been compared.

At 8 A.M. the clinometer had registered a maximum angle for the night of 7° to port and 15° to starboard. But during the hard blowing in the course of the day, the register to starboard had increased to about 25°. The sea was high and cross, but by no means of the height occasionally experienced. A prevalent sea from the west was the heaviest, running at an ordinary height, or frequently so, of 22 to 23 feet waves, with maximum waves of perhaps 25 to 30 feet. At the same time a considerable sea got up from the N. W., the direction of hardest blowing gale. Seas from the S. W., or S. S. W., could also be plainly distinguished at times.

This, as it was now sanguinely hoped, being the last Sunday in which our present society in the saloon was likely to remain together, I was very anxious, feeling, thank God, very much recovered, to have Divine service, and offer to my associates in perils and mercies, a last address—it not being likely, as regarded the greatest proportion, or almost all the passengers, that we should ever again meet in the midst of the congregation.

Feeling, I trust befittingly, the Divine mercy in our happy preservation during the fierce hurricane of Sunday last, I selected for a subject of address Ps. cvii. 23–32 : 23 "They that go down to the sea in ships, that do business in great waters; 24, these see the works of the Lord, and his wonders in the deep. 25 For he commandeth and raiseth the stormy wind, which lifteth up the waves thereof. 26 They mount up to the heaven, they go down again to the depths; their soul is melted because of trouble. 27 They reel to and fro, and stagger like a drunken man, and are at their wit's end. 28 Then they cry unto the Lord in their trouble, and he bringeth them out of their distresses. 29 He maketh the storm a calm, so that the waves thereof are still. 30 Then are they glad because they be quiet; so he bringeth them unto their desired haven. 31 O that men would praise the Lord for his goodness, and for his wonderful works to the children of men! 32 Let them exalt him also in the congregation of the people, and praise him in the assembly of the elders."

The day, unfortunately, was stormy, and the ship unquiet, and the number which assembled far from numerous. I commenced my remarks (mainly extempore) by urging on their attention the important consideration that the religion of the Gospel, in its *saving* comprehensiveness, embraces three essential elements—knowledge, faith, and experience; and that any one of these being lacking, the other must be defective, if not vain. The text led us to consider and elucidate four grand facts in (mainly) experimental religion. 1st, the fact of observation—in respect to those who go down to the sea in ships (v. 23–24). 2nd, the fact of experience—they realise the mightiness of the Divine power and terrors in the storm (25–27); noting, in respect to the expression " their soul is melted because of trouble," that a due apprehension of our peril was needful to the rational and manly development of true courage; that the thoughtless, inconsiderate braving of perils was but a brute quality, in which the ferocious bull-dog or hyena might excel us, whilst the levity sometimes expressed in the perils of life and death was more charac-teristic of the mean cowardice which would shrink under the fatal issue of calamity, than of manly bravery; to be truly brave requires us to see, know, realise, the full measure of the peril, and then with dignified Christian manliness to bear up and discharge our manly duties under it. 3rd, the fact of providential results mediatorially and propitiatorially by Christ Jesus, in respect to praying, God-fearing persons (28–30). The final fact of Christian obligation and duty as to the acknowledgment of providential mercies. Application: *we* had realised these facts, and all the mercies of them ;—had we

M

returned the sacrifice of praise?　We held our lives now for further and better services towards God;—were we apprehending, receiving, applying for life and salvation the incense of intercession and the atonement in sacrifice of Him by whom, standing betwixt the dead and living, destruction is averted?

In the evening, accompanied by two or three ladies, I undertook, with some taxing of my strength, a service in the third-class messroom.　The numbers (the seamen not being able to attend) were smaller than usual, but full of attention, and gratifying indications of interest and feeling.　My address, as usual, was taken from the second lesson for the evening (James ii. 5).

The night closed splendidly—moon brilliant—stars of larger magnitude shining out gloriously.　The ship was still making good though reduced progress, at the average rate of about nine knots.

Monday, April 14 (N. N. W., W., variable, N. W., N. by W.)—A quiet night and a fine bracing morning, but occasional showers of rain during the day.　As I had apprehended, our position at noon was found to be considerably to the southward of the ship's expected place.　On getting careful compass comparisons, the cause was abundantly evident—there now being found, as compared with the compass aloft, an easterly deviation in the steering compass on a N. E. course of a full point, and on a N. N. E. course, on which the deviation was also tried, of $1\frac{1}{2}$ points.　Thence I had verified two inferences I had expressed as being probable, in my remarks under date of the 9th, that the deviation of the adjusted compasses, when steering northward of E., had changed its sign, and that the comparatively small errors we had found in south-easterly or more easterly courses, were probably due, not to the fact of the compasses being generally true, but to the fact of the ship's head on these courses (the reverse of her position on the stocks) being probably on or about the point of change.

The course steered up to noon was E. N. E. *true* (variation 5° E., deviation of steering compass $11\frac{1}{4}$° E.), which gave us a latitude of 42° 50', longitude 138° 45'; for the rest of the day we steered (*true*) about N. E. by E., making from 10 to 6 knots. Clinometer, at 8 A.M., starboard 16°, port 8°. In the evening, moderate breezes; generally fine, with a night brilliant with moon and stars. At the close of the day (midnight) our 58th out, we had approached our destined termination of the voyage within about 330 miles, a distance which, on four or five occasions, we had very nearly reached, or exceeded, in the course of one day's run. Great anxiety and interest had been excited by the expectation of accomplishing our passage within 60 days. But the failure of the wind, our northerly or leewardly position, when a breeze sprung up from the N. W. or N. N. W., prevented the possibility of our accomplishing this nautical triumph; yet still leaving the hope of being privileged to make a shorter voyage than any ship, sailing or steam, had ever accomplished.

Tuesday, April 15.—The wind falling light during the night, the breeze being scant, or partly against us, steam was got up, and the engine put into action at 7 A.M. We were now enabled to steer a higher course, by taking in the square sails, and to direct the ship's head towards Cape Otway, 60 miles from the entrance to Port Philip, which affords a safe and convenient "landfall." Our course by binnacle compass (with a point and a half deviation or more), was now hauled up to about N. by E., N., and N. ½ W. But still, what with so much easterly deviation, easterly variation 8°, and an easterly current, we fell to leeward (with wind north-westerly) of our desired position. The latitude at noon was still 41°, and longitude, reduced backward from 2 P.M. to noon, 142° 4' E., making King's Island 90 miles distant, and Melbourne about 240. A breeze springing up in the afternoon, slightly more favourable, topsails, etc. were set, which advanced our headway about a knot. But all was anxiety on board for wind and progress; the fine day (mild, pleasant temperature, 62° to 64°) and clear brilliant night failed to give repose to the urgent feelings that had been excited in the hope of making the quickest passage ever known.

Wednesday, April 16.—With a splendid sunrise of fine

clear sky, *land* appeared in sight—obviously the high land
(above 2000 feet in elevation) beyond or within and around
Cape Otway. We were now, and had been for a considerable
period, dependent on our auxiliary steam, the weather being
either calm, or the light breeze which occasionally marked
the surface of the water, being against us. Our engine, however, did us timely and admirable service, pushing the ship
ahead at the general speed of 7 to 7½ knots. The sight of
land and gradual approach, slowly because obliquely, to the
coast, was a new and striking novelty. Few on board had
seen any land for 58 or 59 days, or since the evening of our
departure from Plymouth. Nor had we seen sign or indication of human life, beyond our own community, since we
passed the *Thetis* on the 13th of March. But now vessels of
various classes began to heave in sight—coasters, a large ship
to the eastward, then, as we approached the entrance of Port
Philip, a pilot-boat, and subsequently steamers. It was fortunate for the early delivery of the mails beyond Melbourne,
that the first steamer which came out of the port was one of
the regular packets bound for Sydney, into which, within an
hour or two of noon, the Sydney mail was transferred, and
the packet hastened away for the delivery of the earliest
British mail at that port ever before accomplished. But few
minutes elapsed before a second steamer from Melbourne
approached us, bound for Launceston, Port Dalrymple, Tasmania, which likewise sending her boat received the mail
for that colony. Thus in 59½ days, or less, from a British
port, these colonial mails were transferred to their home
packets, and on their way for their final delivery !

Meanwhile a pilot had boarded us—in receiving whom and
transferring the mails, we spent an hour or two ; but yet were
in excellent time for entering the channel into Port Philip
at or near low water, until which, or during the strength of
the ebb, we must necessarily have remained outside.

The information received from the pilot respecting recent
passages and arrivals from England gave us increased satisfaction at the unprecedented performance of the *Royal Charter*,
and occasion of highest thankfulness for the Divine mercies

in our rapid and safe progress to this termination of our outward voyage. From the pilot's information we learnt that for the last five months the shortest passage to Melbourne had extended to 91 days, or nearly one-half longer than that of the *Royal Charter*. In one case, a fine Aberdeen clipper was lost upon or near to Cape Northumberland, only about 250 miles from her destined port. Happily and providentially the crew and passengers escaped safe to land. It appears that the ship had made the land two or three days before, and was beating up against easterly winds along shore, when, having incautiously stood too close in-shore, she missed stays, and not having her cables bent, so as to have afforded her the chance either of bringing up or casting by her anchor, went on shore on the rocks, and was wrecked. The Captain having been charged with blame, a legal inquiry was instituted, and he was acquitted.

Just after we had fairly entered Port Philip, an address, accompanied by a purse of sovereigns, was presented to Captain Boyce, from the saloon passengers. As we entered Port Philip, or rather just after we had passed into the Eastern Channel, the water being now quite smooth, and the accommodation ladders stretching obliquely aft from the gangways to the water's edge, I obtained a hasty but important fact in respect to the ship's external magnetism (previously verified as to the upper plating), of the complete inversion of the original polarity—just as I had anticipated from theory. Both sides of the ship had decidedly northern polarity, from the upper edge of the top plating for some distance downward, and so attracting the south end of the needle nearly right round on the starboard side, and acting vigorously on it on the port side. The magnetic polarity on the starboard side, with the ship's head about east (magnetic), changed at some eight feet below the upper edge of the top plating, whilst that on the port side appeared to continue for at least eighteen feet down—a difference consistent with theory.

The Channel we pursued took us some leagues out of the direct and ordinarily practicable course, but was cleared well before sunset; and notwithstanding this *detour* and our various delays at the entrance and also within the port, in giving up the Geelong mail, etc. we reached our destination at 23h. 44m. Greenwich time, and dropped our anchor in Hobson's Bay —the roadstead of Melbourne, on the outer flank of a large fleet of shipping. And here every devotional heart must have been fervently lifted up in grateful aspirations to the God and Father of our Lord and Saviour Christ, for the abounding mercies, in furtherance and preservation, which we had all so signally experienced—*all*, without exception, for the whole of our numerous fellow voyagers, whilst participating in the same general mercies, had been brought hither without a case of serious illness.

The position of our anchorage, as taken from the Admiralty chart (that is, the position permanently assumed during our stay), was in latitude 37° 52′ S., longitude 144° 56′ E.

Thursday, April 17.—The morning after our arrival being fine, there was a general move among the passengers, though many did not leave till the next day.

Having received offers of kind courtesy from several gentlemen who came on board the preceding evening, and a visit, pretty early, of Lieut. Crawford Pasco, R. N., with his boat, and the timely offer of putting us on shore and escorting us to Melbourne, Mrs. Scoresby and myself gladly availed ourselves of the opportunity, and put ourselves under Mr. Pasco's friendly guidance. We landed at the jetty or pier of a railway station at Sandridge, about ten or twelve minutes row from the ship, and took the first train (the trains running every half-hour) up to Melbourne.

Much having been described to us during the passage out of the enterprise, extraordinary progress, and fine scale and plan of the city, it accorded more with my conceptions of it, except as to being much finer in the width of the streets and the costliness of many of the newer buildings, than almost any strange place in a distant country I remember to have visited. But my familiarity with the progress of prosperous cities and new populations in the United States prepared me for the variety in quality, height, and value of the buildings, and for the irregularities of progress, from wood or corrugated iron to brick or stucco, and fine substantial edifices, of three or four stories high, of cut greenstone or basalt, granitic rock, and well-dressed and executed freestone. All seemed stirring, progressing, and enlivened, by realised or confidently expected prosperity. It seemed almost strange, at our antipodes, and in a city built on a reclaimed desert, and having risen from a population of some 400 or 500 to near 80,000 in less than twenty years, to witness the same dress, manners, and classes of population (except some peculiarities with the working classes), as at home; and it was a surprise to me *not* to witness passers-by in silk and satin dresses, of costly material and incongruous or vulgar combinations of colours, as I had been led to expect. There was no such thing to be seen wherever we walked—many well-dressed and several genteel-looking women in the streets, but all in quiet and sober habiliments.

The streets are of magnificent width—the great streets being about 100 feet broad, with side pavements for foot passengers of five to six yards! Few of the broader streets —for they ran alternately broad and narrow, and are formed by streets intersecting at right angles into square blocks— were completely paved or macadamized; but the best streets were to an extent adequate to the general traffic. Some novelties, in the yoking of numerous bullocks and buffaloes to large wagons, strike one—and something in their smaller carts and carriages; but the gigs, phaetons, and broughams, not a few in number, were—with the addition of a sort of dog-cart structure of much popularity—English, or of the manner of English, with smart, and in many cases handsome horses.

In the course of a visit of about four hours we called at the Government House, where I delivered my letter from the Secretary of State for the Colonies—not indeed for the officer for whom it was intended, the late Sir Charles Hotham, but for Major-General M'Arthur, acting for and administering the duties of Governor. The Governor was not in town, but we had a pleasant introduction to and conversation with the Secretary, Captain Kaye, R. N., who accompanied Sir James

Ross on his antarctic voyage, and for some years had charge
of the Magnetic Observatory at Hobart Town.   He seemed
much interested by the objects of my researches.

We called, too, at the town office of business of the Bishop,
but after two attempts, our time not allowing us to remain,
we were obliged to retire without seeing the admirable, able,
and laborious Diocesan of this important colony, whom I
had formerly known at Cambridge (Dr. Perry).   Whilst at
the Government House we made acquaintance with Mr. Price,
head of the department of goals, a connection of the late
Sir John Franklin, who kindly offered us hospitality at his
country house; but the distance, eight or nine miles, prevented us making any arrangement for the purpose, on
account of my pending magnetical experiments.

Finally we called at the office of Messrs. Bright, Brothers,
and Co., the agents and colonial managers for the Liverpool
and Australian Navigation Company, where, besides Mr.
Charles Bright, whom we had before seen, we were much
gratified with our introduction to Mr. Hart (William Hamilton Hart), the senior partner in the office, whose benevolent
and enlarged views and kind amiability of character interested
us greatly.   He also offered us hospitality whenever we might
be in a condition to accept of it, at his residence of Balmerino,
about three miles from Melbourne.

Lieut. Pasco was called off by some public business, and
obliged to leave us; but his boat, which had waited for him
and us, was put under our direction, and we reached it by
railway at half-past three, after a very gratifying visit and a
charming day.

Friday, April 18.—The next day, Friday, was a singular
contrast to what we had so much enjoyed.   It blew hard,
with heavy squalls, and a deluge of rain.   Some of our associates, who had unfortunately gone up to the city in the
morning, returned at night, drenched with wet, and described
the streets of the city running as rivers of water, and the
lower parts of the streets, running down towards the Yarra,
being two or three feet deep with water—so as to admit of
no crossing but in carts or small carriages, at a cost, each

time, of a shilling. Returning to the ship about 9 P.M., it was found difficult to get off, and a charge of 3*l.* was paid (7*l.* was demanded) for four passengers in one of the large sailing boats, for getting on board. Their condition, though not their spirits, was deplorable.

From the scarcity of labour and the low value of money here, the charges for boats or carriages (except one particular class of the latter) are very high. From Sandridge, the railway terminus at present, to our anchorage is about a mile, but in the present weather a demand is usually made of about 10*s.* for a single passenger, and the lowest ordinarily paid is 7*s.* 6*d.* Two or three years ago the expense was still greater. A Thames boatman, who came out here, told one of our passengers, that some two or three years ago he had made as much as 20*l.* in one day by his boat; and one of the present boatmen stated that he had made 5*l.* on Sunday week by his whale boat. Three mates and youngsters, on Friday evening, were asked 7*l.* to bring them off, but they preferred looking for beds on shore.

Wages, too, though not quite so high as a year or two back, are yet such that a good workman, being an artisan—carpenter, joiner, mason, stonecutter, blacksmith, painter, etc.—can generally earn 20*s.* a day, and some of the classes more. Even more muscular labour in the commonest work—lumpers on board ships and ordinary labourers—may get 10*s.* a day; few classes less.

Lieut. Pasco, at my request, took me on board of H. M. S. *Electra*, Commander Morris, to whom I wished to make application for assistance in swinging the *Royal Charter*, with gangs of men, and boats for kedges, etc. The Captain was on shore, but I received every courtesy from the First-Lieutenant Cholmley, who undertook to deliver my message, and had no doubt of its being cheerfully responded to when needed, as it efficiently was.

Saturday, April 19 (W.)—Weather occasionally fine and bright, but blowing strong, with heavy showers and squalls. This prevented us accepting a kind invitation of the Bishop of Melbourne to dinner, and to be his guests till Monday ;—

this, unfortunately, being the only opportunity that could be offered, as the Bishop was to leave about the middle of the week on a tour of visitation. I accepted, indeed, the invitation conditionally for the next day (Sunday, April 20), purposing to go up to church, and accompany the family out; but the weather was still so unfavourable, in wind and heavy showers, that a lady could not land at Sandridge, which was a lee shore, without getting wet and much additional inconvenience. Our disappointment in not seeing the Bishop and Mrs. Perry we greatly regretted. Weather cold and comfortless, with frequent rain or hail.

On Saturday, the 19th, came in the *Annandale*, which sailed from Liverpool, January 31st, having had a 79 days' passage.

The eclipse of the moon, which occurred April 20th, in the evening, was very well observed. The unshielded part of the moon had the strong silver brilliancy without *yellowness*, peculiar to a transparent atmosphere; and, as usual, the bright part appeared as if belonging to a globe of larger dimensions than the part which was eclipsed. I noticed with some care the termination by the chronometer. The Greenwich time of the last contact of the shadow was 22h. 38m. 7s. The dark part of the moon was plainly visible, with the edge or rim marked by a clearer or lighter line than the rest.

Monday, April 21.—A showery, squally day. Wind from W. S. W. to S. W. Compass observations could avail us nothing, as the distant hills never appeared but once; but a series of observations with the dipping-needle were undertaken—and as to the standard station and three stations on the forecastle, satisfactorily completed. The apparent dip at the standard position was $66\frac{3}{4}°$, and by General Sabine's tracing about $67\frac{1}{2}$; but on the centre of the forecastle the apparent dip was $67°$ $50'$—indicating, with the ship's head about S. $28°$ W., very nearly the true dip.

Tuesday, April 22.—The morning broke fine, and pretty clear, and several ranges of distant hills, pretty well adapted for points of observation, came into view :—

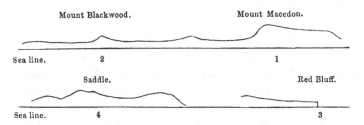

Mount Blackwood.   Mount Macedon.

Sea line.   2   1

Saddle.   Red Bluff.

Sea line.   4   3

No. 1, the western bluff of Mount Macedon, was a good object when it could be seen. No. 2, a conspicuous and lofty elevation—the last remarkable one to the westward of the same ridge. No. 4, also a distant elevation, I named, for my personal convenience, the "Saddle," always observing the depression was not to be mistaken; and "Red Bluff," a conspicuous—though in our view low—point of land, about 9½ miles, constituted my fourth object. This point, which was almost always visible and easily observed, was found a very convenient position for being observed, though not without parallax on the ship's swinging round; but with our short range of cable, about 40 fathoms, + 20 fathoms to the position of the pedestal for the standard compass, this could not produce an error of above 20′ on either side of the mean, and was therefore much made use of.

As there was no chance of being able to swing the ship for some time, whilst discharging cargo from three hatchways, and in a great state of confusion, I satisfied myself meanwhile with taking bearings for the determination of the deviations of the standard compass, incidentally, as the ship's head might lie, or happen to sheer about. In this way, by giving my chief attention to the object, I was enabled to take near 60 bearings of Nos. 1, 2, and 3, with the Admiralty azimuth compass, observing in each case the direction of the ship's head, and ranging from S. 44° 5′ W. to S. 24° 30′ E., comprising about six points of the compass, with deviations at intervals of a degree or less, and never exceeding 8½°. Discrepancies, however, in this first considerable series, occasioned me some anxiety, until I found that the compass, which had acted so

well at Liverpool, required *tapping;* I then obtained excellent results.

Captain Taylor, one of our passengers outward, who was
sent by Messrs. Gibbs, Bright, and Co., to fit out and bring
home an old teak-built East India ship—the *William Monies*
—cordially assisted me in the important object of determining
the true magnetic position of my points of observation.   For
this purpose I contented myself with a set of bearings from
his ship on the Sandridge beach, with his *reverse* bearings
with the common azimuth compass, and connecting this with
simultaneous bearings from the ship, of the Macedon Bluff.
The result of nine bearings from the ship, connected by signal
and the watch with the bearings from the shore, gave the
corrected magnetic bearing of Macedon Bluff, N. 38° 55' W.,
and the deviation of the standard compass (which on the
mean gave the bearing of N. 51° 48' W.), indicating a deviation at the standard position of 12° 53' easterly—the mean
direction of the ship's head by the standard being S. 17° 54' W.
The true magnetic bearing of Red Bluff (No. 3) from the
ship came out at S. 37° 35' E.; but these results, from an
accidental fall of the compass, and the impracticability of
seeing immediately either the shore party or their flag
through the sights of the standard compass, could not be
relied on without verification.

A fearful accident, threatening to be fatal to the life of
one of the labourers called "lumpers," engaged in the discharge of the ship, happened close before my eyes, as I stood
on one of the gangways communicating between the poop
and deck houses.   A block, iron-strapped, was accidentally
dropped from the mainyard, which struck the poor fellow on
the head glancingly; he instantly dropped down, groaning
and senseless.   He was removed on shore, and it was reported
he might possibly recover!

Wednesday, April 23.—The day was hazy, but moderate
as to weather.   The ship's head being mainly in the same
direction, gave but little opportunity of getting deviation
observations; but it gave opportunity for another class of observation of great interest, and calculated further to support

or to contradict my published theoretic deductions. The ship's head turning very nearly south magnetic, the terrestrial induction was equal on both sides. I was thus enabled to get a series of observations on the ship's external magnetism at the accommodation stairs, about 100 feet from the stern, under circumstances for which I had long been earnestly looking. The results were conclusive, and, as to my published anticipations, I may say triumphant. In a small pamphlet printed at Torquay, January 1, 1856, entitled "Illustrations of the Magnetism of Iron Ships," I had remarked, under Illustration No. VI., "The general influence of terrestrial magnetism, as to its action on the magnetism of an iron ship, is to change, in sea-going ships (under a due application of vibratory, straining, or other mechanical violence) the original or organic polarity into some measure of conformity with its own new direction of force. This tendency, in the case of an English-built ship proceeding to the southern coast of Australia, goes *toward* an inversion of the original polarities, or a change of the polarities of our case No. 5 (where southern magnetism is upward) toward this of No. VI.," (having northern polarity above and southern below).

Starboard Side.				Port Side.			
Distance of Compass below the top plating.	Ship's Head by Forward Compass.	Ship's Head by Compass let down outside.	Ship's Attraction on external Compass.	Distance of Compass below the top plating.	Ship's Head by Forward Compass.	Ship's Head by Compass let down outside.	Ship's Attraction on external Compass.
Feet.				Feet.			
4	S. ¾ E.	S. W. ¼ S.	51°	½	E. S. E.	S. ¾ W.	76°
5	S. by E. ¼ E.	S. W. ¾ W.	67	2½	S.E. by E.	. .	64
6	"	S. W. ½ W.	64				
7	"	S. W. ½ S.	53				
8	"	S. W. by S.	48	8½	S. E.	. .	53
10	"	S.W. by S. ¼ S.	45				
12	"	S. S. W.	36	12½	S. E. ¼ S.	S. ½ W.	48
14	"	S. by W.	17				
16	S. by E.	S. ½ E.	0				
17	"	S. by E.		17½	S. 2° W.	S. 1° W.	1
18½	S.	S. by E. ½ E.	17				

Thursday, April 24.—This was a fine, calm day—admirable opportunity for swinging, had we been prepared; though the encumbrance of the ship, with three lighters alongside at once, and the discharging from two or three hatchways, must have rendered such an operation impracticable; but the stillness of the day and the variableness of occasional light breezes brought the ship's head on a great variety of points, and afforded me the opportunity, which I carefully improved, of catching, with the Admiralty standard compass, the bearings of the saddle on the Dandelon hills, or of the Red Bluff, sometimes of both, in 36 positions, ranging from N. 39° W. to S. 36° 43′ W., west-about, and so gave me the data for the subsequent determination of the deviation of the standard compass in these respective positions.

Whilst I was thus engaged, Captain Morris, of the *Electra*, came on board, to reply personally to my message and application, cheerfully agreeing to send me gangs of men and boats whenever it should be convenient to swing the ship— but naming the present week, or early in the next, as the limit of available time, the *Electra* being about to sail for Sydney in the course of a few days. It was arranged that our hoisting the *Electra's* number at the main should be the signal for summoning the required assistance.

Finding no immediate prospect of getting the ship swung, and feeling much hesitation in detaining my wife longer on board, under comparative discomfort and privations, we removed in the afternoon, as I had previously arranged with Mr. Hart, to Balmerino, a pleasantly situated villa, with garden running down to the river Yarra, at a short offset from the Shepherd's Creek-road. Here we found ourselves most pleasantly domesticated in a quiet family circle.

Friday, April 25.—We dined at Toorak, with his Excellency Major-General Macarthur, the acting Governor, where we met Colonel Adjutant-General and Mrs. Neile; the Hon. W. F. Stawell, Attorney-General, and Mrs. Stawell; the Hon. Justice Barry; Captain Boyd, etc. etc. Before we left the house the General kindly asked us " to come out to Toorac, as soon as our visit to Mr. Hart was concluded, and to remain as long as we liked, or found it agreeable;"—an invitation so cordial, that we could have no hesitation in accepting it, reiterated as it was the next day, in the course of a personal call at Mr. Hart's.

Next day, in Mr. Hart's carriage, and escorted by Judge Barry, we visited some of the principal public institutions in the city. We first visited the Public Library, a really fine building and institution, with a considerable and well-selected collection of books. The Judge, I found, had been one of the most active and efficient promoters of this good work. From thence we proceeded to the University, a fine architectural building, which has been but recently brought into operation. Already some talented professors are settled on the premises :—W. Wilson, M. A., Mathematics; Fred. M'Coy, F. G. S., Natural Sciences; Dr. Hearn, Modern History, Logic, etc. The grounds of the University comprise about 40 acres—a grant, I believe, of the Home Government, out of " Royal Park."

We next went round the Royal Park, a pleasant and extensive drive, commanding some picturesque views. The chief timber, however, is the gum tree, a tree having indeed the advantage, like the forest trees of the region generally, of being evergreen, and occasionally growing in good and picturesque forms, but somewhat tiring the eye from their sameness, and displeasing by the prevalence of a white, light-grey surface on the stem by the shedding of the cuticle bark, and by the varieties of irregular shapes of numerous trees in a state of decay.

The Cemetery, comprising a large enclosure out of the park, well laid out and picturesquely arranged and adorned, was also visited. There we had pointed out to us the resting place of the remains of the late Governor, Sir Charles Hotham, whose premature death—hastened by a too acute sensitiveness to public rebuke and censure, and especially to the harshness of a section of the public press, is a most painful incident in that important part of the history of the colony—the substitution under the authority of the British Parliament, of a colonial legislature for the Home government. Amid the struggles of contending parties, the Governor had a very difficult position—a position rendered more difficult

by want of a due apprehension on his part of the state of the
public mind, and a want of adaptation in personal views and
feelings to meet that which would not be controlled. His
death produced a solemn, and in some respects, from the
knowledge of its cause, an awful sensation. It seemed to
act as oil on the waves of tumultuous excitement and passion,
and it was succeeded by an extraordinary calm.

In the course of our day's excursion we saw much, from
certain elevated and commanding points of view, of the
general site and plan of the city. Numerous streets nearly
100 feet wide, with footways in those most advanced to completion of five or six yards or more in width, intersecting each
other at right angles, give the impression of a fine city,
fitted for the chief city of a colony, destined probably in
the order of Providence to be a great and wealthy country;
and considering the recent date of its commencement, the
progress made, notwithstanding the costliness of the work
of artisans and the high rate of wages for labour, presents
something very remarkable in the history of the cities of the
world. It is like a city raised by the power of an Aladdin's
lamp.

The office of the Surveyor-General, the Hon. Captain
Clarke, R.E., was next visited, where we inspected the large
and elaborate maps of the colony; the Post-office, the recent
extensive enlargements of which indicate the rapidly increasing correspondence, and which is characterised by clever
arrangements for facilitating the distribution of the letters;
the Museum, comprising a number of district and colonial
curiosities in mineral and animal objects; and, finally, the
Crystal Palace, a large and imposing structure of wood and
glass mainly, and remarkable for having been erected within
a period of only 73 days.

Sunday, April 27.—The next day, Sunday, we accompanied Mr. Hart's family to St. James's church—the nominal
cathedral—not the most imposing church of the city, but the
oldest, and very commodious in its arrangements. The service was plainly but very well performed—the music pleasing,
and the manner and forms altogether chaste. A stained glass

window in the chancel appeared to me from want of harmony in the colours to be anything but an improvement. The Dean preached—an unwritten expository sermon—the style and doctrine plain and evangelical.

Monday, April 28.—By appointment with Captain Morris of the *Electra*, preparations were made, contingent on favourable weather, for the swinging of the *Royal Charter*—the manual force necessary for the purpose being ordered from the crew of the *Electra*. Going into town early with Mr. Hart, I was enabled, by railway to Sandridge and from thence by a boat in waiting, to reach the ship in good time for our object. Shortly afterwards two boats, one carrying a kedge, with gangs of men and the boatswain, and Lieut. Keene, arrived alongside. Unfortunately, the weather, which for two or three days previously had been calm and fine and the wind light, had changed to a fresh breeze—much too strong for the operation with so large and heavy a ship. The trouble of sending the men, however, was by no means lost; for it being found that the *Electra's* kedge was by far too light for our purpose, the men were set to work to get up two larger kedges of the *Royal Charter*, and a series of hawsers which were coiled on the forecastle ready for use.

Tuesday, April 29.—The swinging of the ship being so essential a measure in my present undertaking and adventure, I now stuck to the ship to be ready to take advantage of any favourable opportunity for the operation. But Tuesday proved as unfitting as the previous day, the wind blowing very fresh. The day, however, was otherwise well employed in a careful series of observations for ascertaining the true magnetic direction, from the ship's anchorage, of the several points or objects on shore which had been fixed on for the determination of the compass deviations.

This determination had been attempted on the 22nd by a set of reciprocal bearings between the standard compass on board and a common azimuth compass on the beach 1¾ miles to the N. E., which gave for the true magnetic direction of the Red Bluff, S. 37° 35′ E. But I had not full confidence in the action of the compass on shore, nor did I feel that my

N

bearings could be perfectly relied on, in consequence of the
flag of the shore party not being visible through the sight
vanes of the standard compass. Hence I was glad to avail
myself of assistance within the command of the Surveyor-
general, Captain Clarke, who had kindly offered me services
in surveyors, instruments, or other appliances, belonging to
his department of the public service. The assistance I
applied for, and which was promptly and effectually rendered
to me on this occasion, was a surveyor, with a theodolite, or
other good instrument for magnetical bearings. He came
very well supplied—having a theodolite and a circumferenter,
to which I added the azimuth compass formerly used. The
place selected for reciprocal bearings was the western shore,
between Sandridge and Williamstown, distant from us about
1⅓ miles. As neither boat nor flag of convenient size could
be seen at such a distance through the sight vanes of my
instrument, I directed him to proceed to a spot on the beach
where I observed a large punt aground, and to place his
instruments in a straight line (one behind the other and
sufficiently distant to avoid mutual disturbance) passing the
punt's stern to the place of the standard compass as defined by
a flag on board—a direction which the well-practised officer
accurately carried out. By this arrangement, having with my
glass ascertained that he was exactly in position, I had only
to observe the direction of the punt's stern and simultaneously
(or nearly so) the bearing of one of my distant objects—the
Red Bluff,—and thus, adjusting the time of our observations
by signals previously arranged and put in writing, I obtained
an ample and very satisfactory series for my object.

The ship's head during the operations had ranged from
S 28° W. to S. 47° W., and the bearings, as affected by
changes in the deviations of the standard compass, differed
to an extent of about 2°; but on taking the angular distances,
now determined betwixt the station on shore and the Red
Bluff, for each of ten observations separately, I found a very
satisfactory measure of agreement. The mean of the whole
gave for the angular difference 152° 44'.

Again, the bearing of the ship's compass from the shore,

taking the mean of the whole of the sets of observations and of the several instruments, was found to be S. 66° 28′ E., or conversely, the true magnetic bearing of the station on shore from the ship, N. 66° 28′ W. To this, adding 152° 44′, also westerly, the true magnetic bearing of the Red Bluff was found to be 219° 12′ W. or S. 39° 12′ E., differing just 1° 37′ from the bearing formerly determined by the common azimuth compass alone.

From this result the positions of the three other places on shore were next ascertained. I had already prepared for it by taking their respective differences in bearings from numerous sets of observations, checked by the angles given by a pocket sextant. The general results casting the bearings into a continuous series from the North round by East, came out as follows:—

True magnetic bearing from ship's anchorage of

Dandelon Hills (the Saddle) . . . N. 76° 28′ E.
Red Bluff . . . . „ 140 48 „
Mount Blackwood . . . . . . „ 293 28 „
Mount Macedon (Western Bluff) . „ 319 28 „

The position of the ship by the Admiralty chart of Port Philip, was lat. 37° 52′ S., long. 144° 58′ E. Distance from Melbourne in a direct line about 13½ miles.

Two improvements in the Admiralty standard-compass, for its better adaptation to delicate observations on deviations, and for determining the true magnetic bearings of distant objects, were suggested by these and other observations made at Port Philip. The first was the adaptation of a much lighter card for being used under circumstances of great stillness and weak horizontal directive force. Before I was aware of the defective traversing of the lightest of the two cards usually supplied, though placed on one of the finest points, I was annoyed by finding differences in observations repeated and in all respects similar or analagous, sometimes amounting to 1½° or 2°, and in two or three instances of still more. Such differences indeed I might have been prepared to expect in a new ship, recently launched, and when being swung in

different directions, especially under the vibration from hawsers
employed in heaving round; but in our case there were no
such influences in operation. On tapping the compass-bowl
with the butt-end of a lead pencil the discrepancies were
sufficiently explained by observing frequently an immediate
change in the direction of the card. This precaution, therefore, was subsequently adopted on each observation. Another
improvement, as it occurred to me, in a fine instrument, designed to be employed as a steering, azimuth or "swinging,"
compass, would be the adaptation of a small telescope, adjusted
to the circularly-moving rim, for giving assistance in observations of objects or positions on shore, when, as in the case of
our reciprocal bearing, the place of the shore instruments
could not be seen through the sights. Such telescope, too,
would be of the greatest use, when, as I repeatedly found,
the shore object (its point, bluff, or other recognisable place)
could only be indistinctly seen. In many cases, whilst pursuing my observations in Hobson's Bay, the saddle on the
Dandelon, or the termination of the Red Bluff, could not be
seen without the help of a glass (I usually employed an opera-
glass), leaving the bearing by the sights to be made out by
various shifts, such as the incidental proximity of some more
visible point or object—a boat passing, a rope on board the
ship, etc., for getting the bearing required.

An attached telescope would have relieved me from all
these embarrassments and risks of inaccuracy. It might possibly be applied to one or other of the shifting rims at present
adopted; but in any convenient form it would be a valuable
addition to the instrument.

[On Tuesday, the 29th April, a dinner was given to Dr.
Scoresby, by the members of the Philosophical Institute of
Victoria, at which there were present the acting Governor,
the Chief Secretary, the Speaker, Mr. Justice Barry, the
Attorney General, the Surveyor General, and other distinguished persons of the colony. The speeches were reported
in the Melbourne newspapers.]

Wednesday, April 30.—The swinging of the ship being
still the object with me of great anxiety, I went down to

Sandridge by an early train, but found the wind, as before, far too strong for the contemplated object, nor indeed, would it have been easily practicable, the ship being encumbered with lighters (small craft), three of them being alongside at the same time receiving cargo.

Thursday, May 1.—This day again failed us, by reason of too much wind; but a good series of bearings, for the deviations of the standard compass, was obtained, extending from S. 13° E. to S. 62° E. Captain Morris came on board with reference to the swinging, stating that he must leave on Monday, or very early next week, for Sydney, rendering, therefore, the object in contemplation the more urgent and important, as it could not, at this time at least, be possibly accomplished by the *Royal Charter's* men, without much detriment to pressing duties about the ship. It was arranged that, if it were possible, operations should be commenced early in the morning. As is not unfrequently the case in the Australian climate, it was a glorious night,—the sky was resplendent with stars, which seemed to stand out of the concave above, as if mysteriously suspended within the aërial vault. This was a phenomenon I had often observed when travelling in the United States of America.

Friday, May 2.—At day break I was called with the agreeable report that it was a fine and still morning, calm, or all but quite calm. I immediately arose, directing that the *Electra's* number should be immediately hoisted at the main. Captain Morris, indeed, had anticipated the signal, by ordering the men designed for our service an early breakfast; and at 7 A.M. Lieut. Keene, followed by the sloop's cutter, with a gang of men and a kedge, in charge of the boatswain, an intelligent and efficient officer, came alongside. Captain Morris soon afterwards also came on board to see that all was satisfactory.

The sun had risen clear, and shortly afterwards threw out his utmost heating influence, unfortunately, to our disadvantage, for it raised up a fog, which rapidly screened from view the Dandelon and Red Bluff, previously visible, and completely hid, indeed, during the whole of the forenoon, every

distant object, so that, whilst every circumstance besides was
most favourable—the day being otherwise beautifully fine
and calm,—all deviation observations were rendered impracticable; for the chance of clearing, however, every external
preparation was made. The ship's head generally lay about
north. Whilst in this position one of the *Royal Charter's*
kedges was carried out towards the W. N. W., with its
hawser leading out of the ship's port-quarter, and brought
in to the capstan on the afterpart of the poop-deck. The
Electra's kedge was dropped a little way from the stern,
tending to the starboard-quarter, as a check kedge for the
swinging.

Whilst we were at luncheon, we got the pleasant report
that the fog was clearing away. The Dandelon hills had
appeared, and the " saddle," my characteristic place thereon,
was faintly discernible. The Red Bluff also had begun to
show itself. By the time I had got my compass in position,
and arranged my staff of observers under Captain Taylor, of
the *William Monies* (who had been a passenger out with us,
and had been living on board the *Royal Charter* ever since
her arrival), the *Electra's* men, who had been prudently sent
to an early dinner, returned, and operations previously commenced were steadily proceeded with, and, just as the day
closed in, were brought, except for a single point or two, to a
satisfactory issue.

The record of the proceedings, ultimately most successful,
—in which I had such valuable aid in the kind attention of
Lieut. Keene, and the active labours of the boatswain and his
men—may be conveniently given by insertion of an extract
from a letter, descriptive of the operation, and summary of
the leading results, which I afterwards furnished to the editor
of the "Argus:"—

" 1. *As to the Compasses adjusted by fixed Magnets.*—Their
errors were found to correspond in nature, though not in
quantity, with the results of ordinary experience. The steering compass, corrected at Liverpool, was found to have now a
maximum error of 19¼ degrees on the S. E. by S. point, and
of about 12 degrees on the N. N. E. In the other compass

the extreme error appeared to be rather less than a point and a half. These deviations, indeed, were unusually small, and for these obvious reasons,—that the steering-compass in the *Royal Charter*, instead of being placed, as ordinarily is the case, near the stern, and close by large masses of disturbing iron, is judiciously placed, in respect of correctness of action, at the distance of 68 feet, and the other adjusted compass 89 feet from the stern, besides being both in a position elevated some feet entirely above the upper plating of the poop-deck. Hence, under circumstances of change in the terrestrial magnetism, not unfrequently producing errors in adjusted compasses amounting to three, four, or even six points, the errors in those of the *Royal Charter* did not appear to exceed at the greatest a point and three-quarters.

" 2. The *standard compass*, unadjusted, having its place on a pedestal, transiently for use on the deck-house, before the mainmast, and not far from the middle of the ship's length, gave results, under my personal observation, of such singular beauty as to consistency and precision as must render them, when mathematically worked out in connection with sets of observations in England, available for very important scientific deductions. The deviations of this compass were found to have diminished greatly since the " swinging" at Liverpool on the 4th of January—the means of the two *maxima* errors having been reduced from near 25 to about 14 degrees. The points on which these *maxima* were found had but slightly changed; but one of the " points of change," or positions in which the compass becomes true in direction, had, contrary to received opinion, shifted considerably,—that is, no less than *four* points. There was also a small anomalous change, which, without considerable extent of description, could not well be explained.

" 3. The *compass aloft* gave results that were admirable. This compass is supported on rods or bars of unmagnetic metal at the height of about 43 feet above the poop-deck, projecting from two to three yards aft, or towards the stern. The present arrangements about it have some awkward inconveniences and defects which, on another voyage, will be easily corrected. But still its indications, when the mean of a set of observations was taken, were highly satisfactory.

"Notwithstanding the unavoidable errors of observation with an instrument unfortunately divided only to quarter points, the deviations on 23 out of 25 points observed were below a quarter of a point, and averaged but little more than a degree! In the other cases there were evidently mistakes of observation, as conclusively shown by the consistency and small errors of other sets of observations in proximate directions of the ship's head. Comparing the apparent deviations of this compass, as observed at Port Philip and at Liverpool, they were found to have been reduced, small as they were, on the voyage; and there could be no doubt (considering the unusual accuracy of the ship's positions by the "deàd reckoning" on the passage out, with the positions proved by observations of the sun, and indications of the chronometers) but that this compass had acted throughout with most reliable accuracy. The deviations, indeed, of the compass aloft, as here determined, it is important to observe, were far less than those of the steering compasses of any ship built of wood in which I ever sailed, and smaller considerably, I doubt not, than the ordinary errors of the compasses in any class of ships afloat!

"As the compass-changes in ships trading betwixt Great Britain and these southern colonies are often strangely and extravagantly great, and the employment of iron shipping of the larger clipper-class for the intercommunication must obviously increase, and that, as is probable, very rapidly, these determinations on the complete success of the plan of a *reference compass aloft* can hardly be overrated, I think, in importance. Different ships, indeed, may require modifications in detail. Some may do with a small, and some demand a higher elevation; but the principle, wherever masts of wood are employed, or a mast of wood for the compass with hempen and wooden rigging and gear, sufficiently away from other magnetically disturbing influences, must prove effective, and, as far as the compass has to do with security, safe. And besides the general advantage of the plan, in destroying the errors ordinarily produced by the ship's magnetism, it has this further distinctive advantage, that sudden changes (the most dangerous of all), which not unfrequently occur in a ship's magnetic condition, and are wont to produce sudden and unexpected errors in the compasses on deck, have no sensible disturbing influence on the compass aloft.

" In regard to other objects of inquiry in respect to the changes of magnetic condition in an iron ship, produced by the inversion of the direction of terrestrial magnetism in this region of the globe, it is gratifying to be able to say that all my anticipations (as plainly indicated in a set of illustrative diagrams, with descriptions, which I had printed before leaving home) have been most strikingly, perhaps without charge of vanity I might venture to say triumphantly, fulfilled !"

This important object being now accomplished, and the great fact of the general inversion of the polarity of the ship's magnetism externally, as I had ventured to predict, determined,—I, for the first time, felt myself free to look into, as far as opportunity might permit, something of the nature and peculiarities of a region, destined in the order of Providence, as it would seem, to become of very high consideration among the countries of the world.

We had hoped to have visited other parts and colonies of this vast island, especially Sydney and Hobart Town. My letters from the Secretary of State in England, included introductions to the Governors of both New South Wales and Tasmania, as well as of the other British colonies of the far south. Mrs. Scoresby received a cordial invitation from an old friend in the former region,—the lady of the Bishop of New South Wales,—to visit them at Sydney; and the kind and considerate Governor acting in Victoria generously offered to place a residence of his, situated a short distance from Sydney, at our disposal. I was anxious, too, to have visited a place and harbour characterised by many objects of interest and beauty; but the 16 days already spent had shown us but little of the vicinity of Melbourne, whilst the time remaining to us, only 20 days more, according to the date at which the *Royal Charter* was advertised to sail, mortifyingly showed us that, with the absorption of five or six days in the intermediate passages, a full week could hardly be spared for our visit there. The project, therefore, was reluctantly abandoned; and we determined to employ the brief remaining period in seeing what we conveniently might of the country and society more immediately accessible.

CHAPTER IV.

After a sojourn of 38 days in the prosperous colony of Victoria, we prepared to rejoin the *Royal Charter* for another adventure on the great oceans of the globe. We had arrived in Port Philip on the 16th of April; on the 24th of May we were required to be on board with the view of sailing, as arranged with the Post-office authorities, on the following morning. The interval, with the exception of the time in which I had to wait for the favourable opportunity of swinging the ship, has been spent in unmixed enjoyment and satisfaction. Our first visit, as has been noticed, was at the hospitable residence and with the friendly household of W. Hamilton Hart, Esq., at Balmerino, on the Gardener's Creek-road, about three miles from Melbourne; our next, and the most continuous was with his Excellency Major-General Macarthur, the Acting Governor of Victoria, at Tourac, three miles from Melbourne, where we were made acquainted with a refined and charming society in an almost continuous series of social dinner-parties. Among the ladies and gentlemen connected with the military staff, the public officers of the city and government, and belonging to the legal and clerical professions, and the mercantile classes, there were many from whom we parted with real regret.

Our luggage being forwarded before us to the station of the Melbourne and Sandridge railway, we left our pleasant abode, accompanied by our kind and considerate host to Melbourne, on the afternoon of Saturday the 24th of May. About 5 P.M. we reached the gallant and imposing ship in which we hoped, under the continued providence of heavenly guidance, in due time to regain our native land. As usual, in such cases, the ship's deck and apartments below presented a scene of much confusion. In every department she was crowded with passengers. Her rapid passage out, and her admirable performance at sea, had gained for her such a popular reputation, that all the saloon berths were in overwhelming demand, numbers of applicants being disappointed, and the accommodation for the second and third classes quite full. The large number of third-class passengers leaving the colony, I felt disposed to regret. But others, and those anxious for the gaining of labouring men to the colony, took a different view of the case. This returning body, chiefly of the working classes, many of whom carried along with them the fruits of their toils in the gold-fields in considerable amounts of treasure, it was considered would be the means of stimulating a far larger number to attempt the same adventure, and thus to yield a large balance, in the way of population, to the rich and promising regions of Australia.

Amongst the matters of interest on arriving on board was a small steamer lying under the port-quarter, and shipping through a large receiving port, communicating with the lower saloon and the treasure depository, a large amount of treasure in boxes of gold. In this and the preceding day nearly 200,000 ounces of gold had been put on board and safely deposited in a strong iron compartment below the lower saloon, and secured by a massive trapdoor of iron and a Bramah's lock, in boxes generally containing 1000 ounces. The total weight of treasure taken on board was estimated at nearly ten tons of gold, which with costly jewellery and other precious things might probably reach to the value of nearly a million sterling.

The shipping of such valuable produce has not always been accomplished with safety. In one instance a daring and successful robbery was carried into effect of a large quantity of gold just put on board a ship bound for England, which was the means of instituting a variety of protective measures which since then have proved safe and effective.

The arrangements for the shipping of treasure are simple,

but satisfactory. Some party appointed by the agents or managers for the ship attends at the Bank during the weighing of the gold, and its being deposited and secured in small square boxes of strong hard wood,—each box being marked with the weight, the sign of the shipper, etc. Each shipment is conveyed, under police escort, from the Bank to a small steamer lying at the quay at Melbourne (on board which no person is allowed except the crew of the vessel and the authorised parties), and from thence is steamed down the river Yarra alongside of the ship destined to receive it. A custom's officer attends the shipment and notes the packages and quantity for the security of the colonial revenue in respect of a duty of 2s. 6d. per ounce, a charge which though small on a value of 4l. per ounce, yields a large annual amount, amounting in the case of the gold in the *Royal Charter* only, to about 25,000l. As the labels on the boxes are respectively read off and put down, they are handed below into the treasure room and there packed for the voyage. The Captain, who becomes the responsible party, keeping the key of the safe, gives a receipt for the whole.

The morning of *Sunday*, the 25th of May, cleared up, after a night of almost incessant rain, and ushered in a beautiful and most enjoyable day. Whilst the cable was being hove in (a light air of wind from the north casting her head that way), I was fortunately enabled to catch sights of the Dandelon Saddle and the Red Bluff, in a variety of positions betwixt north and northwest, which we were not able to complete at the general swinging. The first position I caught was N. 31½° W., which gave me the deviation very proximately of the N.W. by N. point; whilst eight or nine intermediate positions gave the N.N.W. and N. by W., with several repetitions at the north. Fortunately after the ship's head had swung round as far as N. 8° E., it returned westerly as far as N.W. by N.; thus enabling me to check the former observations betwixt that point and the north. The results, however (a phenomenon I had been led to anticipate), of swinging in different directions exhibited very

perceptible discrepancies, extending from one to two degrees, or even more.

At 11 A.M. the anchor was got up, and we immediately
started, not a-head but a-stern (the ship's head lying up the
bay and towards a considerable quantity of shipping), which
retrograde progress we continued to make until we were
fairly clear of the shipping, and beyond the Williamstown
lighthouse. Soon after we had got her head round, Messrs.
Charles and Reginald Bright, and Messrs. William and
Frederick Hart (sons of the senior manager) left us in a
small steamer, the ship having been then or previously
cleared of the visitors from the shore.

The full steam-power being now put on, we proceeded
briskly (from $7\frac{1}{2}$ to 8 knots) down the fine sheet of water
constituting Port Philip,—the day being either calm, or the
breeze exceedingly light. It was a glorious day. The sun
was bright and cheering, with a warmth just agreeable to the
feelings. But the circumstance of putting to sea and the
general excitement on board deprived the Lord's-day of its
characteristic interest and features. We could have no public
service; but in our private room we were enabled to undertake something like the usual service for the day, both
morning and evening, and coming within the limits, at the
least, of two or three gathered together in the name of our
adorable Lord, and seeking to realize something of his
gracious presence. But otherwise, the circumstance of so
sailing was a painful one (resulting, I believe, accidentally,
from the arrangements in respect to the mail), as we could
ardently have wished to have our first Sabbath, on commencing
a new adventure on " the great and wide sea," specially consecrated to the Lord of the Sabbath-day, and in united prayers
for a safe and prosperous voyage.

We passed out of port by the South channel—the deepest
of the two communications with the ocean, in which, however, a clear and continuous depth of water is not ordinarily
to be expected, even on the top of the small tide rise, of
more than 24 or 25 feet, and rarely extending to 28 feet.

At 7 P.M. we were fairly betwixt the heads, steaming

southwesterly until we got the two Port Philip lights in a line, giving the fair way (the night being set in dark) out to sea. Previously, the firing of a gun and the discharge of a rocket notified to the pilot vessel the requirement to take out our pilot, a final batch of letters having previously been made up and sent on shore by a Geelong boat.

The night was quiet and almost calm. We steamed on a S.S.W. course, being our reverse direction on approaching Melbourne, for passing to the westward of King's Island, and rounding the southernmost cape of Tasmania.

The draft of water of the ship, on sailing, was found to be 20 feet aft, and 19 feet 2 in. forward, a far nearer approach to an even keel, in my judgment, than was consistent with the best performance and steering of the ship.

Monday, May 26 (N. E., calm, N. W., variable).—This day was fortunately a fine and enjoyable day, enabling us the more comfortably to settle ourselves in our berths, and arrange our accumulated luggage. The wind, when there was any, continued light during most of the night, but freshened towards sunrise, after a breeze from the N.W. had sprung up. At 8 A.M the power of the sails exceeding that of the engines the screw was detached and taken up. Cape Otway was seen at 7 A.M., bearing N. by E. (true) ; and the north point of King's Island also appeared bearing S.S.E., distant about 20 miles. In the course of the forenoon we coasted the island along on our port side, but it was too distant to present any notable or interesting feature. At noon, our latitude was 40° 1′ S., long. 143° 21′ E. The courses steered throughout the 24 hours (ending at midnight) were S.S.W. to S. by E. ; the rate of going was from $9\frac{1}{2}$ to $4\frac{1}{2}$ knots, and the distance accomplished 168 miles.

Our passengers abaft were now fairly distributed, and, so to speak, domesticated in the saloon. Among our associates were the talented vocalist Catherine Hayes and a Roman-catholic Archbishop.

Tuesday, May 27 (Southeasterly, southerly).—Our progress this day, ending midnight, 120 miles, on a course almost south (true), was almost entirely due to the engines.

But a breeze generally blowing a-head, or nearly so, our average speed was but five knots. We kept further to the westward than our proper course might have suggested from the risk of embarrassment by the coast of Tasmania, in the event of the wind coming from the southward and westward.

The weather was fine and enjoyable, the temperature moderate and beginning to be bracing, with a southwesterly or southerly breeze.

In the afternoon I tried the effect of the change in the terrestrial dip by a series of dipping-needle experiments, our latitude at the time being 42° 20′ S., and long. 143° 15′ east. The dip at the standard or pedestal station was found to be (ship's head S. 17° W., true magnetic—no heeling) 72° 15′ S., the terrestrial dip being (by General Sabine) 71° 15′. The instrument being removed to the forecastle (ship's head the same) gave a mean dip at the midship station of 70° 15′, the needle vibrating in 2·0 seconds. On the starboard side the dip was 74° 50′ (compass-needle there S. 25° E.); and on the port side 71° 10′ (needle at station, S. 40° W.) The vibrations were 2·0 starboard side, and 1·8 port side, and 2·0 midships.

Wednesday, May 28 (E.S.E. to E.), brought us a freshening easterly wind, putting us on a south to S. by E. course, true, and under all sails, causing the ship towards noon to overrun the screw, producing the usual disagreeable vibration. At 6 P.M. the screw was raised, and our speed, close hauled on the port tack, increased to ten knots. The day was generally fine, the air bracing, with a temperature of about 51°. The course made good (till midnight) was about S. by E., and the distance run about 188 miles.

The compass deviations, as determined at Port Philip, were found to apply with great precision, both in steaming down Port Philip, tracing the southern channel, and subsequently proceeding to within sight of Cape Otway, on the starboard side, and King's Island on the port side. But the deviations taken at the swinging referring to the position of the ship's head by the standard compass, were found not to be conveniently applicable, in their present shape, for the use of the Captain and officers, who naturally took the steering compass for the shaping of the ship's course.

To render my observations, therefore, most conveniently and easily applicable for practical use, I cast the deviations of the steering compass into the form of a diagram, which I found to yield excellent curves, and from thence measured off the deviations for each point with reference to the indications of the steering compass itself. The Table of Deviations thus adapted, and made applicable in an easterly or westerly direction, exactly the same as for the variation, was furnished to the Captain and chief officer for use, so long as no particular alteration should be shown by the compass aloft.

The sea was luminous in the ship's track at night, the sky being dark and cloudy. A luminous medusa was passed, when I transiently visited the deck at 10 P.M., probably many others were within view.

On the 29th of May the wind was easterly and south-easterly, nearly right a-head (in respect to our proper course) during the whole day, and blowing a fresh breeze. The ship seldom would lie better than S. to S. by W., sometimes falling off to S.S.W. Hence going generally ten knots we made rapid progress to the southward. At noon we were in 47° 50′ S. by observation, reckoning our meridian at 144° 34′ E. The temperature had fallen to 46° in the forenoon.

The clinometer registered, during the night, a maximum heeling of 22° to starboard, the average being about 12°.

Light showers of rain prevailed in the afternoon, with an occasional break of sunshine out of a generally cloudy sky. A rainbow at 2·30 P.M. appeared of unusual nearness, one of the limbs being plainly visible within the horizon and up to seven or eight yards from the side of the ship! It was a complete arch, well coloured, but not so sharply defined as in the most perfect form of the phenomenon. The night was cloudy and dark, the wind blowing a fresh breeze directly against us, so that, while rapidly advancing towards the southern pole, we made no easting, but rather lost ground.

Friday, May 30 (E. S. E., S. E.)—This day was as little auspicious, in regard to the direction of the wind, which continued, an unusual circumstance in these latitudes and this position, to blow steadily from the E. S. E. and S.E yet we still held on the southerly course, ranging, till night, from S. or S. $\frac{1}{2}$ W., true, to S. by W. $\frac{1}{2}$ W. In order to keep nearer to the wind the steam was got up, and the engine, at 11·30 A.M., put into action; but the wind, though it had much moderated, was yet too strong for us to derive any material benefit from the appliance. The weather, however, was fine, and the air sharp; temperature 43° at 8 A.M., and the sea smooth. Fog showers, very like those of the Arctic regions on approaching the margin of the ice, prevailed in continuous succession, yielding occasionally, under transient gleams of sunshine, small prismatic segments on the horizon, and on one or more occasions the brilliant and complete rainbow. Before daylight we must have passed a little to the westward (about 49' off) of Royal Company's isle.

At 7 P.M., our course having for some time been only S. S. W. $\frac{1}{2}$ S. (true), the ship was tacked to starboard, and to the surprise of most persons, the officers of the ship as well as others, she made exactly the *reverse compass course* of that made on the port tack. This, I showed, was the simple result of the deviation, which, whilst considerable in quantity (—12° when lying little better than S. S. W. by the steering compass, and +11$\frac{1}{2}$° when lying on the other tack N. N. E.), had changed its signs from westerly to easterly. Hence the courses made good, as cleared of deviation and variation (reckoning the variation from Sir James Ross's observations at 11$\frac{1}{2}$° E.), came out proximately S. S. W. or S. by W. $\frac{1}{2}$ W. and N. E., the ship apparently lying scarcely nearer to the wind, with all sails set, than seven points.

The dip of the needle having increased (according to General Sabine's tracing of curves to 76° S.), I took the opportunity of moderate weather, about noon, to get a pretty complete series of dipping-needle observations, where the agreement of the readings of the instrument, with the face both ways, indicated a probable accuracy (supposing the needle to have no index or normal error) within a few minutes of a degree. The ship, however, was not upright, but heeling from six to seven degrees to starboard by the clinometer.

At the "standard" or "pedestal station," the dip was 78° 30′, apparently $2\frac{1}{2}$° in excess of the terrestrial dip. At the midships "forecastle station," the observed dip was 75° 45′, or very nearly what was supposed to be the true inclination. On the "starboard side of the forecastle," the mean observed dip was 76° 45′, and on the port side, 78° 45′. Finally, on the afterpart of the poop, on the two sides of the middle of the after skylight, the results obtained were $72\frac{1}{2}$° on the starboard, and $72\frac{3}{4}$° on the port side, indicating a less dip by $3\frac{1}{2}$ and $3\frac{1}{4}$ degrees than the terrestrial. The time of the oscillations of the needle roughly yielding indications of the magnetic force afforded no conclusive evidence of distinctive change from the force observed at Port Philip. The ship's head, it remains to be noted, was pretty steadily near about the true magnetic S. by E., or ranging, perhaps, from S. by E. to S. $\frac{1}{2}$ E. A set of compass comparisons (of the four compasses) was also taken.

The evening was starlight, numerous luminous medusæ floating like glow-lamps, or will-o'-the-wisps, on the surface of the water. But they were much smaller than those I had observed on our approach to Australia, and only became visible within 100 yards or so of distance.

Saturday, May 31 (Easterly, southerly, south-easterly, variable).—To those anxious and looking for a quick passage home, this day (the 31st of May,) was a disappointing one, in the most unusual continuance of light and adverse winds. Sometimes the ship was steamed, slowly, head to wind, when the breeze was sufficiently light; sometimes taking the wind sharp on the bow with fore and aft sails only set; and sometimes under topsails and other square sails, making some near approach to an easterly course. But the result of the day's work, from noon to noon, only gave an advance eastward, to long. 146° 29′ E., and a total progress for the nautical day of scarcely 120 miles on a course E. $\frac{1}{2}$ S. true. The wind was variable, light, or medium breeze, throughout, and sometimes nearly calm, but always scant or a-head.

The effect of the deviation, at noon, was, to those unaccustomed to iron ships, sufficiently curious and perplexing. The ship steaming and sailing on an easterly course, inclining northerly, broke off about noon to N. N. E. by the steering compass. Disliking so northerly (apparently) a course, the ship was put about, when to the surprise and annoyance of the officers, she now only lay S.W., a difference of 18 points, indicating an angle of nine points from the wind! This new course, indeed, would have led us in a direction of some ten points or more from our course, whilst the course on the starboard tack (apparently N. N. E.) would, with above a point of easterly variation and a point and half of easterly deviation, have given a true course approaching to N. E. by E. being within three points or a little more of the course wished to be steered! The next manœuvre was to take in all sail and, the wind being light, steam on a more favourable course, which at 4 P.M. was found to be about E. by compass or E. by N. true.

The day was cloudy,—a gloomy canopy of dark cloud generally obscuring the heavens; but it was dry, and the air sharp,—the thermometer having fallen to 36°. This, without fires, doors and windows often open, and currents of air flowing through the saloon, made the ship below comfortless enough, especially for those accustomed to much self-indulgence in fires and every appliance of domestic comfort! Nor less so to us, after the enjoyment of our cheerful well-warmed rooms at Tourac, during the latter three weeks of our abode on shore. But no resource was left us now, except appliances of warm clothing and rapidly walking up and down the poop-deck. Often nearly the whole of the saloon passengers might be seen walking rapidly in circuit and ranks, in order by such exercise to gain something by the stimulating of the animal heat. The night was dark and dull. On this day, crossing the meridian of the eastern point of Tasmania, we passed from the hydrographical region of the Indian Ocean into the Pacific.

Sunday, June 1 (Southerly, calm, S. W.)—Scant breezes or calm prevailed during the night, but so as to allow a general progress on an easterly course, and to make good a rate of about six knots. But about 7 A.M. the springing up of a fair breeze from the south-westward soon put all on board in good heart. Sail was immediately made on the ship, so that, in moderate space of time, a goodly quantity of canvass was spread out to catch the propitious breeze. The sky continued however to be covered, as in the preceding days, with a dark canopy of cloud, intercepting the sun's rays and preventing observation. At noon, our estimated position was latitude 51° 31' S., and longitude 150° 26' E., giving us a progress only of about 860 miles, on a true course of S.S.E., in seven days, whilst this small measure of progress, from the defect of easting, could by no means be considered as wholly gain.

The circumstances, happily, were favourable for the renewal of our devotional exercises. An unusually large congregation assembled in the saloon, where much seriousness and apparently interested attention prevailed whilst I addressed them from the second lesson of the day (Mark ii. v. 17.

In the evening I went to the third-class messroom, accompanied by several saloon passengers, where we found a goodly assemblage awaiting us. The uniting interest and crowd of worshippers was a cheering sight. We commenced with the Evening Hymn, which was sung with much animation and good voice. After a selection of prayers, with the psalms and second lesson, from the evening service, and the singing of the 100th Psalm, I addressed the congregation (which might number 150 or more) extemporaneously, from the sublime chapter in 1 Corinthians which we had just read, chap. xv. 51, 52. Alluding to the condition of the assembly, chiefly working men, returning to their homes with a goodly treasure from the gold-fields, or other ample sources of rewarded industry, and congratulating them on their present successful enterprise,— I then pointed out the infinitely momentous revelation before us of a state in reversion beyond the enjoyments and toils of life, where there should be a restoration, in some unexplained and mysterious form, of the present body, after being wasted in the grave, or however dissolved in the ocean or otherwise, a resurrection of the just and the unjust with a spiritual externally abiding body. The attention throughout was fixed, earnest, and solemn, and very encouraging for a continuance, under the Divine blessing, of similar services during the voyage.

A gloomy dark night succeeded to the densely clouded day. But the breeze gradually improving we increased our speed, under all available canvass (after the withdrawing of the steam-power at about 2·30 P.M.), to about 11 knots. A considerable swell from the south-westward.

The courses steered from midnight to midnight were E. or E. by N. by the steering compass, and as on this direction there was *no* deviation, all the compasses very nearly agreed, and the application of about a point of easterly variation became the only correction required on the course steered.

Monday, June 2 (South-westerly, westerly).—The favourable breeze, with some slight increase, prevailed throughout this day, but being right aft for the most part, the sails could not yield their full influence, and our speed was restricted to a general rate of about 11 knots. The rolling of the ship under a strong south-westerly swell and with squared yards was sometimes heavy. At 9 A.M. indeed the maxima registered by the clinometer were only port 10°, and starboard 13°, with a small increase in the registered number (10,245) of the rolls. But in the afternoon the swing from side to side increased so much that a succession of heavy rolls at 7 P.M. registered 24° to port, and 23° to starboard. The day was generally cloudy, and often showery, the cloud-stratum becoming broken so as to give opportunity for good sights for both latitude and time. These gave our meridian latitude 52° 15', and long. 155° 40' E., indicating a distance reckoned on Mercator's sailing (for 137° of longitude) of about 4800 miles from Cape Horn.

A set of compass comparisons was made at 4 P.M. which,

though the ship was rolling heavily (from 10 to 15 degrees each way), were fairly satisfactory. They indicated the fact —pretty nearly agreeing with the observations at Melbourne —of a general approximation of the three fixed compasses and the standard compass on an E. course. The compass aloft in three readings gave the ship's head due east, the steering compass E. 1° S., the companion compass east, and the standard compass E. 1° S. The dipping-needle at the standard station gave about 78° of southerly dip, but the heavy rolling of the ship rendered the result doubtful to the extent perhaps of a degree.

In the course of the day I finished the projecting of the deviations of the three compasses of the *Royal Charter* and the standard compass in diagrams, showing the curves of the deviations, comparatively by inspection, of the several compasses as reduced to the true magnetic directions of the ship's head on each point of the compass, and the differences in the deviations of each compass obtained by the swinging at Liverpool, January 2, and the swinging at Melbourne, May 2. The comparisons present many curious and very instructive results. But the facts which the most strike the eye are these: 1. The considerable changes in the way of error in the *adjusted compasses*. 2. The changes in the *standard compass* and compass aloft in the direction of greater accuracy or improvement in indication; and, chiefly, the trifling quantity of error, or the very close approximation to the truth, in the *compass aloft*.

Tuesday, June 3 (W., variable, N.N.W., N. W.)—The wind was light most of the day, and at noon seeming inclined to calm, though the ship still made progress to the eastward, on an average of about seven knots. But a shift of wind at 2·30 P.M. to the N. N. W. gradually brought up the ship's way to eight or nine knots.

A strong southwesterly swell (in waves of perhaps 15 to 18 feet) and rather short spaces betwixt caused the ship to roll more heavily than we had yet witnessed, I think, during the voyage. The clinometer had registered as the maxima for the night an angle of 34° to port, and 26° to starboard, with a numerical register up to 3066, at 9 A.M., indicating the movement of 2821 rollings in 24 hours. Such was the amount of rolling—the yards being square in the morning and forenoon—that it was very difficult to walk the deck. This unusual rolling motion, it appeared to me, was satisfactorily to be explained in the synchronism of the ship's natural oscillations and the period of the waves successively reaching her on the starboard quarter, about six seconds, as, on the occurrence of a series of waves, in this ratio of progress relative to the ship's position, the rolling was found to be heavy and continuous.

The temperature on deck at 8 A.M. was 41°, and in our cabin below 53°, the former gradually rising to an extent of four or five degrees as the day advanced, and soon after noon the sun, almost for the first time for several days, broke out bright and clear. Good sights for latitude and chronometer were obtained, which gave the latitude 53° 17′ S., and longitude 160° 34′ E. In the afternoon our position was fairly betwixt the Auckland and the Macquarie's islands—course E. by N. by compass (no deviation).

The "variation of the compass," in our line of advance from this position to Cape Horn, became a subject of inquiry of considerable interest and importance, as it does not appear that very many reliable observations are on record within it, beyond the determinations of Sir James Ross in the track incidentally pursued by the *Erebus* and *Terror*. To this inquiry my attention had been particularly directed by the Lords Commissioners of the Admiralty, in their letter by the Secretary, acceding to my application for the loan of instruments and charts, dated December 18, 1855. The Secretary states, " As the variation charts, published at the Admiralty, are now in course of revision, I am to request you will transmit to the Admiralty, by the earliest opportunity, a copy of the variations of the compass which you may observe in the voyage out to Melbourne, in order to the immediate correction of the charts of the Atlantic and Indian Ocean, if necessary." The difficulty in the way of such determinations in the course of my contemplated voyage in an iron ship obviously was the eliminating of the local action in the ship's magnetism. But from the very satisfactory results in ascertaining the deviations of the compasses at Melbourne, it now appeared to me that little as could be done on our outward voyage before the action of the compasses and their relative deviations had become known, that sailing, as we should now frequently be, on an east or E. by N., magnetic course, which fortunately happened to be points of change, or nearly so, of most of the compasses, the actual variation, within a very small amount of probable error, might be very well determined. The first opportunity for the trial occurred in the afternoon of this day, the sun shining clear about an hour before setting. Three sets of azimuths, with simultaneous altitudes, were obtained when the standard compass happened to be tolerably steady. The apparent azimuth, N. 59° 20′ W., corrected for 2° westerly deviation with the ship's head E. ½ N., gave the correct apparent azimuth N. 61° 20′ W., and the true azimuth being found to be N. 42° 36′ W., made the variation 18° 44′ E. (latitude 53° 20′ S., long. 160° 50′ E.), a result which, being afterwards compared with Sir James Ross's determinations for the Auckland islands and vicinity, allowing for the change in the variation during 12 or 13 years, seemed to be very nearly accordant.

Beyond our present position, I do not find any specific information in the charts on board for the guidance of the navigator pursuing our track round Cape Horn. The captains of ships, indeed, proceeding from Australia to the eastward for England, appear generally to be guided in their allowances on the dead reckoning by the agreement or differences discovered from day to day betwixt the positions of the ship derived from supposititious allowances for compass errors and the positions indicated by celestial observations. If they find themselves considerably to the southward of their reckoning on any particular day, they allow more easterly variation or deviation proportionally, or the converse. This practice, when the sun can be regularly seen, and whilst the course lies wide of any land or rocks, does very well; but when, as is not unfrequently the case, celestial observations may not be attainable for days together, or when the ship's course may lie pretty nearly in the direction of land, and especially when a change of wind may oblige a change in the course, and so in the ship's deviation,—a reliance upon such inverse processes may obviously prove not only very embarrassing, but in heavy weather, with fog or darkness, not a little hazardous.

Wednesday, June 4 (N. W., W.)—A fresh breeze gene-rally prevailing throughout the day of nautical time gave us though not a rapid yet a respectable progress. From noon to noon we made 225 miles by the log, which brought us to about 166° 46′ of E. longitude. Yet to our anxious and per-severing Captain the long series of light or contrary winds we have had since sailing was very dispiriting, as, from an experience of about thirteen passages round Cape Horn, from Australia, he had rarely, if ever, found so much calmness. Unceasing rain kept most of the passengers below, but, happily for our comfort, the temperature was warmer, the air being 46° at noon, and our cabin 56° to 60° or upward.

The clinometer noted at 9 P.M. maximum rolling of 15° to starboard, and 13° to port, during the night; and an average heeling of 7° to starboard.

The birds, which for some days had been rather numerous, had now, for a time, almost deserted us; but the phenomenon of the luminous medusæ floating on the surface of the water continued, occasionally at night, to excite our attention. Light or moderate breeze, after sunset, with heavy rain.

Thursday, June 5 (W. N. W.)—The rain continued very heavy till daybreak, and, our breeze deserting us, the screw was lowered and the engines started at 8 A.M., which in-creased the ship's rate of going from 2½ to 7½ knots. In the afternoon the breeze from the westward (very gentle) returned, and so far became steady that at 6 P.M. the engine was stopped and the screw raised, but our speed became considerably reduced. The day was throughout fine, cloudy, occasionally clear, and the temperature more mild, 46°. A strong swell from the S.E. indicated the probable occasion of calm or light wind just experienced; we apparently being betwixt conflicting currents of the atmosphere.

The clinometer at 9 A.M. marked maximum rolling or heeling for the night $5\frac{1}{2}°$ to starboard and 4° to port.

Several cetaceous animals were seen during the day apparently of medium or small species. I saw several "blowing" (three or four at a time), but they were too distant for the discrimination of the species.

The day being fine I took the opportunity of making magnetical observations.

1. *Azimuths.*—There being a swell from the southward, and the compass-card but rarely steady, it was a matter of great difficulty getting good and satisfactory azimuths of the sun. In order to test the observations, and obtain a better result, I took advantage of three opportunities presented to me in the course of the day—two in the forenoon and one in the afternoon. The first of these gave me two azimuths with corresponding altitudes; the second, three; and the third, one. The data for the variation were these (the latitude was 54° 26′ S. for all):—

Polar. Dist.	True Alt.	Obsd. Azim.	Ship's Head.	Deviation.	True Azim.	Variation.
° ′	° ′	° ′	°	° ′	° ′	° ′
112·33	12· 3	N. 6·25 W.	E. 5· N.	2·10 W.	13·24	=21·59 E.
· ·	12·15	N. 8·13 W.	E. 9· N.	2· 0 W.	11·57	=22· 8
112·34	5·44	N. 56·20 W.	E. 17· N.	1·40 W.	36·32	=21·28

This measure of agreement was as gratifying as the general result, giving a mean of 21° 52′ for the variation separate from deviation, was satisfactory.

Where the deviations of the standard compass on the point steered were so small, the determinations at Port Philip so recent, and the agreement of the compass aloft likewise corroborative of the process for eliminating the ship's magnetic action on the needle, it appeared to me that the result might be fairly relied on as abundantly accurate for practical application in navigation, and probably as near as any determinations at sea can be expected to come. The ship was without heeling.

2. The *comparison of the four compasses* agreed very well with their previously determined deviations on the course, about E. by N. ½ N., then steered, the difference in the four being very small.   No heeling.

3. *The dipping-needle* was placed on four stations, and its indications for inclination and force carefully taken, with the face of the instrument both east and west at each station. At the standard station the apparent dip was 77° 20'; on the forecastle midships, 78° 45'; on the afterpart of the poop, starboard side, 70°; port side, 82°.   The great difference in the last two stations was obviously due to terrestrial induction, the ship's head being E. 17° N., the polar axis must be inclining greatly over to the port side, and the equatorial plane running obliquely across the ship's breadth.   Hence, in experiments made May 30, with the ship's head lying nearly south, the dip on both sides was very nearly the same, being 72½° and 72¾°.   Correspondingly with the observations of this day, and their great difference, I find under date of April 11, ship's head S. 82° E., the dip on the starboard side 68¾°, and that on the port side 78¾°.   In both these cases the ship was upright.

Friday, June 6.—During the whole of the night and day the wind was very light, sometimes it was quite calm.   Our progress was therefore but small, for we were without the engine, in constant expectation of wind, until 8 P.M., when the breeze having entirely died away the screw was once more put in action and most of the sails taken in.   The weather was fine and generally fair, the air mild (as it was considered) 41°, but bracing.   A very heavy swell prevailed all day from the S. S. E., the waves sometimes rising to 20 or even 24 feet in elevation, indicating conclusively not only the present or very recent blowing of a heavy gale to the eastward or southeast of us, but the cause of the long continuance, with a rather low barometer (29° 50'), of light winds or calms.   We were obviously in the quiescent atmospheric position betwixt conflicting winds, and we should have been well satisfied with our dallying breezes rather than to have encountered a heavy adverse gale.

The clinometer at 8 A.M. marked 8° port and 14½° starboard, as the maxima rolling for the night, with a register of 4426. Many albatrosses were about the ship, with numerous Cape pigeons, gulls, etc.

A numerous set of sights for the variation were taken about 2·30 P.M., the heavy sea prevailing rendering good azimuths difficult to catch. The compass error, ship upright and head E. 7° N., with deviation eliminated, gave 21° 41' E. for the variation, the latitude being 54° 20' S., long. 175° 58' E.

Saturday, June 7 (Easterly, calm, S. S. E., S. W.)—Light airs, chiefly a-head, prevailed during the night, so that our headway, which was but very small, was entirely due to the engine. This long continuance of light winds in the season of winter in these far southern latitudes has been occasion of as much surprise, being so exceedingly unwonted, as it has been the occasion of anxiety and disappointment to our active and persevering Captain. The unprecedented success in our outward passage, bringing us to our ultimate destination in a shorter space of time than any similar ship had ever succeeded in accomplishing, shorter too, with but one exception, I believe, of a full-power steamer which made the passage in the same time, than had ever been made by any ship whatever, beating at the same time every clipper ship or steamer arriving from England at Melbourne, from January 1 to the date of our sailing (or within the last five months, I may say) by 13 days on the fastest, and 69 days on the slowest, and 41½ days on the average—everything seemed, under Providence, to encourage the hope that the return voyage of the *Royal Charter,* at a season when strong favourable gales for the passage by Cape Horn were with much confidence expected, might likewise be successful, as to shortness, or, at all events, might vindicate the high expectations formed of the capabilities of this experimental ship of a somewhat new class, in establishing the principle of clipper ship, large tonnage, and auxiliary screw steam-power, as the type for the best and most popular intercommunication of England and her great colonies of the south.

About 7 A.M. a light breeze sprung up from the S.S.E.,
with rain, which soon seemed to promise—connected as external appearances were with a rapid and considerable fall in
the barometer (from 29·150 the preceding evening to 28·870
at 8 A.M. of this day), an abundance of wind in a good direction for progress.   The increase was such that by 9·30 A.M.
the steam-power was stopped, and as soon as practicable the
screw raised up.   This work, however, was not yet completed, and the double topsails and courses, fore and aft,
being set to the wind, when a very awkward accident happened to the gear of the upper topsail, which occasioned
the withdrawal of all sail from the mainmast, for a space of
54 hours, before the damage was corrected, and the yards
and sails placed in effective position.

This accident consisted in the breaking of the iron-work,
the swivel suspension of the gin or topsail tie-block of the
upper topsail-yard, by which the ponderous yard, weighing
—without sail and ropes—about five tons, fell suddenly from
its position when fully hoisted upon the lower topsail-yard,
which, in its turn, was broken away from its supports, by
the carrying away of its slings, bursting some of its hoops,
and otherwise damaging the yard and upper part of the
lower topsail, and this lower topsail-yard descending from
its position and threatening by the concussion further de-
rangement of the mainyard.   This was the most serious mischief which had yet proceeded from the breaking of iron-
work about the masts, sails, and yards, unfortunately a too
common occurrence.   Providentially the mainyard held fast,
and the fall on deck of several pieces of the broken iron,
etc., did no essential injury, the rain which prevailed at
the time having happily induced the great mass of passengers, generally crowding every part of the deck, to remain
below.

All hands were forthwith summoned for the clearing away
of the crippled yards and sails, and the reparation of the
damage.   But it was not until about sunset that the securing of the upper topsail was effected, and the lower topsail
first and then its crippled yard got down upon deck.

Happily these important preliminaries to repair and replacement were accomplished, under increasing wind up to a gale and much rain, without further accident.

By 3 P.M. the gale was such that with topsails and courses only forward and aft, our progress reached to about 11 or 12 knots.

The wind continued but for a short time in the southern quarter, though it blew hard from the south for a little while, but in the evening it began to veer to the westward, so as by 8 or 9 P.M. to have come right aft, being due west. With squared yards the mainsail was then set, as also the upper topsail lowered down to its greatest practicable nearness to the mainyard.

Sunday, June 8 (W., W. N. W., to W. and S. W.)—The night was wet, and the wind being right aft, the ship rolled considerably, as much at a maximum as 20° to port and 17° to starboard, and an increase of the register, at 9 A.M., to 6400.

A favourable morning, very different from what we had expected with the barometer at 28·720 with a reduction of the gale to a strong breeze, gave convenient opportunity for our usual morning service in the saloon. There was again a very considerable attendance, and great fixedness of attention whilst I addressed them from 1 Peter v. 8, 9—a portion of the epistle for the day.

This was anything but a Sabbath of rest to the ship's company, officers and men, artizans and mechanics, were engaged from daylight to sunset in endeavours to replace the damaged topsail-yard by a spare one lying on the spar-deck or deck-house. Soon after the conclusion of dinner the capstan began to be put in motion, but the affixing of guys for security in raising so heavy a mass of timber and iron (some five or six tons weight), and the obtaining of a safe and sufficient "purchase" for heaving it up, and the adjusting of some disordered iron-work, so retarded the operation that the day closed in before any considerable progress had been made, and the operation was obliged to be abandoned for the chance and opportunity of another day. Meanwhile a

serious abstraction from the spread of canvass was necessarily occasioned, and a loss of perhaps two knots on an
average was suffered in speed.

The latter part of the day was rendered very uncomfortable by frequent and almost continuous showers of rain
and sleet with a fall in the thermometer to 39°. The wind
was very variable both in force and direction, sometimes
being so moderate that the courses flapped against their
masts, and at other times blowing a brisk gale; at the same
time veering about in direction, from S. W. to W. and
W. N. W., and conversely. Heavy seas from different
quarters, occasionally from the south-eastward, but principally from the westward, made the ship roll considerably,
and indicated the otherwise probable fact that the breeze was
kept down in force by reason of conflicting currents of the
atmosphere. Waves of 20 to 24 feet in height, however,
marked the general maximum.

At the appointed time in the evening (7·30 P.M.) I renewed
my visit, accompanied by six or seven saloon passengers, to
the third-class messroom, for the purpose of a devotional
service. The place was again crowded with probably 150
persons, mainly third-class passengers. After a selection
from the prayers, I took verse 2 of 2 Corinth. vi., the second
lesson for the day, as the subject of discourse for the en-
forcement of the divine monition and call, "Behold, now is
the accepted time! behold, now is the day of salvation!"
All seemed earnestly attentive, and many indicated considerable feeling and emotion. It was again an encouraging
assemblage and service.

Our course pursued throughout the day was E.N.E. magnetic, giving, with about two points of easterly variation, an
east course, true. The rate made good from midnight to
midnight varied from 6 to 11 knots. Our latitude at noon
was 54° 2' S., and longitude 175° 30' W., we having passed
the antipodal meridian of Greenwich at about 6 P.M. of yesterday, and at 4 P.M. our position both in latitude and longitude was very near to the antipodes of Whitby, the place
where I spent my early life at home from earliest childhood.

The snow and sleet which occurred during the day became occasion of surprise and interest to several different persons on board to whom atmospheric moisture in the form of snow was a previously unrealized phenomenon.

Monday, June 9 (W. N. W., W., S. and W., etc.)—We held on our way making, under frequently shifting winds betwixt W. N. W., W., and S. S. W., a course very nearly of due east, and for our defective canvass pretty good progress, except at times, when the generally brisk gale failed us greatly for some hours in the morning. Heavy swells from the north-westward and south-westward prevailed.

Daylight brought us a fine bracing and drying air, thermometer 36°, with patches of frost on the decks and a considerable quantity of snow laid over the windward side of the boats and masts. A heavy swell at night from the northward caused the ship to roll greatly, the maximum angles registered being 21° to port, and 24° to starboard, registering, at 10 A.M., 8350 on the clinometer. Another swell from the S.W or W. S. W. was still heavier, and prevailed without abatement during the day. The waves frequently rose above 24 feet, and were estimated (the highest) at 26 to 28 feet.

The great and critical work of getting up the spare lower topsail-yard was recommenced with the early day. The difficulty arose chiefly from the heavy swell, with a less apprehended but not overlooked risk, in the sufficiency of the purchase and appliances for raising up so heavy a mass of timber and iron (about six tons) to the requisite height for its being secured. The purchase consisted of a pair of treble blocks of large diameter, and a new rope of five inches in circumference. I had myself watched the progress upward with much anxiety, and the various appliances of tackles and guys for steadying it. These were admirably applied, and it was soon obvious that little was to be apprehended from the rolling of the ship, unless something should give way.

Just as the yard had risen to within a very short distance of its position, but whilst yet entirely dependent on the purchase worked at the fore-capstan, both yard, men, and

P

ship had a most providential escape. Though the rope employed for the purchase was deemed sufficient for more than twice or even thrice the weight of the yard, Captain Boyce, happily, had been somewhat anxious, if not distrustful, but there was no other fitting and adequate appliance on board, and had so far made provision for any accident which might afford a moment's warning as to get placed in position sufficient "stoppers" for holding the "purchase fall" in certain places betwixt the capstan and the "leading block" on deck. It was, I say, a very providential provision; for just at the point referred to, when the yard was suspended many feet above the spar deck, one of the men employed in heaving, suddenly gave warning that a strand of the rope at the capstan had given way! It was the business of a moment, under the feeling of urgent danger, to stop the capstan, and apply the stoppers, a security which had hardly been completed when the rope at the warning place broke. Thank God! the stoppers were effective and preserved us from the threatened calamity. The fall of such a ponderous spar from such a height would probably have carried away the mainstay, or, if not, might, from its resistance, have canted on an end, and have penetrated the decks; and where or how its force should be arrested, where the side of the ship could scarcely have sufficed for the object, it was impossible to conjecture.

With certain cautionary measures, in the application of a hawser in the form of a "messenger" to the "tackle fall," and so avoiding the *nip* of the other rope from the whelps or ribs of the iron capstan, and as much as possible the heavy "swinging," the yard was at length got into position and properly secured. The men worked bravely, and by 4 P.M., whilst yet the twilight remained in the sky, the sail was spread to the wind and once more commenced its important service. I examined the rope afterwards, and the capstan—the former was without fault, being all but new, well made, and of sound good hemp; but the capstan of iron with rigid and protruding whelps of iron, was wholly unsuited (though the only mechanical power on board of sufficient power) for a heavy lift, not only on account of the nip at the whelps, but the formation of the barrel which expanding rapidly towards the base occasioned frequent heavy surges on the flitting upward of numerous turns of the rope placed round it.

On the first sight of the accident, with the obvious crippling of the lower topsail-yard, rendering its replacement indispensable, I had ventured to predict that the accident would occasion us the loss of 100 miles in distance eastward. The result more than verified the prediction; for, from the loss of a portion, sometimes the whole of the sails on the mainmast, during a period of 54 hours, with the loss of studding sails during a fair wind and generally brisk gale for the time (there being no fitting opportunity of taking the men off their work to set these extra sails), the loss was frequently three knots or more in the ship's speed, and under any of the intervening circumstances not less than two; so that the ship's retardation by the breaking of a piece of faulty iron could scarcely have been less than four degrees of longitude.

Tuesday, June 10 (South-westerly, W., calm, E. N. E., E., N., N.W.)—We kept our favourable breeze, varying to different quarters, all night, and, though it fell to little wind towards noon, and calm in the afternoon, we had made progress, at mid-day, to longitude 161° 50′ W., in latitude 54° 25′. Showers of snow prevailed occasionally at night, and at 6 A.M. the decks were covered to a depth of some inches, and the thermometer indicated 31°. It dried up, however, in the forenoon, and the temperature rose to 36°. The sharp bracing and dry air brought most of the saloon passengers on the poop-deck, where various modes of exercise, walking, running, hopping, gallopading, were resorted to for the acquiring of some comfortable measure of heat. The sea being high, however, and the ship rolling much with squared yards and light wind, there were considerably difficulties in carrying out the different modes of exercise.

My observations for the variation, which I had been very anxious to carry out regularly by means of azimuths, eliminating the deviation by the method noticed under date of June 3, had unfortunately been prevented for three days by

the encumbering of the spar-deck by the side of the standard
compass, besides the want of help for taking altitudes and
compass comparisons, which I could not reasonably apply for
when the officers as well as men were so engaged.

But the ship's affairs being in a fair measure of adjustment
and the cumbrous yard out of the way, I became solicitous,
heavy as the rolling of the ship was, of catching sights for
azimuths, when the sun incidentally broke out on this day.
But the swing of the azimuth compass induced me again to
resort, in the first instance, in the forenoon, to Captain
Becher's repeating card, comparing it with the foremost or
companion compass, which, being on the floating principle,
was less disturbed and not unfrequently steady.

On the mean of three sights, the true altitude was 12° 3';
the angle of the sun with the ship's head, by the repeating
card, was E. 81° 40' N., and the ship's head by the companion
compass, E. 13° 50' N., which, being additive, gave the
apparent azimuth by compass 95° 30' westerly from the magnetic east; a correction of 1° easterly deviation indicated
94° 30' as the deviated or true magnetic azimuth. The latitude was 54° 10', longitude, 161° 50'; polar distance of sun,
113° 1'. Hence the true azimuth appeared to be N. 11° 54' E.,
to which the magnetic azimuth of (94° 30'—90°) 4° 30' W.
being applied, indicated, proximately, a variation 16° 24' E.
As, however, the deviation of this adjusted companion compass could not be so well relied upon as that of the standard
compass, which had been taken with such elaborate precision,
I was anxious, if possible, to test it, and improve it by sights
with the azimuth standard in the afternoon.

Happily, an opportunity—embracing, however, but one
good sight—occurred about 2 P.M., and though the compass
card was swinging considerably, the azimuth appeared to be
a tolerably reliable one. On working it out with a true altitude at 7° 6', latitude 54° 25', and longitude 161° 30' W.,
apparent azimuth, corrected for deviation, 1° 45' W., ship's
head by standard, E. 12° N., and the ship upright, it gave
18° 5' for the variation (easterly), which I considered preferable, for the reason stated, to the former.

The dip at the standard station, ship's head by compass E. 10° S., appeared to be 75° 45', and the force oscillations, 2''.

The calm of the afternoon continued until about 6 P.M., when a breeze sprung up from the E. N. E. to E. by N., rapidly increasing to a fresh gale, and putting us on the unpleasant course of S. E. to S. E. by S.—unpleasant because urging us farther into the severe climate of the south, and endangering the more the falling in with ice. Rain, squalls, and threatening gale, brought us under a very low sail, but, fortunately, only for a short period; for at 8 P.M. the wind suddenly chopped round to the N., and thence to N.W., enabling us to make good our desired course of true east.

Wednesday, June 11 (N.W., W. to N.W., and W.)—The wind, though variable, continued moderate and fair, giving a progress throughout this day of seven to eight knots on the required course of almost due east. This gave our latitude, at noon, 54° 32', longitude 159° W. The morning was bright and dry; for the season of winter in this region, the weather was singularly fine and mild—temperature 41° to 43°. In the afternoon we had showers of rain; swell moderating; clinometer at 9 A.M., maximum 8° to port, 28° to starboard, for the night previous, and registered number 1560.

Thursday, June 12 (W., W. S. W. to N.W. and W. etc.)— Light or moderate sometimes fresh breezes, constantly vary- ing betwixt N. and W. S. W., still prevailed, yet giving us a fair run, on a course pretty nearly east, of 197 miles for the nautical day. Our rate of going was from 7 or 8 up to 10½ knots. With occasional showers, the weather continued fine ; the morning was splendid, and, for the season and region, far more enjoyable than had been anticipated. The temperature was about 41°.

The want of a steady breeze gave much and almost con- tinuous work to our seamen, in trimming and retrimming yards, hauling up and setting canvass, and in shifting studding sails.

At the same time that we for such a long period had sailed under light or moderate or occasionally fresh breezes,— scarcely 24 hours having yet occurred since our departure from Port Philip in which royals could not be carried,—heavy gales had been almost perpetually blowing at no great distance from us. Of this we had evidence in the heavy swells, reaching us from various quarters, and not unfrequently from two or more quarters at once. Latterly, these swells had been from the S.W, W., N.W., or N., and reached a maximum height of 20 to 24 feet. This afternoon, a fresh onset of seas from the west overtook us, when waves of 24 feet, or upward, were observed.

Hence it appeared probable that a vessel leaving Australia a week or ten days after us,—as the fine and fast ship the *Kent* had been advertised to do (her date for leaving being announced for the 5th of June)—might be expected, especially if pursuing a more northern parallel, to fall in with these strong and favourable gales, and so to gain on us in the passage round "the Horn" a week or ten days, and possibly to beat us.

The hopes of a quick passage home, which had been in some degree revived by the apparent setting in of smart and commanding gales from the prevailing quarter on the 7th, were now generally abandoned. There was now obviously no chance of making a passage of extraordinary shortness; and all that could be reasonably hoped for was, under the blessing of Providence, that we might still make a fair or good progress in respect to the general average of clipper ships.

The fineness of the weather and gentleness of the winds hitherto experienced, in a region generally boisterous, and at a season approaching the shortest day, may not indeed appear very extraordinary, if compared with the corresponding month of December in similar latitudes in our own country and neighbouring seas. November and January are for the most part characterised by severe or boisterous weather. But it is not unusual to find December mild, fine, and by no means tempestuous. And such Captain Boyce's experience indicates as to this region, corroborated by many instances of long passages by reason of light winds round the Horn, by vessels sailing from Australia in May or June; on this special occasion, he seemed to reckon on a less severely tempestuous passage than at seasons which might be thought favourable for moderate weather. The risk too of meeting with ice—a formidable peril in the long dark nights—is greatly less in the winter season than in the summer, when the ice ordinarily sets more into the track of ships from Australia.

Friday, June 13 (Westerly, N.W., N.)—A continuance of fine weather,—beautiful sunrise, enjoyable and bracing day; and, as we had for several days in succession, glorious moonlight. The sky with moon and stars, a blaze of nocturnal light. Towards evening, indeed, the wind freshened to something like a gale, and became so scant, being north or more easterly, that the ship with full sail and the yards faithfully squared would not quite lie on her purposed course. But at the cost of getting farther to the southward than in this meridian was thought desirable,—having regard to the possible risk of falling in with ice, and an unnecessary exposure to the severities of a high southern latitude in winter, —the sails were kept well full, and in effective action, so as to urge the ship on her way at the rate of 10 to 12 knots.

Observations for latitude and longitude gave our position at noon 54° 71′ N. and 148° 26′ W. The clinometer at 9 A.M. registered maximum heeling for the night at 9° to port, 16° to starboard; registered number 2400.

Observations for the variation of the compass were continued each day when opportunity favoured. The results indicated a gradual reduction of the easterly variation. We had hitherto been going on an important parallel, in respect to the ordinary navigation from Australia to England; yet, a parallel, as already indicated, in which we have no reliable determinations, so far as I can make out. Hence the obtaining of a complete and tolerably continuous series was obviously an important desideratum.

Saturday, June 14 (N., N. N. E.)—In the night the northerly wind still further freshened, causing royals and spanker to be taken in. But the direction, though continuously scant and too much tending to easterly for our wishes,

yet availed us advantageously. At noon we were in longitude 141° 22′ W., and in latitude 55° 36′, having been pushed
near 40 miles more southerly than we had wished, with a
further easting of the wind in the afternoon, and so rendering
the otherwise effective breeze or gale less satisfactory.

The thermometer at 8 A.M. was 43°. The temperature of
the sea, at the surface, had undergone some remarkable
changes recently. At noon, yesterday, it was as low as 37°,
indicating a northerly set of a south polar current; this
morning, at 8 A.M., it had risen to 42°.

Heavy seas from the W. S. W. and N.W. again prevailed,
whilst no wind of any force had yet been experienced by us
from either quarter. The " sky rose," as the phrase is, both
this morning, yesterday, and on several other recent occasions
from the westerly quarter—a circumstance, with the corresponding drift of the upper or rather mid-region of clouds,
which proved the prevalence of a westerly wind at no great
elevation above us; but no corresponding favourable gale, as
general experience led us to expect, reached us.

The day was throughout fine; decks generally quite dry.
The barometer high, for a northerly wind—a quarter in which
the mercury generally stands *low*. Captain Boyce had rarely
if ever seen the barometer keeping permanently so high in
this region, where the average height of the column is very
considerably lower than in the corresponding parallels of the
northern hemisphere.

Sunday, June 15 (E. N. E. to E. by S.)—The wind throughout the night continued to get increasingly unfavourable—
the ship breaking off to E. S. E. (true), and, finally, by daylight, to S. E. or S. E. by S. Going rather fast at the same time,
and rapidly advancing into the region of frost, and more and
more limited daylight, the result was being rapidly produced,
which I thought to be most of all undesirable and disadvantageous. Thus, at noon, whilst the gain in easting had been
by no means considerable, we found ourselves, exactly as
I had supposed, almost two degrees in latitude nearer the
southern pole. Happily, to my feeling, the ship was then put
on the other tack, and, with reduced canvass and increasing

gale, made a progress, as slow as could be desired, on a
N. N. W. compass course, which, on application of 16° E.
variation, and 11° easterly deviation, placed her *head* at least
to the eastward of north—but with the leeway made her real
progress about N. by W. ½ W.   The barometer now fell
rapidly, and the wind (with dry, sharp, hazy weather) pro-
gressively increased up to a strong and hard gale.   Sail reduced to lower topsails and fore-topmast staysails.

The temperature in the morning was 34°.   The clinometer
marked 5° to port and 21½° to starboard, with a register of
3026.   The sea temperature was 37½°.

Divine service was conducted as usual in the saloon in
the morning, and in the forward messroom in the evening.
The subject of the address in the morning was from (Mark
xv. 34) the touching words of deeply felt abandonment by
heaven, " My God, My God, why hast thou forsaken me ! "
The subject for the evening was also from the second lesson
(2 Cor. xii. 7—9).

Monday, June 16 (E. N. E. to N., W. S. W.)—The gale
continued directly against us all the night, moderating somewhat after daylight, and, as the day advanced, gradually
falling.   The sea got up to an extent, so far as could be
observed after 8 A.M., of waves of about 24 to 28 feet maximum elevation.   In the night, however, there was a cross
sea from the N.W., which caused the ship to roll heavily.
The clinometer marked 16° to starboard and 33° to port;
number of register 4937.   The air had become milder.   The
day was singularly dark, generally hazy, and rain set in in
the afternoon, the usual and almost essential precursor, as
the experience of our navigators notified, to a favourable
change.   The ship was kept on the starboard tack from noon
of yesterday till 3 P.M. of this day, but made very little progress.   The day's work of the log gave a N. 19° W. course,
and distance of 59 miles up to noon.   No sooner had the
ship been got round on the port tack, than the wind began
to shift round to the northward, and by 6 P.M. the ship's
head was directed pretty nearly towards the east, our required
course, and at 8 P.M. the wind had become fair.

Tuesday, June 17 (Variable, N.W. to N., or N. by E.)—
Singularly variable winds, both in strength and direction,
prevailed during the night, with occasional rain, and much
head sea. The men were in almost continual and harassing
employment, bracing round the yards, setting and taking in
sail, and shifting studding-sails. The yards and sails being
very heavy—nearly as heavy as those of a line-of-battle ship
—the work on this and various other preceding occasions,
in which we have had an extraordinary prevalence of variable
winds, became not a little severe. On many of these occasions the wind would seldom settle in any particular point
for an hour together; and sometimes would range, whilst
blowing a five to seven knot breeze, from abeam to astern, or
to the opposite quarter, several times during a few hours.

At noon, notwithstanding the head sea and recent adverse
gale, we found ourselves farther to the eastward than our
reckoning, which we ascribed to a current setting in our
favour, to the extent of a degree of longitude. Our position
by the sun and chronometer (two sets of sights) was found to
be 56° 18′ S., and 133° W. longitude.

There was heavy rain in the previous night; but the sun
rose brilliantly clear, and the day continued fine with a brisk
freshening gale from the N. to N. by E. The course made
good was mainly easting, but still involved a southerly tendency. The clinometer marked angles of 12° to starboard,
and 7° to port, for the night, with a registry (10 A.M.) of
6000. The temperature of the air was then 37°.

For the first time since our present outset, I obtained a
capital amplitude,—that is as good, where the rising is so
slow, as could ever be expected. Corrected for deviation,
ship's head E. 17° N., in the standard compass, 1° 37′ W.,
it gave the variation 17° 56′ E.,—indicating an increase by
reason, probably, of an advance of about 80 miles towards
the south since the last observation. Wishing to verify or
modify this result, I took sights again in the forenoon, about
10 A.M., which were again capital (the ship's head then being
E. 14¼° N.) making the apparent azimuth (N. 9° 10′ E.—
correction for deviation 1° 40′) 7° 30′ E. The true azimuth

came out 26° 8′ E., and the variation 18° 38′ E., differing only from that derived from the amplitude by 41′, and giving a mean result, which I prefer to adopt, of 18° 20′ E.

Dip experiments were also made (ship's head about E. 11° N. true magnetic, and heeling 6° to starboard) at the standard station, and at the three stations on the forecastle. The former gave 76° 45′ dip, whilst General Sabine's chart indicated only about 71°. On the midship station of the forecastle, the dip was 78° 30′, on the starboard side, 74°, and on the port side, 89½°, the needle being all but vertical. These latter great differences were probably due to the heeling to starboard, and the broad intersection of the magnetic equatorial plane of the earth with the transverse level section of the ship, carrying the polar axis of terrestrial induction quite over to the port side. The excess of apparent dip over the supposed true dip, 5¾° and 7½°, might possibly have had some reference to the plunging and labouring of the ship, with her head nearly on the magnetic north, in the recent gale.

The freshening gale in the evening did not, by reason of the continuance of a considerable head sea, give us the advantage in speed which ordinarily we should have gained. In fact, the plunging of the ship, and her obvious straining, which recently had become painfully apparent, induced the Captain to take in the outer jib, one of the most important of the sails, with the top-gallant sails, spanker, etc., which reduced her rate of sailing, going a point free or off the wind, from about 10½ or 10 knots, as it might have been, to 8 or 8½.

An unpleasant circumstance, which had for some time been anxiously watched by Captain Boyce and the chief engineer, Mr. Wigsall, became known to me several days ago. It was found that a leakage in the foremost compartment, at the bows, supposed to come in by an open rivet-hole, had of late greatly increased, so as to cause a small fire to be kept continually under one of the boilers for the purpose of pumping by means of the auxiliary or "donkey engine." This requisite work, which was speedily done by it on the outward passage, required during the gale its constant application, and even that was scarcely sufficient. But the source of anxiety and alarm was the discovery of weakness and too perceptible working of some of the plates and angle-irons near the junction of the stem with the keel. Some support had been applied within to the upper part of the weak place; but the apprehension of a risk of forcing off the plates had hitherto prevented any appliance to the lower part,—where, as the external pressure varied, as the ship plunged, the working was most considerable. The compartment itself, happily, was but small, and it was thought to be securely watertight, having in the outset of our voyage been filled, as a tank, with fresh water. But in the case of plates working loose, or by working cutting off the heads of the rivets, as was feared, it was impossible to calculate on what the effect of the entire displacement of a plate or plates in the bow might be or where the mischief might end. One measure, at all events, became absolutely necessary, the prevention, as far as the weather would admit, of the ship's deeper plunging; and this, whilst availing ourselves of a tolerably fair wind, could only be done by going under moderate or easy sail— no small sacrifice in our natural anxiety for progress in good weather in this ordinarily boisterous region.

Wednesday, June 18 (N. N. W. to N.)—The cause referred to kept us under rather scanty sail during the night, keeping our rate of sailing, with a fresh gale of wind nearly abeam, down to 8 or 9 knots, instead, as it might have been, 2 or $2\frac{1}{2}$ knots faster. Nevertheless we made 193 miles, or near six degrees of easting, for the nautical day ending at noon. The progress brought our distance from Cape Horn something within 2000 miles.

Satisfactory, however, as this amount of progress was, it involved some 40 to 50 miles of southing, a tendency which, in our already short days and high latitude, I greatly regretted.

The morning was cloudy, but fine; thermometer 40°; clinometer marking 1° port, and 18° to starboard, registered number 6067.

No particular change being found below, a little addition

was made to our sail, and our progress, with some increase of wind, advanced to 10 and 11½ knots.

The head sea getting down, the ship made a capital progress, a point or two off the wind, to the eastward, making a little southing. Her rate of going from noon to noon was 10 to 12 knots, and the distance accomplished was 252 miles, making nearly eight degrees of easting. This brought us to longitude 120° 4′ W. at noon, in latitude 57° 26′. The wind continued a fresh gale in like direction for the rest of the day, and our course, as before, pretty nearly due east.

The ship being out of trim, as I had noticed on our departure from Port Philip (May 26th), by not being sufficiently by the stern, which now, in the working of the bowplates and in the steering, proved very disadvantageous, a quantity of cargo (copper ore) was transferred from a forward compartment to the storeroom abaft, with excellent effect—the quantity removed amounting to about 45 tons. The third-class passengers, well appreciating the importance of the operation, for a rumour of increase of leakage had got abroad, assisted efficiently in the transference.

Friday, June 20 (N. by E. to N.W.)—We had a gale at N. by W. on the beam, with an abatement of the head sea to an inconsiderable quantity, enabling us to make a progress of 10 to 12½ knots, on a course about E. ½ S., and giving a distance up to noon for the 24 hours of 264 miles. The sun, indeed, never appeared, and the day was extremely dark—heavy clouds with rain occupying the whole heavens. We were obliged to have a light in our cabin, without one for the most part neither reading nor writing was practicable. The barometer, usually standing low in these regions, had been as high as 30·55 inches, an elevation which Captain Boyce never before witnessed. It now, however, began gradually to fall.

Saturday, June 21 (W., W. S.W., S.)—About midnight, the wind, after heavy rain, began to moderate and draw aft; and at 6·30 A.M. it suddenly shifted to the starboard side, and soon blew a gale at south. The diverse action of the north-west swell, and of a fresh southerly sea which soon got up, made the ship labour much. The clinometer, however, did not mark

a greater heeling for the night than 18° to port, and 28° to
starboard; the register at 10 A.M. had gone forward to 8720.
Forthwith, on this change of wind, the temperature began of
course to decline; at first so slowly that remarks were made
on the surprising mildness.　But little consideration would
have shown that, after such a long prevalency of mild winds
from the northward, it must require some time to drive back
the warmer air and give due effect to the antarctic frost.
About mid-day, the temperature of the air was 32½°, but it
sank in the advance of the night to a sharp frost.　The ship,
on the increase of the southerly sea, rolled more and more
heavily.　Barometer falling.

Our compass course, to ease the ship and facilitate her
speed, was now made about N. E. by E., which, with easterly
deviation and variation, was supposed to make good E. ½ N.
The distance accomplished, according to the log, was 253
miles.　Strong gales, squally, with snow showers, prevailed
during the night.

Sunday, June 22 (S., S. by E., S. by W.)—It blew hard all
night from the southward, and the gale continued all day,
increasing in force in the evening—in the squalls it was a
heavy gale.　The barometer was not low for this region,
being at 10 A.M. about 29·40, but sail was reduced to single
fore-topsail and foresail, with jib and stay-sail forward, and
double topsails on the mainmast,—mainsail and all after sail
being taken in.

Yesterday was cold, with frost and snow beginning; but this
was a regular stormy antarctic winter's day!　Thick showers
of snow were frequent, generally accompanied by heavy
squalls.　The thermometer throughout the night had been
at 25°, and there was but little alteration during the day.
The decks were covered with snow or ice; and the interior
of the cabin's porthole lights indicated the severity of the
weather without by repeated incrustations from condensed
vapour on the inside.　The saloon was unpleasantly cold.
Our cabin, warmed by our continued presence there, and
after 3 P.M. assisted by the flame of the candles, yet retained a
moderate and bearable temperature,—the thermometer rising

in the course of the evening to 57° or 58°, and keeping a very agreeable temperature during the day of about 47°. The sea, on the night of 21st–22nd, was cross and rather high, which made the ship roll heavily, extending, at a maximum, to an angle of 33° to port, and 28° to starboard; the registry of the clinometer was now (at noon) 10,028.

Steering with the wind on or slightly abaft the beam, on the starboard tack, and carrying double topsails (until evening), reefed mainsail and foresail, etc., we made capital progress from 11 to 16 knots; completing in the 24 hours from noon to noon about 317 miles of distance nearly, or ten degrees of longitude, on a compass course of about N. E. by E. ½ E., making, with allowance for the swinging off by the sea, about an E. by N. course, true. This was a more scant direction for doubling Cape Horn than could have been wished; but the advantage of making such good way, whilst yet near 900 miles distant, was believed to be more than a compensation for the loss of a little of our southing.

This was the *shortest day* for these regions; we had but very little more than six hours daylight, with a declining moon, the absence of which for the earlier hours after sunset begun to be felt. The risk of falling in with ice in our present high latitude, though less than in the summer, when the ice advances more to the northward, had been contemplated as very possible, especially whilst we were traversing the meridian from 140° or 150° W. to 100° W., where ice at those seasons most usually is understood to be met with. In the early morning, indeed, Captain Boyce had been a little startled by an appearance, during a partial intercepting of the moonlight, as of ice on the sea to windward. It happily appeared to have been a deception; but still we were by no means safe.

This Sabbath, though to us a day of rest, was not a day of assembling ourselves together. A severe hurt, which I had received in my knee by a fall, had greatly incapacitated me for exertion, as I could not walk without support, nor stand even for a short period without inconvenience or risk of damage to the strained muscles or tendons about the knee.

All we could manage was our social and domestic gathering of two or three together, in our cabin, in the Lord's name,—encouraged by the gracious promise of His presence in the midst.

A solemn incident of the day—the burial of one of our number—made me still more regret my inability to perform the last sad sublime service over the dead. The departed one had been a stonecutter in Australia. He was ill of a consumptive disorder when he came on board, and by no means in a safe state for a winter's voyage. A few days ago he was taken with dysentery. Unfortunately, I never heard of his case till after my accident, when visiting him could not be safely attempted. He died, rather suddenly, yesterday; and his body being enclosed in sea-form, in a hammock, was committed to the deep about noon, in the form prescribed and arranged for the burial of the dead at sea in the Book of Common Prayer. The ship's surgeon, Dr. Elliott, at the Captain's request, officiated at the solemn service.

Monday, June 23 (S. to S.W.)—The gale still continued, with fierce squalls in showers of snow, and a high irregular sea. The ship was put under low sail,—ultimately only two lower (or close-reefed) topsails and inner jib. The wind veered in our favour, going round to S.W., which, with beam sea and low sail, caused the ship to roll very heavily. The clinometer at 10·30 A.M. marked the maxima for the night 33° to port, increasing to 37° soon afterwards, and 28° to starboard; the registry was 3104, indicating the number of rollings sufficient to relieve the escapement in $22\frac{1}{2}$ hours = to 3076. In about three hours from this the registry had advanced to 3428, or 324 additional.

The risk of falling in with the ice had been a frequent subject of discussion on-board. Captain Boyce's experience, as to winter passages, was in favour of a clear sea in the line of our present advance. He had never met with ice in any passage at this season. My own apprehensions, in steering on a parallel so far south as 58 or 59 degrees, were different. The night's run, however, revealed to the watch on deck the unpleasant apparition. Betwixt 4 and 6 A.M. four of these floating masses were passed, one or more evidently of large dimensions. Favourably for our passage amid them, the moonlight was effective to a useful extent even in the snow showers. Their appearance was admonitory—as no security could now be felt against the meeting with others. For twelve hours we saw no more, but about six in the evening we passed another, the fifth; it was on our port side; and, shortly afterwards, a sixth. One of these last was turretted, as with lofty spires; also one of the four first seen. Another was plainly tabular.

Everything in and about the deck retained the same polar aspect as yesterday, with an increase in the frequency of the snow showers. The temperature had been down as low as 21°. A considerable number of the men were knocked up and gave in. Captain Boyce's hardihood surprised me. He wore only occasionally a moderately warm overcoat at night, which, with the breaking of seas over the ship abaft, got stiff with the frost. But his hands were altogether uncovered. With the temperature at 25° the previous night, he had stood on deck for long periods together, sometimes grasping an iron railing for support in the heavy rolling of the ship and on the slippery decks, or getting hold of handfuls of snow! I had never met with an instance, among the numerous sailors I had been associated with in a 21 years' experience in the Greenland seas, of such insusceptibility to cold and cold substances of the bare hands.

No part of the Southern Ocean, beyond the 40th to 45th parallel of south latitude, it would appear, possesses an absolute immunity from occasional visitations of icebergs. According to the useful and intelligent inquiries of Mr. Towson, of Liverpool, on "Ice Impediments in Australian Voyages,"—a region determined, doubtless, by two great currents descending from the Austral regions, and their modifications of direction by the continent of South America, and its proximate currents,—there is a large space of the Southern Ocean which may be represented by a spherical triangle, having one of its angles near the latitude of 40° S. (30° according to Mr. Towson), and longitude 30° W., within the lines of which icebergs and other icy impediments to navigation specially abound. The western arc or side of this triangular area, as described by Mr. Towson, stretches from the apex here defined, in a sweep, taking a south-westerly direction, about two or three degrees to the southward of Cape Horn, and extending in the direction of about 61° S. latitude, in 100° W. longitude. To the south-eastward of this arc ice is understood to be dangerously prevalent. The eastern side or arc of the same area seems to extend from the same apex in a south-easterly direction, but in a form not yet sufficiently defined. Hence, though ice may be met with, and frequently is, outside of this area,—and though the lines or area of it may vary with the season of the year,—the prudence of avoiding navigation within it seems established. And further, the suggestions to pass round Cape Horn, so as to sight or nearly to sight it, and then to keep well to the westward, so as to pass to the eastward of and near the Falkland islands, seem equally judicious and well supported by experience; ice, it would appear, being seldom if ever seen in the route thus described.

With Mr. Towson's view as to the security against falling in with ice to be derived from watchful attention to the temperature of the sea, and as to the thermometer being capable of ensuring safety by giving indications of approach towards ice, my personal experience and consideration of the question have not brought me to agree. Not that in the vicinity of large bodies of packed ice, or of numerous icebergs near together, the temperature of both the air and the sea, in certain relations to the direction or drift of the ice, may not be sensibly affected and sometimes to considerable distances; nor that the unusual lowness of the temperature of the sea water will not indicate a descending polar current, and therefore a greater probability of ice being met with (as indeed our present and recent experience characteristically showed); but that the rule of alteration in the sea temperature, or in that of the air, can apply with any sort of certainty to the determination of approach to ice, much less to detached masses, is, I am satisfied, a mistake. The practice of noting the temperature of the sea is an excellent one, and *may* happen, as in cases cited by Mr. Towson, to be admonitory, but, if relied on, it would be found deceptive. The chilling or radiating influence of an isolated iceberg cannot sensibly affect the temperature of the air to windward of it, or on the sides, to any considerable distance; nor could the sea be altered in its general temperature for miles around except in the track perhaps of its drifting.

The attention to the thermometer recommended is good, and may be useful; but the only effective safeguards, under Providence, will, I believe, be found to be these two: First, a perpetual and strict attention to the look-out day and night by men made aware of the signs of ice and the danger of encountering it, within the possible ice limits; the look-out to extend to a watch for fragments of bergs, which may be discerned by the roll of the sea over them as if breaking on a small sunken rock; and, secondly, in giving such attention to the adjustment and quantity of sail during the night or thick weather, when ice has been seen or may be reasonably expected, as may allow for hauling up, or bearing away, in the event of sudden necessity. To these may be added the suggestions given under date of April 1, with the importance of a look-out aloft in critical cases.

Important, however, as the admonitory indications of the surface temperature are, the terms in which confidence in such indications are frequently expressed need limitation and modification. The existence of a descending current from the polar regions, as I have said, will generally be pointed out by a temperature in the surface water, below that of the prevalent temperature; and still more characteristically by a sudden considerable reduction. The general inference within the ice region is *not* that ice *will* be seen, but *may* be seen; not that the actual approach towards an iceberg will be pointed out, but that an iceberg or icebergs may not improbably be near. The importance of observing the temperature of the sea, therefore, as a cautionary and admonitory sign of ice, can hardly be overrated. But it may not be trusted as either infallibly certain or as a sign always to be rightly

interpreted. Sometimes the transition from the waters of
temperate regions to those of polar currents is so sudden
that the sailing only for a few minutes may take a ship
from one to the other; whilst a mile or two of distance may
change the temperature some degrees. In such cases, the ice
brought down by the currents from the polar regions, especially *drift* or *packed* ice which drifts rapidly in the direction
of the wind, may be swept into the strange waters of another
climate, where the surface temperature, except in the track
over which the ice may have passed, can give no intimation
of its proximity or existence.

As icebergs, from the great depth at which they float, have
but small motion under the action of the winds, their principal drift is obviously due, first of all, to descending polar
currents, and then to modifications in the polar drift produced by the influence of laterally running or opposite currents with which the former in its equatorial progress comes
into contact. Thus the antarctic drift current, descending
(probably in a curvilinear form) into the western part of the
Pacific Ocean, eastward of New Zealand, and running in a
north-easterly then easterly direction towards the southern
part of South America, brings down the icebergs annually
detached from their parent source, and scatters them about
within portions of the track of ships from Australia round
Cape Horn. The eastern edge of this current appears to
have brought down the icebergs which we have recently met.
Whilst the main body of the arctic drift current is supposed
to proceed by a northward deflection from the coast of Chili,
a branch of it, probably a more superficial one, leaving the
icebergs generally behind, and passing round Cape Horn,
with increased velocity, proceeds, apparently, north-eastward,
past the Falkland islands up to the outset from the river
Plate, or not improbably until it meets with the " Brazilian
current " running in an opposite direction. The meeting of
these oceanic streams probably produces, by their mutual deflections and coalescing, the current known by the name of the
" Southern connecting current," passing betwixt the parallels
of 30° and 40° S. latitude, in an easterly or south-easterly

direction towards the Cape of Good Hope. Supposing the southern edge of the connecting current to bulge out towards the south-east as an effect of the recent deflection, it would obviously form another (the eastern) arc of the spherical triangle described by Mr. Towson, and so, in the comparatively still water betwixt it and the north-easterly current from Cape Horn, might constitute a convenient lodgment for the accumulation and condensation of icebergs in the general descent, northward, from the antarctic lands and shores.

Another serious accident, from the failure of the ironwork of the *Royal Charter*, occurred this evening, again crippling (at first it was feared almost irremediably) the lower main-topsail-yard. Whilst blowing strong in the evening, but by no means heavily, the "truss-band" of this yard, or hoop connecting the yard with the truss binding it to the main-mast head, suddenly gave way; the yard then sustained on a crutch or iron pillar, as on a pivot, rising beneath it from the maintop, swung out forward on the ship's pitching, carrying the pillar along with it, which, becoming disengaged, came down on the starboard side, and, penetrating through the range of rattlings in the rigging, unfortunately fell overboard. The first officer, Mr. Kirby, who, on the first alarm of the breaking adrift of the yard aloft, had sprung on the rigging, and was ascending the rattlings, had a narrow and very providential escape from this pillar of iron of about half-a-ton weight, which descended so close to him, as to carry away one of the two rattlings of the rigging which he had laid hold of.

The reduction of sail to two lower topsails and foretop-mast staysail (the inner jib having unfortunately given way at the head from the bolt rope), by reason of the heavy labouring of the ship, and the falling in with icebergs, greatly reduced our speed and progress. We continued, however, to make pretty good way on a course about E. ½ N. to E. by N., true. Our position, at noon, from imperfect observations, (for the weather was generally cloudy and frequently thick with snow) appeared to be—latitude 57° 32′ S., and longitude 86° 22′ W., distance from Cape Horn about 630 miles.

The sea, during this and the two preceding days, had, in
ordinary estimation, been high as well as irregular or cross,
and, when coming on the beam, which for the previous days
it did, frequently struck the ship with smart or rather heavy
blows, causing the whole after-frame to quiver in sensible vibrations. In more than one instance, the slap of a topping crest,
extending several yards along the ship's quarter, and striking her in parallelism with her broadside, produced the effect
and impression as of some concussion with a hard substance,
such as a small lump of ice! On these occasions, sprays or
seas frequently blew over the poop as well as on the main
deck—coating the former, under the hard frost, with ice. The
ordinary higher waves, as far as I could estimate their height
from the top of the companion staircase, appeared to be
about 24 to 26 feet in height, with a probable maximum, in
unbroken crests, of about 30 feet. The seas, obviously, were
by no means so high as those we had repeatedly observed in
the Indian Ocean. Of this, indeed, their shorter intervals in
time of passing gave sufficient evidence. Thus their period,
as notified by the rolling of the ship in a series of the higher
waves, appeared on an average to be about 12 seconds. In
the highest waves of the Indian Ocean, I had found the
intervals, when coming astern of the ship, 18 seconds, from
which, 9 seconds for the speed of the ship being deducted,
gave the true intervals about 9 seconds.

Tuesday, June 24 (S. S. W., S. W., westerly.)—Dark
stormy weather—though with some moderating in the wind
in the day—prevailed throughout. The ship (because, perhaps, of her being somewhat light in draught of water, and
very heavily masted, etc.) rolled heavily. For the *night*,
indeed, the angles by the clinometer were (at 10·15 A.M.)
27° to port and 22° to starboard, with a registry of 6100,
indicating an advance, for 24 hours, of 2996. But in the
course of the day our extremest inclination was, I believe,
repeated.

By the giving way, the second time, of the lower main-
topsail-yard, our progress was again impeded. What our actual
position might be, had, after the want of observations for four

days, whilst making no small progress in gales of wind, become very uncertain, and our reckoning under the circumstances could only be taken as a probably wide approximation.
Such as it was it gave our position, latitude 57° 29′ S., longitude 79° 58′ W., and seemed to note an advance now within
about 480 miles of the long-looked-to Cape Horn, which we
were so anxious to double. Our compass course continued
from N. E. to N. E. by E., and true course, as was assumed,
about E. by N. or E. ½ N.

The occurrence of icebergs in our track,—always an occasion of anxiety, and in dark nights or thick weather of no
inconsiderable risk,—I now hoped had ceased to be probable.
The ground of this hope was derived not merely from the
less frequency of their appearance in the position into which
we were advancing, but specially from the rise of the surface
temperature of the sea to about 40°, which for several days
had been at 36° to 37½°,—indicating our having passed
through the descending antarctic current, and, if so, probably
beyond the ordinary drift of this class of polar ices.

Wednesday, June 25 (variable, south-easterly.)—The gale
abated or ceased during the night. The heavy rolling of the
ship, of 37° to starboard and 35° to port, registry (10 A.M.) 7580,
rapidly diminished ; and, on the whole, except the changing
of the position of the yards, we had a tolerably quiet night.
But previous to this condition of things in the preceding evening we had by far the heaviest blow from the sea hitherto
experienced. It struck the ship on the starboard quarter, in
parallelism with the side, and gave so sudden and heavy a
blow—causing a violent tremor and vibration of the entire
hull—that the instant impression with many of the passengers
was that the ship had received a blow from a mass of ice,—
but the shipping of a heavy sea which immediately ensued,
sweeping over a large portion of the poop-deck, and a considerable quantity of water entering the saloon by apertures
in the skylights and a partially open companion, explained at
once the cause of the violent shock.

The mainmast, in respect of setting sail, being crippled
by the recent accident to the topsail-yard, scarcely any sail,

generally none, could conveniently be spread on it. The
effect was a great loss of distance, and with the light or variable winds of the night and scant or unfavourable wind during the day we made but a poor day's work—reckoned up to
noon at only 169 miles of distance.

Not having been able to get observations now for five days,
there was much anxiety in nearly approaching Cape Horn to
ascertain correctly our position, which no one could positively
calculate with confidence within 100 miles or more. Sights
for the time were obtained, but failing in the meridian altitude—for which a proximate altitude was attempted to be
reduced and employed—the longitude could not be relied on
with any degree of confidence. What was done appeared
to give the latitude about 56° 39′ and longitude 75° 2′ W.
making our distance from Cape Horn, at noon, about 250
miles.

The temperature of the air rose at night above the freezing,
but a strong southerly breeze coming on in the forenoon, the
decks became speedily coated with snow and ice—heavy
snow showers still occasionally prevailed.

The suspension of my series of variations and other magnetical observations had become a subject of much concern to
me; but the weakness and injury of the knee prevented my
attempting to go on deck. This morning, however, the weather being finer and the motion of the ship inconsiderable,
I made the attempt to catch an azimuth: only one sight—
a pretty good one—was obtained.

Towards night the wind went down, and we made but
little progress. The topsail-yard, which had been crippled,
being slung by chains from the mainmast-head, was considered secure notwithstanding the loss of the crutch, and in
the afternoon the sail was again spread to the wind—now,
indeed, of little or no avail.

Thursday, June 26 (Southerly, easterly, variable, calm,
etc.)—Calms or light variable winds prevailed the whole of
this day, which led to the putting into action of the screw
about 11 A.M. After that, with most of the sails furled,—
sometimes with a breeze from the eastward, sometimes from

the southward or south-west, and sometimes calm,—the engine gave us a pretty general average of about six knots. Snow fell towards morning so as to cover the decks to a depth of several inches; it then cleared up, and happily, after six days want of the meridian sun, capital sights were obtained for the determination of our exact position. These gave our latitude, at noon, 56° 28′ S., and longitude 72° 32′ W. This made our distance from Cape Horn about 190 miles, and from the nearest land 130.

The day being fine, and the ship having very little motion, I had the great satisfaction of renewing my magnetical observations. Two excellent sets of azimuths were obtained, which, corrected for the small deviation of the standard compass, gave the variation respectively 27° 17′ E. and 26° 47′ E., differing only half a degree. The mean, which I prefer to adopt, gives the variation, at very nearly our noon position, 27° 2′ E.

The crippled state of my knee allowed me only to use the dipping-needle at the standard station where all parts of the apparatus were at hand, and where I could have secure footing on the generally slippery decks. The dip was now found (ship quite upright, with her head E. by N. ½ N.) to be 65°, —a reduction since the last observation (June 17) of 11° 45′. The terrestrial dip, according to General Sabine's chart, had also diminished by about the same quantity, and in both cases the ship's dip was greatest to the amount of about 5°.

A set of compass comparisons was also obtained with the ship's head E. N. E., which showed, as at Port Philip on this direction, a very general agreement.

So long a suspension of my magnetical experiments and azimuths had been the occasion of much concern and disappointment. In regard to azimuths, however, not above two opportunities, or rather, I believe, but one, had been really missed, as for the seven intervening days the want of the *correct* latitude—an essential element for azimuths in this region of very low solar altitude—would have rendered any occasional observations of the sun's magnetic azimuth unsatisfactory, if not useless. And in regard to dip experiments, the labouring of the ship during the large majority of these days would have rendered them particularly difficult. On the whole, therefore, I found I had lost less, much less, than might have been expected.

So quiet had been the preceding day and night, that the clinometer only gave a maximum of 7° to starboard and 8° to port for the heeling; the register (10·45 A.M.) had advanced to 7603. Two ships were seen.

Friday, June 27.—Another night and day of light variable winds often ahead, or calms, till near midnight—so that we depended wholly on the engine. The general average from the prevalency of head wind was only about 5 or 5½ knots. The day was fine, the air sharp and bracing.

As the day came on we had the pleasant revelation of land —portions, as it proved, of Diego Ramirez islands, very proximate to Cape Horn. Their appearance at 9 A.M., when I first saw them, was as two cliffs, under much atmospheric refraction, rising out of the sea. The bearing was N. by E. ½ E. true, distance about 25 miles.

The anxiously looked for passage of "the Horn" was now, under the blessing of a gracious Providence, to be accomplished. At 7 P.M. I reckoned we crossed the meridian of the Cape, and at the distance of about 45 to 50 miles— high land, beyond it probably, having been seen in the afternoon, but we were more than doubtful as to the Horn having been sighted. This gave us the second transit, since leaving Port Philip, across the conventional hydrographical divisions of the great oceans of the globe. On the 31st of May we passed from the Indian Ocean into the Pacific; and on this evening, after being 33 days out, from the Pacific into the Atlantic Ocean.

Our progress for a winter's passage appeared to us to have been very remarkable. We have with difficulty and much retardation from contrary and light winds made our way mainly on or near a parallel of latitude said to be characterised by boisterous *westerly winds*. But we have had no westerly wind beyond the strength of light breezes, and these generally very transient in continuance, during the whole passage. We have often had light variable winds—experiencing heavy crossing seas for days together. And during the whole passage thus far, we have only had two gales blowing really hard, that from the E. N. E., N. E. by E., of June 16–17, and on the 23–25, from the southward. Of such a series of light winds and prevalency of fine and till recently of enjoyable weather, Captain Boyce has had no previous knowledge, though with an experience extending to 13 passages from the westward round Cape Horn. With the weakness of the port bow-plates of the *Royal Charter*, this circumstance, which has so much delayed our advance, may have had a providential bearing on the safety of our progress.

So passed we, under our moderate auxiliary steam-power, this famous insular termination of the South American continent and associate islands, without anything but showers of sleet or snow to remind us of our proximity to the regions of frost and ice of the south, or to the extent yet experienced, of the hard gales and mountainous seas of the Southern Pacific. Everything went on, except for the discomfort from a temperature about freezing, as it might have done in genial and moderate climates. Our breakfast—not an unusual thing indeed even in boisterous weather—was served up with all the varieties of provisions which so well furnished our table on coming out, with rolls hot from the oven. The dinners, too, were as good and satisfactory and as handsomely served up as before, and the provision generally as ample and varied and mostly as good. Our fresh beef from Australia was, so long as it lasted (about three weeks), excellent. The flour and bread had somewhat deteriorated in quality. The poultry, Australian market poultry, was excessively poor. Poultry, indeed, was not to be had in sufficient quantity in Melbourne market; Captain Boyce had obtained several dozen fowls from Sydney at a high price. Poultry is dear, and market poultry generally poor, because the rearing of fowls is but moderately remunerating. Seven shillings is a usual price for a pair of ordinary fowls; turkeys are often a guinea, and three guineas have been paid for a good one! Poultry, too, is by no means a secure property. After the "hot winds" which occasionally prevail, the birds are frequently carried off in large numbers.

The circumstance, however, of experiencing fine or moderate weather in the winter season, is, I find, by no means so extraordinary as we naturally inferred it to be. On the contrary, the experience of Captains Beechey, Fitzroy, King, and others, go, I think, in favour of expecting fine or moderate weather in the winter season near to Cape Horn. Indeed the opinions of these and other observant officers go far to prove, that in winter the passage of Cape Horn, from the Atlantic to the Pacific, is most easily effected, and with more favourable winds and better weather, than at other seasons. And the opinion seems to be general with the most experienced, that there is no necessity for going far to the southward for this object; better passages in June and July being likely to be made without proceeding further to the southward than for obtaining a safe and commanding distance and position whilst rounding the southern islands of Diego Ramirez. This favourable condition of the winds, blowing frequently from the eastward for the passage of the Horn to the westward, is, of course, as it proved to us, a disadvantage for the eastward passage,* besides the loss of the strong westerly winds to which the navigators from Australia homeward bound look for a rapid passage.

The passage by the Cape of Good Hope at this season is characteristically stormy; the gales frequently being very severe. But with a sound, buoyant and well-found ship, my impression is, that time would be gained by that route being taken in winter by sailing clippers, rather than the passage by the Horn. On the other hand, for steamers with full appliances of power, the route we took would have been particularly advantageous; as during the whole passage to the Horn and eastward of it, and, indeed, up to our present position, we have experienced only about two days in our progress in which such a steamer need have been stayed or

---

* Much useful information on this subject has been collected and published in Purdy's "Sailing Directory for the Southern Atlantic." *See* pages 57, 60, 62, etc. (Third Edition).

materially retarded—which occurred in the southerly gale of June 23-25.

Saturday, June 28 (North-easterly, N., N.W., var., calm.) —On occasion of snow setting in, and a moderate squall of wind from the north-westward, at 11 P.M. the engine was stopped and the screw raised up. This proved unfortunate to our progress, as the breeze did not long continue, and for near half the 24 hours it was either calm or extremely light winds, hardly giving steerage way.

There were rather frequent showers of snow or sleet, with dark cloudy weather during the whole of daylight. The sea was smooth, and, except as to temperature, not unlike that of regions approaching the tropics.

A set of compass comparisons (excepting the standard) was obtained on, to us, a new course, nearly N. E. true. They were nearly identical as to their respective deviations with what they were at Port Philip.

Compass Aloft.	Steering Compass.	Companion Compass.	—
N. 42° E.	N. 31° E.	N. 37° E.	June 28th.
N. 42° E.	N. 31¾° E.	N. 35¾° E.	May 2nd.

These results seem to indicate that their deviations are yet unchanged. The main unsettlement of the adjusted compasses appeared to have been produced whilst we were steering on an easterly course, in a high southern dip, and with heavy seas. Our subsequent course has generally been the same; and the general tendency of the heavy blows from seas recently received whilst we were steering on or near the general point of change, should, according to theory, be only to confirm or keep up the change which had taken place (chiefly in the Indian Ocean) in the general character of the ship's magnetism. It now remains to be ascertained under what circumstances and in what ratio of progress the *Royal Charter's* original magnetic condition may be again approximated or restored.

The night set in dark, and frequently thick with sleet or snow—fortunately attended with a breeze (a gentle one) from the westward. Clinometer (two days register) 12° starboard and 7° to port, with a registry of 8620.

Sunday, June 29 (variable, calm, S. E., S., S.W.)—Light winds or calms continued with us during the whole night, which (after 36 hours cessation—the Captain constantly expecting, from the low state of the barometer, the sudden rise of a gale) brought the engine again into operation. But its commencement was just as unfortunate as its withdrawal had been. At 10·30 A.M. the screw began to revolve, and within an hour a fresh gale came on from the south-eastward, at once causing the sails to overrun the engine. It was discontinued after only a two-hours application.

The remainder of the day was favourable to our progress, though the wind veered about from S. E. to S. S.W. etc., and occasionally became rather light. We steered an easterly course—that is about E. N. E. $\frac{1}{2}$ N. true, to prevent our becoming embarrassed with a scant wind about the Falkland islands—though to my mind, with some danger of getting into the drift of the antarctic ice—especially as the temperature of the sea had gone down to 37° or 36½°. Showers of sleet or snow still occasionally prevailed, with bright starlight sky. Our distance, at noon, from the Beauchesne islands, or rocks, the southernmost attachment of the Falkland islands, was about 240 miles; the bearing by compass almost north, or with two points variation and one point deviation both easterly, about N. E. by N. true.

An awkward accident, which might have been attended with serious results, happened at the lowering of the screw in the morning. The capstan around which the rope connecting with the screw-chain was passed for heaving and lowering was very slippery, and the rope very rigid, which, from some incaution, caused the rope suddenly to run off, and letting the screw fall, or all but fall, with alarming velocity into its place. The Captain alarmed at the risk, by a momentary impulse, attempted to stop the running chain by his feet; he had a providential escape, for his feet had only been thrown off an instant when the hook and appliance for attaching the rope to the chain flew past; if it had struck him it must have caught him about the ankles. Providentially, too, the screw did no damage, or apparently so, to the ironwork against which it must have struck so formidable a blow.

It was with much regret that I was obliged again to omit our usual Sabbath-day services, both in the saloon and forward. The injury to my knee causing an entire suspension of exercise, with confinement below, had become so disordering, that for two or three days I had been really ill, and much incapacitated for making exertion. In our cabin, however, we were enabled to gather ourselves together as aforetime, and to have the privilege of at least an abridged service both morning and evening.

Clinometer at 3 P.M., port 10°, starboard 7°.

Monday, June 30 (S. E., S., S. W. to N. W.)—Under a fresh breeze or brisk gale generally in a favourable direction, we made 224 miles of distance from noon to noon, on about a N. E. by E. course, true,—the log giving from $7\frac{1}{2}$ to 11 knots; but in the afternoon and evening the ship went from 11 to $12\frac{1}{2}$ knots. Our course being so much easterly we did not make the quantity of northing which could have been desired,—our latitude at noon being about 54° 5′, and longitude 57° 46′ W. The weather on the whole was fine—the sea smooth.

Tuesday, July 1 (W. N. W., N. W.)—Blowing a fresh gale all night, the ship made a rapid progress, sailing about wind abeam on the port side, sometimes making as much as 14 knots. But it was not altogether so satisfactory as could be desired, our course, N. E. by E. to N. E. by E. $\frac{1}{2}$ E. true, giving us far more easting than was desirable, and much less northing. For again we risked—as was further indicated by the lowness of the sea temperature through which we passed, which was 35° to $33\frac{1}{2}$°,—being pushed into the region of ice.

In the morning the gale subsided and became very light, then, about 11 A.M., shifted suddenly to the south-westward, snow and sleet falling at the time. But though the barometer had gone down very low, no gale ensued. We were obviously again betwixt conflicting atmospheric strata, and happily enjoyed a brisk favourable wind, whilst the currents of wind were spending their strength elsewhere.

Good sights for the chronometer were obtained, but the latitude only by an extra-meridian altitude. The results yielded were latitude 51° 42′ S. longitude 52° 7′ W. for noon. Our run from noon to noon (occasionally rising to 15 or 16 knots) was about 270 miles.

The clinometer at 11 A.M. marked 3° to port and 28° to starboard, with a registry of 7885.

The night was fine and moderate, occasional showers of sleet or snow, with much fine clear starlight. Course steered N. E. by N. ½ N., which with half a point variation E. and one point deviation, gave N. E. by N. true.

Wednesday, July 2.—Our progress, which during the night had been moderate, was reduced by the failure of the breeze (though the barometer was still low, being about 29·4) to a rate of from seven to two knots during the day. For the 24 hours ending at noon we had gained about 200 miles of distance, being, by good observations, in latitude 48° 49′ S., longitude 49° 11′ W. The clinometer at noon gave angle of maximum heeling to port 4°, to starboard 28°, with a registry of 7887.

As regards progress the day was unsatisfactory, but otherwise pleasant, enjoyable, and advantageous. The air was nearly calm. The sea had but little motion, and the sun shone among small occasional clouds nearly the whole day; and for the first time for some weeks shed a cheering and moderately warming influence abroad. The sun's amplitude had increased to 53½ degrees; and the day had lengthened from 6 hours to 8 h. 10 m.; and our time had gained on Greenwich of this date 20 h. 56 m.

For my personal investigations, after, from indisposition, having made no magnetical observations (except a set of compass comparisons) for six days, the opportunity was as fully improved as possible, and results of no inconsiderable interest obtained. These may be thus summarily given.

1. *As to the Ship's external Magnetism in the upper edge of the top plating.*—This interesting inquiry was made with a delicate pocket-compass, held two or three inches from the top plating, externally, on the level of the upper edge. The ship's head, fortunately, was near the magnetic meridian, being N. by E. On the poop, from the forepart or break, to the position of the wheel, on both sides, the polarity of the plating had northern polarity,—becoming weak at 63 feet from the stern. Opposite the middle of the after skylight, about 40 feet from the stern, the iron appeared to be neutral; and abaft that position, say from about 30 feet from aft, began to show *southern* polarity; and round the stern and near it, decided and vigorous polarity, the converse of the rest of the ship and of the condition at Port Phillip.

2. *The Dip.*—The needle which had acted very well at Port Phillip, and along the whole of the region of strong magnetic intensity, had, under the reduction of the magnetic force, become again- sluggish,—I may say annoyingly so. Still the results seemed to be reliable within the extent of from half a degree to a degree—perhaps less. The changes with the ship's head nearly north (N. by E.) were curious and instructive. The leading results were as follows :— [Terrestrial Dip about 48°; Force 1·215.]

STANDARD STATION.		FORECASTLE.						AFTERPART OF POOP.			
Midship.		Midship.		Starboard.		Port.		Starboard.		Port.	
Dip.	Oscillation.	Dip.	Oscil.	Dip.	Oscill.	Dip.	Oscil.	Dip.	Oscill.	Dip.	Oscil.
45⅓°	2ˢ·8	64°	2ˢ·4	61°	2ˢ·35	66°	2ˢ·4	26¼°	3ˢ·33	29½°	3ˢ·0

Here, it is observable, how plainly the differences have reference to terrestrial induction, considered with reference to the ship's head being in nearly a meridional position, or pointing in the N. by E. magnetic direction. Thus, on the poop, where the ship's previous polarity was beginning to change, the dip was about 28° (the mean), or 20° below that of the earth. At the standard station, near the ship's centre, it was 45⅓°, or nearly 3° less than the terrestrial. But on the forecastle, in midships, it was 64°, or 16° more than the ter-

R

restrial dip. These differences are, it may be seen, quite consistent with the apparent terrestrial action.

3. *Compass Comparisons.*—The changes in the ship's magnetic condition were found to have produced, consistently with theory, analogous changes in the relative indications of the compasses. The following were the apparent relations of the several compasses with each other, and proximately, their respective relations, in very nearly the same direction of the ship's head at Melbourne.

—	Compass Aloft.	Steering Compass.	Companion Compass.	Standard Compass.
At sea, S. Atlantic, July 2nd	N. $16\frac{3}{4}°$ E.	N. $8\frac{3}{4}°$ E.	N. 11° E.	N. 18° E.
At Melbourne, May 2nd	N. 14° E.	N. $3\frac{1}{4}°$ E.	N. $7\frac{1}{4}°$ E.	N. $22\frac{1}{2}°$ E.

Hence it appears that the compass differences at Melbourne, with the ship's head within about a quarter of a point of the same direction, had already become greatly altered and reduced. The change in the adjusted compasses is obviously towards a return to a better performance, and towards their original relations to the compass aloft and to each other. The approach of the standard compass, however, to the compass aloft, from $8\frac{1}{2}°$ of difference at Melbourne to about $1\frac{1}{4}°$ on this day, presents an anomaly—being the reverse of an approach to its original relations—which would have needed examination and repetition before it could be satisfactorily received, had not the results of the next section verified the inference.

4. *The Variation and Errors of the standard Compass.*— The entire errors of the standard compass were very satisfactorily obtained by means both of a set of azimuths and an excellent amplitude at sunsetting. The former gave the true azimuth for latitude 48° 45', N. 28° 4' W., whilst the mean observed azimuth (ship's head N. 17° E.) was 37° 53' W., indicating an error in deviation and variation combined of 9° 49' E. The amplitude observed was N. 64° 43' W.; the true amplitude for latitude 48° 40' was 53° 44' W.; giving

the compass error with the ship's head N. 19½° E., equal to 10° 59' E. Assuming the deviations taken at Melbourne to have remained unaltered, the corrections respectively for deviations would have been 8° 16' W. and 7° 30' W., and the resulting variations would be 18° 5' E. and 18° 29' E. By the Admiralty chart of the South Atlantic, the variation is 11½° E., indicating (if this were correct) a change in the standard compass deviation on the course specified of ½ (6° 59' + 6° 35') = 6° 47', or say 7°. And this accords very nearly with the change indicated by the observations in § 3. Thus applying the new deviations of § 3, of 1¼° westerly, we should have, on the mean, for the variation 11° 43' E. This is without allowance for deviation of compass aloft.

In the evening a breeze sprung up from the north-westward, veering quickly to north (compass), by which we were placed on a course about due east, and making more speed than perhaps was to be cared for.

Thursday, July 3 (N., N. by W.; lat. 48° 25' S., long. 44° 51' W., dead reckoning.)—The breeze which sprung up in the evening went on increasing to a gale. The ship was soon put under her double topsails, reefed foresail, and inner jib. Blowing harder in the morning, the upper topsails were lowered down. The ship lay only E. N. E. by compass, making about an east course true—though from noon to noon it was found we had gained about 24 miles of northing, our latitude by observation being 48° 25'.

The clinometer at 9 A.M. marked maximum to port 4°, starboard 28°, mean heeling 13° to starboard ; register 7900.

Strongly as it blew the whole day—sometimes a really hard gale—it was remarkable in how slight a degree the sea got up. The highest waves were of inconsiderable elevation considering the strength and continuance of the wind. In the Indian Ocean the sea seems always ready to get up on the most moderate gale ; here, with the wind from the northward, it seems the very reverse. It is indeed a fact of popular observation that this is a characteristic of northerly gales in this region, " they never produce high seas."

The air was dry throughout and temperate, thermometer 42°, sea water 46°. R 2

The gale continued all the night, which was unusually dark. A heavy blow from the sea on the port quarter was experienced in the evening, producing a violent shock, and resulting in the shipping of some water.

Friday, July 4 (N. to N. N. W. to N. E.)—The gale abated towards morning and enabled the ship to carry all her principal sails up to topgallant sails; but the direction was still greatly against us, the ship never lying within five points of her course. The sea, though low in elevation for such a gale, caused, with the hard squalls of the night, some heavy lurches, occasioning a maximum register of heeling to starboard of 44°, the greatest I remember as yet being indicated. The port maximum was $6\frac{1}{2}$°; the registry (9 A.M.) was 9367, and mean heeling at 9 A.M. 7° to starboard.

At the breakfast hour, eight guns were fired in compliment to the Americans on board, four gentlemen of that country being among the saloon passengers. A delicate and delicious punch, of which the basis was sparkling Moselle, instead of water, was liberally distributed by those gentlemen among the saloon passengers at luncheon, and a most liberal provision of champagne was placed by them on the table at dinner. Toasts were drunk, and speeches with snatches of songs or choruses were made after dinner,—all, as the good habit of the Americans is, being in the greatest good humour and pleasantry.

The day though cloudy was fine and dry, and the temperature mild, being as high as 46´. The sea had ranged from 45° to $38\frac{1}{2}$°, and back to a like maximum.

Compass comparisons being made with the ship's head (true) about N. 64° E., indicated on this course a return to, or at least no variation from, the deviations obtained at Melbourne. The comparative results were as follows (angle of heeling 7° to starboard) :—

—	Compass Aloft.	Steering Compass.	Companion Compass.	Standard Compass.
July 4th	N. $64\frac{3}{4}$° E.	N. 59° E.	N. 61° E.	N. $64\frac{1}{4}$° E.
Melbourne	N. $67\frac{1}{2}$ E.	N. $61\frac{1}{2}$ E.	N. 64 E.	N. $67\frac{1}{4}$ E.

Whether this close agreement had been restored by the ship's rolling and straining on an E. by N. to E. N. E. course, or whether it might be due to fixidity of magnetic action in the line of direction, could not be determined; though my impression went, in considerable part, to the first of these causes. For it was found that the ship's external magnetic condition in the upper plating of the poop had made some retrograde movement; the southern polarity noticed on the 2nd, as extending to about 30 feet or upwards from the stern, being now limited to a space of scarcely 15 feet from the stern on the port side, and something more extended on the starboard or leeward side.

The fine dry day with moderate wind disappointed us in the issue—the wind freshening in the evening, with dark hazy weather, and before many hours increasing to a hard gale directly against us, or if possible, worse, as it led us on an easterly course greatly tending to the southward.

Saturday, July 5 (N., N. N.W. to N. E. by E., N.N.W.) —It blew a hard gale all night, which continued with but little abatement until near noon. Its direction shifting from N. N.W. was prevalently from the N. E., turning the ship's head, as we continued to stand on, on the port tack, to the S. E. Fortunately, being under three lower topsails and foretopmast staysail only, the ship made but little headway, but the result was a course (true) for the whole night and forenoon of about S. by E. A strangely tumultuous sea got up—not very high but irregular and lumpy—the waves being mainly separate peaks with little lateral extension, with short intervals, and deep hollows like pits between. For about half an hour in the forenoon, when there had been some slight reduction, perhaps, in their height, I found the higher waves seldom rising more than 24 feet above the hollow, the average being still lower; and only one wave was noticed during that time, near to the ship, which rose a little higher —perhaps to 28 feet. These features of the waves produced in a severe gale from the northward or north-eastward in this region were again conspicuous and remarkable, and formed a great contrast with the waves of the next night, the storm producing which, evidently from the W. N. W., did not reach us, but only a moderate breeze from the same direction.

The ship's external magnetism about the top plating of the poop was found (course N. E.) to have still further retrograded.

The clinometer at 10 A.M. marked the maxima of heeling as 6° to port and 35° to starboard (increasing soon afterwards), with a mean heeling of 12° to starboard, and a registry of 9736.

In the afternoon the gale subsided to comparative moderateness, and the wind shifted to N. N. W., and then near N. W. The turbulent sea now taking the ship ahead, when she began to lie up about N. E., prevented the setting of much sail; but as the night advanced, the mainsail, foresail, and then the upper topsails being set, we made a litle headway about five to six knots.

Retrograding in latitude, as we had recently done, till come again within the 50th parallel, making at the same time and for many days so much easting, and passing as at different times we had done through portions of sea water of a low temperature ($38\frac{1}{2}°$ to 45°), evidently branches of the descending antarctic current, I felt some anxiety lest in the dark and stormy nights we should fall in with ice. Captain Boyce, on account of its being the winter season, felt great confidence that we should not meet with ice, and happily thus far his confidence has been justified.

Sunday, July 6 (N. N. W., N. W., W. N. W., S. W. southerly.)—Though we had but a moderate gale with rain during the night, and that chiefly from the N. N. W., a heavy sea arose and set in with increasing violence from the W. N. W., which caused the ship to roll very heavily (in frequent series of five or six rolls, at intervals or rate of 5' each), the angle of heeling sometimes extending to 40°, and at a maximum to 45°, to starboard, and reversely, 15° to 20° to port. The numerical register at 1 P.M. was 1027. The ship, meanwhile, lying up N. E. to N. N. E., and making three to seven knots headway, ceased to loose ground, and gained a few miles before noon,—when our latitude was reckoned at

47° 26', and the longitude at 35° 9' W., no opportunity, because of the dense atmosphere, occurring for celestial observations.

Suddenly, at noon, the wind shifted to the southward, and soon began to blow fresh. Heavy as the sea was coming now, on steering a N. N. E. course, before the beam, and cross and tumultuous too, by reason of a strong head sea, the ship gradually began to make progress under increase of sail so as to reach nine knots by 3 P.M., in the course of the evening advancing to 9 and 10½ knots. The sea from the westward appeared to be the result of the same gale which we had had from the southward, travelling and revolving, and giving to us striking indications of its severity in these turbulent and elevated waves. For in regard to character and elevation they were strikingly different from those produced by the N. and N. E. winds. Thus as to character they were of the nature of the general oceanic storm-waves— coming in frequent parallel series extending to a quarter or half a mile in breadth, and in elevation reaching to from 24 to 25 feet, with not unfrequently waves of 30 feet and upwards. The height being a question of some interest, I contrived with difficulty to get into the mizen rigging, and in a position where my eye had an elevation of about 32 feet, found many waves intercepting the horizon. Some of broad and extended crests of unbroken water rose so much above my position, that I could not estimate them (the highest) at less than 34 or 35 feet from the depression to the summit. The estimation, it should observed, was always taken when the ship, lying in the hollow or trough of the sea, was quite upright.

It was with much satisfaction that I found myself able to return, partially at least, to my clerical duties—performing Divine service in the saloon in the morning, and addressing the congregation as usual from a portion of Scripture (Rom. vi. 23) comprised in the Epistle for the day, " The wages of sin is death," etc.

Monday, July 7 (W. S.W., S. W., N. W., N. N.W.)—We had but short advantage from our cheering fair wind. By 8 A.M it had fallen to a light breeze, and then became scant, veering to the N.W. and N.N.W., so that the ship for a considerable time could not lie within two to four points of the desired course. The day, however, was fine and enjoyable—the sun shining from rising to setting, and giving all our passengers the advantage of enjoying the fresh and pleasant air. Though the head sea of the previous morning and the cross W.N.W. sea of the day had greatly subsided, a new and high sea from the S.W. ranged up on our port quarter, with waves often of 20 to 24 feet in height. This probably was a continuation of our N.E. gale, in its westward progress and revolving character, from N.E. to N.W., westerly, to the southward. This characteristic of the direction in which the winter gales of this southern region revolve, travel, and subside, is the converse of what happens in the winter gales of high northern latitudes—where, as in the Spitzbergen seas, the heaviest gales of winter and spring generally commence at or about S.E., shift in the same regions to the southward and westward, and never subside till they reach the N.W.

The rolling at night of 6th-7th was again heavy, giving maxima on the clinometer at noon of 40° to starboard and 21° to port—a mean heeling of $3\frac{1}{2}$° to starboard at noon, and a register of 2554. The barometer had made in 24 hours a very remarkable movement—rising from about 29·00 to 30·20.

Good observations were obtained after a lapse of four days, for the determination of our position, which gave the ship's latitude, at noon, 44° 58′ S., longitude, 32° 56′ W.

A series of compass and magnetical observations, to me of much interest, were made in the course of the day, which gave results confirmative of, and analogous to, those obtained on the 2nd. The leading results, with an angle of heeling of from 3° to 4° to starboard, were as follows:—

1. The compass comparisons, with the ship's head N. 25° E. true, gave these comparative and actual relations:—

——	Standard Compass.	Compass Aloft.	Steering Compass.	Companion Compass.	Standard Compass.
July 7 .	N. $25\frac{1}{2}$° E.	N. $24\frac{1}{4}$°E.	N. $13\frac{1}{4}$°E.	N. $18\frac{3}{4}$° E.	N. $25\frac{1}{2}$° E.
May 2 .	N. $30\frac{1}{4}$ E.	N. $24\frac{1}{2}$ E.	N. $10\frac{1}{2}$ E.	N. $16\frac{1}{4}$ E.	N. $30\frac{1}{4}$ E.

The agreement, as I have observed, with the observations of the 2nd July is found to be quite beautiful, especially with reference to the changes in the compass differences since leaving Melbourne. The changes in the compass differences, as relatively given by the observations of the 2nd July and those of this day, run thus (the difference of the ship's head being about 10°) :—

Differences.	Standard and Compass Aloft.	Compass Aloft and Steering.	Steering and Companion.	Companion and Standard.
July 2nd .	$1\frac{1}{4}°$	$8°$	$2\frac{1}{4}°$	$7°$
July 7th .	$1\frac{1}{4}°$	$11°$	$5\frac{1}{2}°$	$6\frac{3}{4}°$

Or comparing the *deviations* at Melbourne with those of July 2nd and July 7th, we obtain,—

Ship's Head true Magnetic.	—	Standard Compass.	Compass Aloft.	Steering Compass	Companion Compass.
N. $24\frac{1}{4}°$ E. {	Melbourne,} May 2 ..}	6 W.	. .	$13\frac{3}{4}$ E.	8 E.
N. $16\frac{3}{4}°$ E. {	At sea, } July 2 ..}	$1\frac{1}{4}$ W.	. .	8 E.	$5\frac{3}{4}$ E.
N. $24\frac{1}{4}°$ E. {	At sea, } July 7 ..}	$\frac{3}{4}$ W.	$\frac{1}{2}$ E.	11 E.	$5\frac{1}{4}$ E.

These sets of comparative deviations plainly show that, on a course from N. $16\frac{3}{4}°$ E. to N. $24\frac{1}{4}°$ E., the standard compass has changed its deviations from 6° W. to about 1° W.; that the deviation of the steering compass has been reduced to an extent of probably about 3°—there being some inaccuracy, I think, in the 8° observed July 2 ; that the companion compass has had its deviation reduced about $2\frac{1}{2}°$; and that the compass aloft has remained unchanged.

2. *The Variations of the Compass.*—A set of azimuths, excellent steady sights, were obtained in the afternoon, and a capital azimuth at sunsetting. The former gave the total compass errors for deviation and variation, 17′ W. by the standard compass, the ship's head being N. 29° 40′ E., angle

of heeling 4° to starboard ; mean latitude 44° 40′, mean longitude 32° 42′ W.; the amplitude gave the amount of errors
= 53′ W. Correcting these results for *deviation* on the
determinations now made,—that is, 46′ W. for the azimuth,
and 50′ W. for the amplitude, we obtain the variation of the
compass by the first = 23′ E., and by the second 3′ W., or
the mean 10′ E. On examining General Sabine's Chart of
Magnetic Declination for the South Atlantic, after these
calculations were made, I found the variation for the above
position about 45′ W.; an accordance, especially taking into
account the secular change, particularly satisfactory.

3. The force had become reduced from that at Port Phillip
in the ratio inversely of $(2·04)^2 : (3)^2$ — the times in seconds
of an oscillation of the dipping-needle, or to about one-half.

4. The ship's external magnetism (of the top plating) on
the poop was nearly the same as on the last experiment, but
the change of polarity abaft had, I think, become a little
more extended, especially on the lee side—the starboard.

Tuesday, July 8 (variable, calm, E. N. E., E. to S. E.)—
The night was fine and pleasant—the sky lit up with a glorious blaze of stars. The Magellanic clouds were conspicuous.
But there was little wind and variable, and sometimes calm,
so that in the absence of our steam-power we made until
towards noon very small progress. The steam indeed was
got up and the engine started about 9 A.M., but, as on a
former occasion, the period of calm happened to have ended,
and, in 2½ hours, a fresh breeze from the eastward having
sprung up, the ship outran the screw; at 11·30 A.M. the
engine ceased its work. Up to noon we had scarcely made
100 miles by the 24 hours log—making our latitude about
43° 28′ S. and longitude 31° 51′ W.

So still and calm was the night referred to, that the clinometer at 10 A.M. marked only 4° to starboard and 1° to port,
with a registry of 2567, indicating only 13 movements acting
on the pendulum escapement.

In accordance with the indications of the barometer, which
had now begun to fall as rapidly as the column of mercury had
recently risen, our new breeze soon freshened into a gale, in

the evening shifting gradually round to the southward, with rain, from which quarter a considerable sea (renewing the ship's unpleasant rolling when the wind was aft) very speedily got up. It was amply compensated for, however, by the rapid progress it yielded,—the log which at noon had given $8\frac{1}{2}$ knots, by 7 P.M. indicating $12\frac{1}{2}$ and incidentally more, which, to the extent of at least 12 knots, continued to urge us on our way throughout the day. The course steered was N. by E. true.

The calmness of the forenoon and early afternoon, as to the sea, enabled me to undertake and carry through a series of observations on the ship's external magnetism.

The results were beautifully conclusive and of singular interest. The instrument employed was the small gimbal-compass, before used, of $3\frac{1}{2}$ inches card, which was securely attached to a frame with a crossbar of wood for steadiness and for keeping it square in respect to the ship's side whilst being let down from the ship's railing. In the arrangement adopted, the centre of the compass was kept uniformly at the distance of 14 inches from the ship's side. The frame was suspended by cords, to the chief of which was attached a measuring tape for indicating the depth of the instrument below the upper edge of the iron plating.

The stations selected for the experiments, to which I was partly guided by the rough experiments recently noticed, were (taking both sides of the ship at the same distance from the stern): No. 1, opposite the after-capstan, or $23\frac{1}{2}$ feet from the taffrail. No. 2, opposite the steering wheel, or 65 feet from aft. No. 3, opposite the middle of the main companion, or 100 feet from the stern. (All these were on the poop.) No. 4 was on the forecastle opposite the great capstan. The first position as to elevation was 6 inches below the gunwale or upper edge of the plating. The second and third usually were at the depth of 3 ft. 9 in. and 8 ft., downward.

Incidental trials, as the steadiness of the compass would allow, were also made. The *angle of heeling*, during the experiments, extended from 2° to port to 7° or 8°, as to the forecastle experiments. The ship's head by the steering compass was pretty steadily at north, being N. by E. true (magnetic).

July 8.	STARBOARD SIDE.						Head N. 11° E.	
Distance of Compass below the top plating.	Station No. 1, 23½ feet from Stern.		Station No. 2, 65 feet from Stern.		Station No. 3, 100 feet from Stern.		Station No. 4, — feet from Stern.	
—	Apparent direction of Ship's Head.	Devi-ation.	Ship's Head.	Devi-ation.	Ship's Head.	Devi-ation.	Ship's Head.	Devi-ation.
ft. in.								
0 6	N. 62 E.	51 W.	N. 21 E.	10 W.	N. 17 W.	28 E.	N. 57 W.	68 E.
3 9	N. 56 E.	45 W.	28	17	11	22	47	58
8 0	N. 67 E.	56 W.	45	34	N. 22½ E.	11½	34	45
12 6	..	..	57	46				
12 9	..	..	..	..	34	23		
PORT SIDE.								
ft. in.								
0 6	N. 56 W.	67 E.	N. 11 E.	..	N. 50 E.	39 W.	N. 67½ E.	56½ W.
3 9	50	61 E.	16	5 W.	50	39	56	45
5 4	..	..	..	..	46	35		
8 0	67½	78½ E.	2	9 E.	39	28		
11 0	..	..	..	..	28	17		
12 0	..	..	½	10½ E.				
14 6	..	..	..	..	17	16		

These particulars afford very conclusively, I think, the following results:—

1. That the external magnetism of the *Royal Charter*, which, at Port Phillip, exhibited strong northern polarity in the upper plating from stem to stern, had now, under a greatly diminished dip, and the inductive terrestrial action (whilst the ship with her head towards the north had of late received much mechanical straining from heavy rolling), begun consistently with theory to change, and that most obviously abaft.

2. That the ship's position relatively to the earth's inductive force with a southern magnetic dip, and her head whilst subjected to violent action of the sea in a northerly direction, corresponds with the diagram, fig. 1, of " Illustrations of the

Magnetism of Iron-ships," (No. ii.), where the stern or after-part from the gunwale downwards obtains an entire southern polarity.

3. That the new or southern polarity has up to this time extended on the *port* side, at the upper edge of the top plating, to about 65 feet from the stern, and on the starboard side, as approximated by casting the upper line of deviations into a curve, to about 75 feet from the stern,—whilst the intensity of the northern polarity appears to have correspondently increased.

It should be noted in respect to the deviations of the compass let over the side, that these must not be taken as very exact, as, being compared with the direction of the ship's head that was taken at the general course steered, from which there would not infrequently be some degrees of departure because of the defects of steering; an *apparent discrepancy*, too, in the ratios of change in the deviations in certain cases is deserving of notice. Thus, in the first position in the table where the southern polarity of the ship's stern or afterpart should as it might seem *increase* with the descent of the compass, we find a *decrease* in passing from a depth of 6 inches to that of 3 ft. 9 in. The same occurs in the first position of the *port* side, when the ship's deviating action first decreases and then consistently advances. This circumstance, which I have in various instances noticed, may possibly arise from a tendency to condensation, so to speak, of the magnetic energy on the terminal edge of the plate.

Wednesday, July 9 (S. W. to W. N. W.)—It blew a gale all night, which, though veering round from the southward to W. N. W., continued to urge us rapidly forward on a N. by E. to N. E. by N. (true) course. At noon we had made a distance of about 300 miles for the 24 hours, and by excellent observations had reduced our latitude to 38° 49′ S., the longitude being 30° 48′ W.

During the night of 8th–9th, the sea being very irregular and "lumpy," the ship rolled very heavily, and, whilst the wind was aft, continuously. The clinometer at 9 A.M. had registered the maxima 39° to starboard and 33° to port—the mean heel-

ing was then 7°, increasing in the day to 12° to starboard.
The number registered was 3522. It may here be observed,
and should have been before noted, that the maxima of the
clinometer in many of the larger angles appear to be in
excess of the actual heeling, in consequence of the resistency
of the mechanism, which, though beautifully constructed,
occasionally *detains* the pendulum for a certain extent of
angle, and then by virtue of the swing necessarily produced
causes it to go beyond the real deviation of gravitation, and
so to register too much. The extent of error when it does
occur I have not yet been able to ascertain.

The temperature of the atmosphere had now under winds
favourable for increase become mild and agreeable. The thermometer was at 53° in the afternoon, and had been higher.

It may be noticed, that ever since leaving Australia we
have had a fair and often profuse accompaniment of birds.
The most general and persevering of the species following
the ship was the Cape pigeon, which has never been missed
from our wake from the Australian coasts to our present position. These pretty birds come boldly and closely near, and
like the fulmars of the Arctic seas, may easily be caught
when the ship moves slowly by hook and line. Two were
thus taken up (one escaping) by the first officer, and one of
them preserved as a specimen for Mrs. Scoresby. The most
numerous of other species which have accompanied us—
especially in our more recent progress—were Cape hens,
with occasional petrels, gulls, and for a while albatrosses.

The fresh gale mainly prevailing moderated towards night,
and became more scant in direction for our desired course;
but still, with a considerable proportion of easting, yielding a
preponderance of northing. The moon of eight days old was
brilliant and beautiful till its setting—the sky generally being
clear and full of stars.

The ship's having made 300 miles in the day's log, was the
occasion of a supply of champagne at dinner, and a progress
of 12 knots produced cheerfulness and excellent spirits with
the whole party. It is difficult, without the experience of the
fact in a long sea voyage, to enter into the exhilaration of

spirits and feeling which, especially after long continuance of light or baffling or contrary winds, a brisk swiftly-propelling fair gale universally produces.

Another source of enjoyment of a different order was afforded to the saloon passengers by Miss Catherine Hayes, who sang popular English and Scotch songs, concluding with " God save the Queen."

Thursday, July 10 (N.W. to N. to N.W.; lat. 35° 44' S., long. 28° 8' W. Variation, by General Sabine, 7° W. at noon.) —About four in the morning, the wind, which had for several hours been moderate, again increased to a brisk or strong gale—blowing hard in occasional showery squalls. Lightning was seen in the western quarter up to the middle of the morning watch. The ship lay up within two or three points of her course, but then broke off,—the braces being slightly checked to E. N. E. or even E. by N. until noon. Having, however, gone on at a rapid rate we made, up to noon, about 185 miles of northing, with little more than $2\frac{1}{2}°$ of longitude to the eastward.

The day was showery, squally, and the wind occasionally blew very hard. Considerable seas prevailed at the same time from the south-westward, rising to 20 feet, and similar waves from the northward or in the direction of the wind. The temperature became increasingly mild, 59°. The clinometer (at noon) marked 40° rolling to starboard and 3° to port, with registry of 4026, and occasionally regular heeling to starboard of 12° to 13°.

Being the season and latitude for the fierce gales of the South Atlantic known by the name of " Pamperos," as coming from the pampas or plains over which they pass, we were in strong expectation, caused by a considerable fall of the barometer and the squally weather with lightning which now prevailed, of encountering one; but happily we escaped. These formidable storms sometimes rising suddenly, ships unprepared for them not unfrequently suffer loss of spars or sails. A passenger with us was in a ship thus unexpectedly assailed at night. The ship was thrown on her beam ends and only righted on the giving way of the canvass, almost

every stitch of which was blown away! Frequently, if the
wind has commenced at N. E. or varied from N. to N.W., the
pampero will burst in by a sudden shift to the south-westward, which, if taking the sails aback, may be very dangerous. Admiral Beechey, when in the *Blossom*, and making a
winter's passage to the eastward round Cape Horn, having
encountered N. E. and N.W. gales, in July, as he was advancing towards 35° of latitude, had several pamperos; and
from the experience and observation of this intelligent officer,
it is inferred that, " ships may generally reckon upon encountering at least one pampero between 33° and 37° S."*

Friday, July 11 (W. N. W., W., W. S. W.)—Though the
wind was rather scant, yet, with the braces checked and a
fresh gale blowing, we made good way in· direction about
N. N. E ½ E., true—averaging in speed about ten knots.

This brought us at 6 A.M. to an interesting position in our
voyage—an epoch in our adventure,—the crossing of the
Royal Charter's track in her outward voyage, and thus completing the circuit of the globe. This position of our intersecting tracks was in latitude 33° 30′ S., and longitude
26° 50′ W., where we were on the 16th of March at the same
hour in the morning, being an interval altogether of 117
days by nautical time, but subtracting the day gained in time,
116 days. The time occupied in making the circuit has of
course to be reduced by 39 days nearly, spent in Port Phillip,
and by one day more occupied in going into port and steaming out. The remainder, or 76 days, indicates the time of
sailing during which the ship performed the feat of going
round the world.

The morning was charming, mild (temperature 61°), and
bright; and the day, though we had many showers of rain,
was enjoyable. We continued to make a good progress as
to speed, with an improvement in our course by the wind
drawing more to the northward and south-westward in the
evening. The night was brilliant with clear moonlight and
the bright shining of the principal stars.

---

* From Purdy's " Sailing Directory for the Southern Atlantic," pp. 316, 270.
Third Edition, 1845, 275, 6.

Considering that the heavy rolling of the ship since the magnetical experiments of the 8th, on a curve inclining to the northward, would probably have advanced the changes in the ship's magnetic condition, I made a complete series of trials with the dipping-needle and the compass at almost all the stations previously arranged. The following were the most observable results (ship heeling 10° to starboard) :—

1. *As to the Dipping-needle.*—Experiments at the various stations fore and aft afforded some curious and interesting results,—in striking accordance with previous theoretic deductions. Though the terrestrial dip was yet, according to General Sabine, about 30° (perhaps 27½°) S., the dip at the station nearest the stern (23½ feet from the taffrail) was found to have changed its signs, being on the starboard side 2½° N., and on the port side 10° N. On the forward part of the poop, it was 21¼° on the starboard and 29° on the port side, southerly! At the standard station, it was 28° S.; and on the forecastle (amidships), 49° S.

2. *The Ship's external Magnetism.*—Though the great heeling of the ship, the rolling, and strength of the wind, rendered these experiments difficult, yet the advance of change—the pushing forward of the new or southern polarity of the stern—was abundantly evident. Thus on the 8th, on a course only about a point different (being in the present instance about N. 26° E., true magnetic), the westerly deviation at the second station from aft, on the starboard side, had increased to about double aloft, and the easterly deviation at No. 3 had gone down from 28° and 22° to about neutral,— showing an extension of the new or southern polarity from about 75 to about 100 feet, the extent of the third station.

Saturday, July 12 (W. S. W., S. W., variable.)—Blowing a fresh gale all night, with strong squalls, we continued to make good progress on a course about N. N. E. true, so that at noon we found ourselves in latitude, by observation, 28° 36′ S., and longitude, by chronometer, 24° 57′ W. The sun rose cheeringly, and we had a day of real and exhilarating enjoyment. The air being mild and soft and dry, most

s

of the ladies spent the day on deck, walking or sitting whilst
reading and working. The wind indeed abated sometimes to
a very light breeze, but still, though variable, always enabling
us to hold on our course, and, generally being far aft in its
direction on the sails, allowing a large spread of canvass on
every mast. Though we had heeling and rolling in the early
part of the day to the maximum extent of 28° to starboard
and 3° to port, yet the sea greatly subsided before noon,
allowing all sorts of industrial and recreative occupations to
be pleasantly carried on.

The birds, I observed, which hitherto had been generally
numerous, in great measure deserted us—even the persevering Cape pigeon. A few albatrosses, after a long cessation,
appeared in small numbers on the 9th and 10th, but to-day
they had again deserted us.

The progress of the ship's magnetic changes being of much
scientific and no inconsiderable practical interest, further experiments on the *outside* polarity were made in the evening.
On this occasion, I adopted a somewhat different plan—
applying the compass only at one particular level. And
having found that the highest level previously adopted (with
the compass-frame resting on the uppermost moulding outside, or 6 inches below the upper edge of the top plating,)
was not so satisfactory as the next moulding, which gave
the compass a depression of 3 ft. 9 in.; this latter level was
followed throughout. Several additional places were taken
into the series, as admitting of the convenient letting down
of the compass. This series, with the exception of one or
two discrepancies occasioned by incidental bars or masses of
iron—particularly the cranes or davits of the boats—was on
the whole fairly consecutive and satisfactory.

The series with the ship's head generally very steady at
N. 27° E. true magnetic, and the angle of heeling 3° to
starboard, ran as follows :—

July 12. Deviations of the upper plating externally of
the *Royal Charter*, taken with compass 3 ft. 9 in. down the
side.

Head N. 27° E. true.	I.	STATION 1. II.	III.	IV.
Dist. from Taffrail .	18 ft.	23½ ft.		59 ft.
Starboard side Deviations . .	N. 76° E.	N. 73° E.	N. 56° E.	. .
Port side	N. 75° W.	N. 67½° W.	N. 11° W.	N. 2° E.

	STATION 2. V.	VI.	VII.	STATION 3. VIII.
Dist. from Taffrail .	65 ft.	68 ft.	94 ft.	100 ft.
Starboard side Deviations . .	N. 62° E.	. .	. .	N. 45° E.
Port side	N. 22½° E.	N. 17° E.	N. 50° E.	N. 67° E.

Here a discrepancy is observable in col. v.,—a position which was very near to a massive iron crane or davit sustaining the after ends of two life-boats, resting on the poop rails on the two sides of the ship. But the main facts of change in the magnetism of the ship's stern and " after end," and of its gradual extension, whilst a course approximating the true magnetic north was being steadily pursued, were, consistently with theory, specially evident.

Thus the observations on this level made yesterday, indicating a general *southern* polarity of the ship to the extent of about 65 feet from the stern on the port side, and about 100 feet on the starboard side,—now indicated an advance of this new polarity to about 76 feet on the port side, and far beyond (to an extent that could not conveniently be determined) the 100 feet of the previous day on the starboard side ; the deviation at this level, which had been only about 2° W., being now about 18° attractive of the north pole of the compass.

Sunday, July 13 (W., W. S. W., W. N. W., S. W.)— Urged by moderate or fair breezes with occasional squalls we continued, under all available sails, our favourable though not very rapid course towards the N. N. E. (true) direction. This brought us, at noon, to latitude 26° 9′ S., and longitude

23° 59' W.   As on the outward voyage, we found the
approach to the tropics marked by a striking absence of the
birds wont in other regions to follow our track.   Life began
to wane and solitude to grow upon us.   On first going on
deck in the morning, two birds only were seen, and these
keeping at a great distance—a contrast to the close and in-
timate association of the Cape pigeons, Cape hens, gulls, etc.,
which had hitherto accompanied us.   These appeared as
chance strangers, seeking no intercourse with or benefit from
the presence of the *Royal Charter*.   Later in the day several
petrels (a large species) made close advances; but these recognised no abiding companionship and soon like the others
deserted us.

The clinometer at 9 A.M., after a quiet night, shows maximum angles of heeling of only $3\frac{1}{2}°$ to port and 11° to starboard
—with an advance of the numerical register of rolling from
4486 to 4500.

The sea was very blue—the water warm (70°), and the air
65°.   To antarctic voyagers it was a glorious day!   How
refreshing, cheering, beautiful, by recent and long contrast!
In the evening there was a brilliant display of sheet lightning in the distant horizon and in various quarters; but no
thunder was heard.   Heavy showers of rain and strong
squalls became frequent after the close of the day, alternating
with brilliant moon and starlight—the stars of the principal
magnitude shining out and overcoming the power of the
lunar illumination.

Our usual Sunday service was held in the saloon in the
morning, with an address from the first lesson for the day,
the 20th–22nd verses of the chapter (1 Kings xiii.)   In the
evening I had the satisfaction of renewing my ministrations
in the third-class messroom, where as usual we had a goodly
attendance, and a most attentive congregation.   The subject
of address was taken from the second morning lesson (John i,
12–13).

Monday, July 14.—With a fair but moderate wind we
entered the southern tropic.   At 7 A.M. it was, when crossing
the tropic of Capricorn, we had the satisfaction of finding

ourselves beyond the ordinary region of storms and cold, and enjoying moderate temperature and beautiful weather in the torrid zone.

The change was marked by the gradual change of costume of the passengers—the antarctic clothing had for three or four days been diminishing till it disappeared, and tropical habiliments began now to take their place; the change was rapidly progressive, and exceedingly enjoyable. At noon of the 9th the thermometer first reached 50°, at noon of the 11th it rose to 60°, and about noon of the 12th we had the first registration of 65°. To-day it made a further advance to about 68°. This region was thus pleasantly temperate because of the high northern declination of the sun—the sun's meridian altitude having yet only reached about 45 degrees. And for the same reason, probably, we had thus entered within the tropics without yet meeting with any token of the usual south-easterly trade wind. With all the general consistency of these specific belts of atmospheric currents, the strength, breadth, commencement, and termination, with the intermediate space of calms, are all subject to considerable variations, mostly not calculable, but in certain respects referable to conditions, such as that of the sun's position, on which some measure of reasonable conjecture may be founded. One of the most specific of these conditions, perhaps, is the tendency of the trade wind belts to follow the course, in declination, of the sun, so as to shift môre to northward in the European summer, or with high northern declination, and further to the southward at the converse season.

Indicating, perhaps, our approach to the regular trade winds, the long continued westerly breeze was observed to shift to the southward in the afternoon.

It was my privilege this evening to witness for the first time a vertical moon. Somewhere about 8h. 29m. P.M., a phenomenon which I had for some time been anticipating occurred. The moon, 13 days old, shining gloriously in the tropical sky over our heads. Personal shadows were absorbed within the outline of the thickest part of the body or dress.

Though the ship had made capital progress for the fluctuating strength of the wind, we could not but remark that
her sailing in light and moderate breezes was, for a ship of
her build, rather heavy. This was ascribed to a disadvantage
not hitherto overcome in iron-ships—the rapid and extensive fouling of their bottoms by the adhesion and growth of
weeds, barnacles, etc., which necessarily produce a deteriorating effect on a ship's sailing qualities. Various contri-
vances in paint and chemical colouring have been resorted
to for the overcoming of this evil, but hitherto without much
success. Nor has any process yet been devised for the clearing away of the fouling attachments without the costly and
often impracticable operation of docking the ship. It has
occurred to me, however, that much of the disadvantage
might be overcome by a process of scraping, with no very
complex arrangements for the guidance of the apparatus, by
means of a framework of changeable curvature guided by
the keel; and an essential element in such arrangement,
it seems to me, would be the attachment to the scraping
apparatus designed for the bottom of the ship of buoyant or
water-tight air vessels, by which any degree of upward
pressure on the scraper might be given. To the details of
such an apparatus,—which might be moved by ropes forward
and aft as well as at right angles with the longitudinal
motion by similar means,—my attention has not been
directed, nor might my habits of contrivance at all compete
with the ingenuity and invention of practical and ingenious
engineers or artisans.

The breeze hitherto for some time enjoyed began in the
evening to fail us—reducing the ship's rate of going, at 9
and 10 o'clock, to about five knots or less. The screw was
therefore lowered—there being no appearance of squalls—
and the engine was put in motion about 11 P.M. The night
was so beautiful, and the air so soft and pleasant, that numbers
of the passengers, ladies as well as others, remained on deck
until a rather late hour.

In our present position, relative to the earth's magnetical
force, and whilst sailing very nearly on the magnetic meridian,

the changes rapidly going on in the ship's magnetic condition had assumed so high a degree of interest and importance, that I again undertook a very extensive series of magnetical observations with the dipping-needle, on compass differences, and on the ship's action on the compass, externally near the gunwale of the poop deck.

1. *As to the Dip.*—The recent changes in the ship's magnetic condition may conveniently be elucidated by a comparison of the experiments of this day with those already described of the 11th, under a change from $32°\ 33'$ S. to $22°\ 52'$ S., or of $9°\ 40'$ in latitude towards the north, and of $2°\ 39'$ of longitude towards the east, and of a reduction in the terrestrial southern dip from $29\frac{1}{2}°$ to $16\frac{1}{2}°$, or of about $13°$:—

	Station I. Poop.		Station II. Poop.		Station III Standard.
	Starboard Dip.	Port Dip.	Starboard Dip.	Port Dip.	Midship.
July 14	$20°$ N.	$26\frac{1}{4}°$ N.	$9\frac{1}{6}°$ S.	$4°$ S.	$14°$
July 11	$2\frac{1}{2}°$ N.	$10°$ N.	$21\frac{1}{4}°$ S.	$29°$ S.	$28°$

	Station IV. Forecastle.			General Angle of Heeling.
	Starboard.	Midship.	Port Dip.	
July 14	$41°$ S.	$49°$ S.	$42°$ S.	$3°$ Starboard.
July 11	. .	$39\frac{1}{4}°$	. .	$10°$ Starboard.

Here observing the line of mean differences, we mark the most interesting feature of the ship's magnetic changes,—showing most decidedly the same fact as that obtained from observations on the ship's external magnetism,—viz., the transition of the polarity in the poop plating from northerly to southerly and the rapid progress of the change. Thus we notice nearly twice the change in the dip on the poop to that on the forecastle, the former being from 4 to $5\frac{1}{2}$ degrees in advance, apparently, of the change in the terrestrial dip, and the latter 3 to 4 degrees in the rear, whilst the change at the standard station appears to agree very nearly with the actual change in the direction of the earth's magnetic force. All these results, it is interesting to observe, are strictly accordant with the theoretic deductions.

2. The change in the external magnetism of the upper poop plating (3 ft. 9 in. below the edge) was found to have still further extended on the port side. But the observations presented otherwise no particular results.

3. The compass comparisons, too, exhibited nothing worthy of being noticed.

Tuesday, July 15.—The winds being generally light—varying from the southward to westward in the night—gave opportunity for the engine to do useful service; but in consequence of the prevalence of a rather high south-westerly swell, not without very considerable vibration or rather shaking of the afterpart of the ship. Sometimes this shaking attended by a disagreeable noise occurred under different circumstances,—where, under incidental freshening of the breeze, the action of the sails overran the speed of the screw, and where in the violent lifting or heaving of the stern-frame the proper and ordinary revolutions of the screw were interfered with. The disadvantage was ascribed to the fact of the direct action of the engine with nothing of intermediate and yielding contacts to break the force of incongruous contacts and operation of the water. The effect was often such as to cause the ship perceptibly to shake in the manner of an elastic flexure as, apparently to sensation, of some inches up and down, giving correspondent vibration in the spanker boom, as supported at the two ends, amounting to a spring-like movement of the intermediate timber up and down.

We had another cheering and beautiful day; the sun, except when occasionally screened by one of the small detached clouds which floated somewhat numerously in the sky, shone brilliant and hot, and yielded a rich treat of enjoyment as contrasted with so much recent experience of unpleasant weather; one of the awnings was for the first time spread. At noon we found that an advance had been made of $3\frac{1}{2}$ degrees of latitude within the tropic, our latitude being 19° 59' S., and longitude 23° 54' W.

As usual, as far as my personal experience goes, the scene abroad – except as to some whales which from time to time appeared—was one of entire desolation! Not a single bird, nor flying-fish, nor dolphin, boneto, nor even shark, was observed during the whole day. The whales, therefore, were more notable, as breaking the solitariness and monotony of the external scene. They were probably of more than one species; two which appeared very near to the ship were of a large kind, apparently 40 to 50 feet in length; the dorsal fin was small and hooked backward. The blowing was scarcely more dense but a little more forcible than that of the mysticetus, or Greenland whale, but the colour was more of a slate or bluish grey. One of them came within 100 to 150 yards of the ship's stern, and after blowing could be traced sometimes till its next appearance a minute or two afterwards pursuing a devious course just beneath the surface of the water—appearing under the penetrating action of the solar rays, sometimes as a dark patch in the partially illuminated sea, and at others as a greenish-white object shining through the intermediate stratum of the waves. The whale was obviously following and examining the strange and huge object rapidly swimming as it might seem on the sea—sheering first towards one quarter then towards the other of the ship, and for convenience of inspection, as I have often noticed in the Greenland whale, occasionally turning on its side or on its back whilst urging its onward course, and so exhibiting the white surface of the under part of its body. Its manner or intervals of breathing appeared to be in sets of two or three respirations of ten to twelve seconds betwixt, and then a disappearance for a space of about two minutes. The ship was going about eight knots, and the whale easily kept pace with her, and when its curiosity appeared to have been satisfied, it sheered off on a different course, and soon ceased to be visible to us.

In a former part of my journal I have remarked on the prevalent error and persuasion of the spouting of water by whales—a persuasion which in one of my earliest essays on natural history I showed was founded in mistake, for nothing like water, much less fountains and jets of water, is to be seen in the respiration of this order of animals. Yet the error is not only a popular one, but even to this day is supported by many of our eminent naturalists. In again noticing the subject, it is with the purpose of adding the experience gained in this voyage of the blowing of several kinds of whales previously unknown to me. In the whales of the Indian ocean, of the South Pacific, of the South Atlantic, as well as of the north and of tropical regions, the "spouting" is but the respiration common to animals breathing the atmospheric air, except in being more dense and steam-like in its expirations!

The south-westerly swell continued to follow us the whole of the day—with waves sometimes rising to about 20 feet, and causing the clinometer to register maximum rolling to starboard of 29° and 6° to port. This, I believe, according to general experience, is uncommon,—a sort of intrusion of South Atlantic waves into the ordinarily smooth and quiet sea of a tropical region,—continuing as I have said for the whole day, and thus extending at least 300 miles within the torrid zone.

Soon after noon the breeze shifted gradually and softly to the starboard quarter (the ship sailing on a true course of about N. 20° E.), and gave us the impression of the commencement of the south-eastern trade wind.

Among the magnetical and compass observations made in the course of the day, I made further trial (having before had the instrument in action) of an ingenious patented invention of Mr. Andrew Small, maker and adjuster of compasses, of Glasgow, for finding by direct observation the true meridian, and consequently "the errors, in deviation and variation, of the mariner's compass." The instrument consists of an attachment to a compass on the ordinary principle of certain moveable circles in brass—a true meridian circle, an hour circle, and an equatorial circle, on the adjustment of which, in reference to the latitude, declination of the sun and apparent time, proximately, a stream of light from the sun's rays passing through a slit in the moveable hour circle is made, by the turning of the interior framework of the apparatus to fall on the middle of the interior of the opposite side of the same circle (or to be put into position by using the slit as sight vanes) by which the meridional circle takes the position of the true north and south. The difference betwixt this and the direction of the compass-needle gives of course the errors of the compass observed with, in whatever position it may happen to be placed, and by comparison with the steering or other compass, determines also its errors. My object being simply to test by experiment the performance of the instrument, to this object I confined myself, by a series of trials carried on from about 2 to 3 P.M., and frequently repeated. The results were tested by the foremost of our adjusted compasses, the deviation of which combined with the variation I had ascertained to be 7 degrees westerly, and appeared to agree very well.

Wednesday, July 16 (S. E. to E., E. N. E., N. E., E.)— The breeze which shifted yesterday to the south-eastward has since proved to be the regular trade wind. It blew with some degree of briskness all night and most of the day—frequently, by the quantity of sail spread, causing the ship to over-run the screw, and giving an average speed of 8 to 8½ knots on a course making due north. The result of the day ending at noon was about 200 miles of distance, and latitude, bringing us by observation to 16° 40′ S. and 23° 52′ W. Variation (by General Sabine), 13½° W.; deviation of steering compass about 7° easterly. At 9·30 A.M. the engine was stopped and the screw raised up. There was some loss of speed, perhaps a knot, by the suspension of it.

The day throughout till after sunset was fine, pleasant, and enjoyable—though a larger quantity of the small detached cumulus clouds were prevalent with an entirely clouded sky at sunset. The temperature rose to about 72° in the open air. The swell from the S. W. or S. S. W. of yesterday had diminished in the forenoon considerably, and shifted its direction to nearly the south. In the course of the day it went down to a very inconsiderable measure of elevation. Sea blue. No living thing seen beyond the ship. Temperature 72°; sea higher.

In the evening and night we had repeated showers of rain with strong squalls, in which the ship invariably broke off from her N. N. E. (compass) course to N., N. N. W., and even N. W.; but the deviation of course towards the west seldom lasted long, whilst the breeze improving with us gave us a pretty good progress.

Thursday, July 17 (Easterly, N. E., N. N. E. to E.)—Showers of rain with strong squalls of wind continued to prevail throughout the night. After sunrise, however, the rain ceased, and the wind, which always veered in a northerly direction in the showers, became more steady.

On deck, where we always had a fine, cheering (because fair) and fresh breeze, the weather was most enjoyable. We had crept by a gentle progress into warm weather; and now the sun having been so long in the northern hemisphere, this southern tropic was far from hot—the temperature which at 8 A.M. was 73° not rising throughout the day higher than 75°. So comparatively quiet had our position now become, that the clinometer, at 8 A.M., marked only 8° to port and 2° to starboard as the maxima for the night. In our room the temperature was 73° in the morning and 76° at night.

In the evening after a rather cloudy day the rain showers were renewed, but with less disadvantage as to the direction of the wind, the ship generally being able to lie up, or very nearly, on a true north course, and ordinarily steering a quarter point to the eastward of the true meridian. The squalls and general increase of the breeze gave us a good progress, averaging for the afternoon and evening above ten knots.

The position reached by the *Royal Charter* at noon—latitude 13° 28′ S., longitude 24° 8′ W.—was one of peculiar interest with reference to my magnetical researches; for at that time, or close upon it, we crossed, according to General Sabine's chart, a part of the flexuous line of the earth's magnetic equator. The terrestrial induction now being about to reverse its direction of action on upright iron, it became a subject of much interest, and perhaps scientific importance, to determine the then existing condition of the

ship's magnetism. With this view I instituted a regular and
complete series of my usual observations with the dipping-
needle, the compass placed outside the ship, and of compass
comparisons.

1. The experiments for the dip gave results bearing out
the inferences suggested by those of July 11 and 14, and
confirming, strikingly and beautifully, former theoretic deductions. The results being tabulated, and after the plan
adopted July 14, being placed in juxtaposition with a few of
the recent series, became more characteristic and instructive.
As the ship was steering very steadily, I give only the
general direction of the ship's head (true magnetic); and
the noon latitude and longitude for all (see pp. 270–271).

Friday, July 18 (E.)—Throughout the night we had a
strong favourable breeze (the ship with the wind abeam
making a north course, inclining to the eastward,) and we
made good way, averaging from noon to noon just ten knots.
Showers with hard squalls prevailed during the night,
obliging us to keep our little portlights shut, which rendered
the berths warm and close. But the day was beautifully fine
and the breeze cool and refreshing (the temperature not
rising above $76\frac{1}{2}°$), without a single interruption to the enjoyableness of this fine climate. Our distance run up to noon
was just 240 miles, or 4° of latitude, bringing us to 9° 30′ S.,
and 23° 57′ W. longitude. The clinometer marked for the
night 12° to port, and 2° to starboard, and an ordinary heeling of from 7° to 10° to port; the number registered was
4706. The sea, considering the freshness of the breeze, was
beautiful and characteristically (for the tropics) smooth.
Deadness in external nature still reigned; I never saw a
living thing beyond us the whole day. At dinner-time there
was an unwonted stir from a cry of a sail on the port bow.
The table was in a minute almost deserted. It was a ship
sailing in a direction opposite to ours or more westerly.

The scene around soon after sunset became gorgeous in
the tints and splendour of the still strongly illuminated sky.
As the sun disappeared the full moon rose on the opposite
side of the horizon, exhibiting that broad surface and sur-

OUTWARD PASSAGE.

Date.	Latitude.	Ship's Head (true Magnetic).	Terrestrial Dip.	Station I. Poop. 40 ft. from Stern.		Station II. Poop. 100 ft. from Stern.		Station III. Standard. 181 ft.	Station IV. Forecastle. 273 ft. from Stern.		Station V. 284 ft.	Ship's Heeling.
				Starb. Dip.	Port Dip.	Starb. Dip.	Port Dip.	Midships Dip.	Starb. Dip.	Port Dip.	Midships Dip.	
	° ′	°	°	°	°	°	°	°	°	°	°	°
Mar. 8	6·58 S.	S. 43 W.	10 N.	. .	. .	. .	. .	34½ N.	. .	. .	. .	7 Starb.
11	16·46	S. 11 W.	3½ S.	. .	. .	.	. .	16	. .	. .	. .	7 Starb.
12	19·30	S. 8 E.	10 .	. .	. .	(15 in.	abaft)	6¾	. .	. .	. .	7½ Starb.
13	23·30	S.	15	5¾ S.	4 N.	. .	. .	½	. .	. .	6 S.	9 Starb.
14	{28·32 / 29·20 W.}	S. 5 E.	23	. .	. .	. .	. .	8¼ S.	25¼ S.	15½ S.	15¾	9 Starb.
15	{32·29 / 27·32 W.}	S. 21 E.	28½	16¼ S.	8¼ S.	29 S.	25 S.	13¼				
17	36·15	S. 24 E.	36	. .	. .	31½	35½	23¼	33	41½	28	5 Starb.
19	41·13	S. 47 E.	44	27⅓	26	38½	53½	33	. .	. .	. .	10 Starb.
22	42· 9	S. 47 E.	50½	33½	35¼	41¾	59¾	43	. .	. .	. .	4 Port.
25	43·52	S. 37 E.	57¾	42¾	44¼	53¼	65¾	52	. .	. .	. .	2 Port.
28	45·49	. .	62½	. .	. .	. .	. .	62				
Apr. 3	47·31	S. 50 E.	70¼	63	70	67½	79½	70½	. .	. .	. .	6 Starb.
11	44·24	S. 82 E.	73¾	68¾	78¾	. .	. .	70				
12	44· 7	E. 3 N.	73½	. .	. .	71½	87½	75½	72	85½	75½	1 Port.
15	40·48	N. 10 E.	. .	. .	. .	. .	. .	70½	. .	. .	. .	6 Starb.
	Port	Phillip.										
21	37·52 S.	. .	. .	. .	. .	. .	. .	66¾	73⅙	69	68	

HOMEWARD PASSAGE.

Date.	Latitude.	Ship's Head (true Magnetic).	Terrestrial Dip.	Station I. Poop. 40 ft. from Aft. Starb. Dip.	Station I. Poop. 40 ft. from Aft. Port Dip.	Station II. Poop. 100 ft. from Aft. Starb. Dip.	Station II. Poop. 100 ft. from Aft. Port Dip.	Station III. Standard. Mean Dip.	Station IV. Forecastle. Midships Dip.	Station IV. Forecastle. Starb. Dip.	Station IV. Forecastle. Port Dip.	Ship's Heeling.
	°			°	°	°	°	°	°	°	°	°
May 30	50·45 S.	S. 5 E.	76° S.	72½ S.	72¾ S.	. .	. .	78½ S.	75¾ S.	76¾ S.	78¾ S.	7 Starb.
June 5	54·26	E. 17 N.	74½	70	82	. .	. .	77⅓	78¾	. .	. .	0
17	56·18	E. 11 N.	71	. .	. .	. .	. .	76¾	78½	74	89½	6 Starb.
July 2	48·45	N. 20 E.	48	26¼	29½	. .	. .	45⅓	64	61	66	1 Starb.
11	32·33	N. 25 E.	29½	2½ N.	10 N.	21¼ S.	29 S.	28	49	. .	. .	10 Starb.
14	22·52	N. 7 E.	16½	20	26¼	9⅙	4	14	39¼	41	42	3 Starb.
17	13·28	N. 18 E.	0	34½	38⅓	15½ N.	7¾ N.	2¾ N.	21¼	19	24½	3 Port.
21	0·36 N.	N. 22 E.	23½ N.	50	54	43	40	27½	12½ N.	15 N.	6 N.	4 Port.
24	9·17	N. 28 E.	35	56¼	59½	52	53¼	35	29¾	. .	. .	0
29	18·38	N. 25 W.	49	67¼	68¼	62¼	65¼	54½	52	. .	. .	0
Aug. 2	27·11	N. 10 E.	58¼	72	73	71¼	72½	60¾	60¼	. .	. .	0
6	36· 5	N. 40 E.	64½	75½	75¼	74¾	78	68¼	68¼	. .	. .	0

prising apparent magnitude so often observable at the rising
or setting. Whilst the sky on that side exhibited a clear
cold blue passing into grey—that of the sun's quarter was
gorgeous in warmth and colouring. At some elevation (about
a quarter of an hour after sunset) the colour of the atmosphere was still blue; about 10° above the horizon it passed
by insensible gradations into a rich rose-pink, then downward
into a warm orange, yellow, and greenish-yellow,—the colours
changing and merging in the most perfect effects of the
softest tinting. The intensity and warmth of the tints were
of course the strongest in the quarter in which the sun
had descended, but spread right and left with diminishing
richness about 60° each way, and then on disappearing rendered the cold grey of the sky more cold and striking. The
sky though generally clear was yet not cloudless. A number
of small cumulus-formed detached patches (the characteristic
tropical cloud) being seen floating in all directions, some
tinted and illuminated with admirable picturesque effects.
But the sea alive in the stirrings of a fresh breeze, whilst
dead as regards visible animal life, was an important and rich
feature in the general scene. The general surface of the
strongly marked circular area bounded by the horizon to
leeward of the ship or in the direction of the sunset (up to
within half a mile of the ship), was tinted in a manner I
never before observed, with a rich purple, and then continued up to the ship in splendid reflections from the millions
of undulating surfaces of the richly coloured western sky!
It was a vision of glory—soft, impressive, and of exquisite
harmonies—and well serving to elicit admiring impressions
of the glory of the great Creator, who in the first review of
his wondrous handiwork saw, and caused it to be put on
record, that it was very good.

The great apparent magnitude of the moon at its rising is
a phenomenon with which most persons are familiar, but on
the occasion now referred to the moon seemed to exceed her
usual apparent magnitude in such position.

Saturday, July 19 (Easterly, E. S. E.)—For some time in
the night the breeze was so fresh that the ship made 11 knots

and upward; but towards morning it declined so as to reduce our speed to 7½ to 9 knots for the day. The course
steered (the wind drawing from free) was generally N. N. E.
by compass, or about N. by E. ¼ E. true. The day broke
beautifully, and the weather, though somewhat warmer, was
very enjoyable on deck, soliciting all to turn out, and bringing around us on the poop not only passengers but Australian birds of varied plumage and capabilities of imitation of
language and other sounds of the human voice. The poop
deck, its fixed seats and places available for chairs, was quite
alive with human beings,—ladies and gentlemen occupied in
reading, working, conversing, or games, from morning till
night. Cheerfulness and enjoyment were universal.

At noon, we observed, in latitude 6° 3′ S., and longitude 23°
33′ W. Another vessel,—a stout ship,—was seen on a course
crossing at a large angle our track; but she fell far astern of
us, and did not come nearer than five or six miles. The
temperature of the air was 77° at 8 A.M.

Some compass experiments on the ship's external magnetism were again made, showing a further extension in the
southern polarity of the ship's stern and a tendency to
change the polarity of the upper plating, especially on the
starboard side of the forecastle, where the compass was very
slightly affected by the ship's magnetism.

Repeated trials of the ingenious contrivance of Mr. A.
Small, of Glasgow, for determining by inspection the direction of the true meridian, and from thence the compass errors
in deviation and variation, enabled me at length to form a
decided estimate of its capabilities. The result was, that
for moderate horary angles from noon the instrument acted
well, pointing out the true meridian generally to within a
degree or two, where the latitude and apparent time were
pretty well known; but in the case of considerable intervals
from noon, extending, as in our circumstances, to upwards of
three hours from noon, the results became obviously erroneous to the extent, as at 7·30 A.M., of about 10 or 12 degrees
to the westward of the truth, and at 5 P.M. to nearly as much
to the eastward. This tendency to error in large hour angles

T

I had suspected at an early period of my acquaintance with
the instrument, on account of the results being obtained in
dependence of a plane only proximately coinciding with the
plane or cone of the sun's motion in the heavens. For observations, however, within moderate limits, the instrument
appeared to work well and readily gave the required information.

In consequence of the absence of the moon for an hour
after sunset the effects were more varied than last night, and,
if possible, more magnificent. Soft fleecy clouds of the small
detached cumulus form, suspended in the quarter where the
sun had disappeared, were embossed, as upon the rich glowing tints of the clear sky, in complimentary tints of olive-
green, brownish-red, and faint purple, in succession. The
sky next to the horizon glowed at first in rich and delicate
greenish-yellow, rising into orange and delicate purple, till
lost at some 20° elevation in the blue ether of the upper sky.
The tints deepened and changed as the sun continued to sink
lower and lower below the horizon, and the general breadth
of the glorious display for awhile seemed gradually to sink
to a lower altitude like a transparent screen disappearing in
the process of its descent, but becoming richer in the orange
and red as it seemed to be departing. As this resplendent
scene seemed about to pass away the sky was again over-
spread with a glowing pink.

The appearance of the surface of the sea was repeated
with the same richness of purple colouring, but changing
with the change of the atmospheric tints and being replaced
by a beautiful blue. Opposite to the region of magnificence
all the sky was of a cold hazy-looking grey till the moon arose,
and in its turn became the mistress of the ascendant.

Sunday, July 20 (E. by S. to E. by N.)—We were surprised as well as disappointed by the failure of our breeze in
the mid-watch. For a time it was nearly calm, and preparations were made for putting the screw into action; but just
as the steam was got up and all in readiness to lower the
screw, the breeze, a little more scant in direction, sprang up,
again rendering the intended appliance of engine unnecessary.

The morning was showery, with strong squalls on the starboard beam—the ship then lying only about north, or rather west of north, true. It had been proposed to have Divine service on deck; but the frequency of the squalls and showers rendered the arrangement inexpedient, and disappointed us greatly. We, therefore, though in much smaller numbers than otherwise might have been gathered together, resorted to the saloon. After the second lesson, I baptised an infant who had been born on board, the son of a gold digger. The subject of discourse was taken from the first morning lesson, 1 Kings, xviii. v. 21, "And Elijah came unto all the people, and said, How long halt ye between two opinions? If the Lord be God, follow him; but if Baal, then follow him. And the people answered him not a word." And v. 39, "And when all the people saw it they fell on their faces; and they said, The Lord he is the God! the Lord he is the God!"

As the day advanced the weather became more settled, and evening service was held on the poop deck at 8 o'clock. It was an interesting and impressive scene, some hundreds of persons gathered together on the deck. The text was 1 Tim. vi. 11-12. "But thou, O man of God, flee those things; follow after righteousness, godliness, faith, love, patience, meekness. Fight the good fight of faith, lay hold on eternal life."

During the nautical day, though the wind was frequently light and never strong, we made out three degrees of latitude, and found our meridian position to be 3° 1' S., and longitude 23° 30' W. The showers and squalls which had prevailed during the morning ceased about noon, and the rest of the day, whilst cloudy, was fine, and the air temperate, (76° at 3 P.M. on deck) refreshing, and for the equatorial region very cool. The sea was somewhat warmer.

The course steered was N. N. E. by compass, making, by 6° deviation E., and 15° variation W., nearly a N. by E. course true. After the moon rose, the sky, as usual became clear, and the atmosphere beautifully serene. No living creature externally to be seen.

Monday, July 21 (Easterly, E. by S., E. S. E.)—Pretty good work was done by the ship, for a moderate trade wind, during the night. We reached in the forenoon another interesting position of the voyage—the equator, crossing the line at 9·30 A.M. This event, of universal interest to those on board—especially to the passengers and officers— was to myself, with reference to magnetical investigations, important, as I had fixed on this position for conducting a series of observations on the ship's magnetic condition for particular comparison with those made on crossing the line of the magnetic equator.

Terrestrial magnetism having now, within four days, not only changed its sign as to the inclination of the needle, but advanced up to 23° or 24° of northerly dip, must, it was expected, have produced great changes in the ship's magnetic condition,—changes, the nature and extent of which it was important to determine by varied and careful observations. Three processes—those adopted on former occasions—were practically available for this end, besides observations of the sun's azimuth or amplitude, or both. Thus, first, by observations with the dipping-needle made near the two extremities of the ship, in intermediate positions, and on the two sides of the deck, results should be obtained showing by inspection the general character of the changes which might have taken place, and likewise supplying data for more profound scientific and mathematical inquiries. Then, secondly, by trial of the ship's deviating action on a compass let down on the outside of the iron plating, the actual disturbing force on the horizontal needle at the place of experiment would be determined. And thirdly, by compass comparisons, the effects of these changes in the ship's magnetism on the several fixed compasses, the practical result, as to the particular position to which the ship's head might be directed at the time, would be directly ascertained.

1. *As to the Results of the Experiments with the Dipping-needle.*—Carrying back these results to the table placed under the date of the 17th instant, we may perceive at a glance, not only the direction of the magnetic force on board in eight different parts of the deck, but the extent of the

change which has taken place in that direction at each of those stations. It should be remembered that the needle was in each case placed upon its gimbal table on a stand having a sufficient aperture in the top for the free action of the balancing rod and weight, and adjusted to the height at which the needle of the standard compass was placed, and that at all the stations the needle had very nearly the same elevation as measured from the keel.

The most observable changes in the dip, as appears from the inspection of the table, are the following:—first, that the negative or southerly inclination of the needle, which existed to the general extent of 70° to 80° when the terrestrial magnetism was 70° or upward, diminished in a different ratio from that of the terrestrial dip; secondly, that the ratios of change were different at every station on the deck, those aft being the most rapid; thirdly, that casting the observations at each of the five principal stations into curves, with reference to an uniform scale of degrees for terrestrial dip—the ship's dip at the several stations being set off as ordinates—all the curves, except that of Station I., came out of singular regularity, approximating very nearly to straight lines; fourthly, that in those curves representing the changes in the ship's dip we find the sign of the inclination of the several stations (measured from the taffrail) changing from negative to positive, or from southerly to northerly, in the following positions and order:—

Places of no dip in Fox's Needle at the several Stations.

	STATION I. 40 ft.		STATION II. 100 ft.		STATION III. 181 ft.	
	Latitude.	Terr. Dip.	Latitude.	Terr. Dip.	Latitude.	Terr. Dip.
Homeward	30° S.	27° S.	20° S.	11° S.	12° S.	2$\frac{1}{2}$° S.
Outward .	. .	14° S.	. .	. .	. .	15$\frac{1}{2}$° S.

	STATION IV. 273 ft.		STATION V. 287 ft.	
	Latitude.	Terr. Dip.	Latitude.	Terr. Dip.
Homeward	4$\frac{1}{2}$° S.	15° N.	4° S.	15$\frac{1}{2}$° N.
Outward .	. .	. .	. .	8° S.

Thus, fifthly, that the sign of the dip changed *first* in the aftermost stations, and in regular succession (the ship's head always being in the northerly or north-easterly direction) as to time, latitude, and dip, as the distance of the stations from the stern increased; sixthly, that the position of change from northern to southern dip on the outward voyage, with the ship's head generally directed to the southward (south-westward or south-eastward) was not coincident, not even approximately, with the homeward positions of change, but were slower in their attainment or greatly behind the changes in the terrestrial dip; whilst, in the transition from southerly to northerly dip, the magnetic forces in the ship anticipated, in all the stations on deck, the transition in the earth as marked by the crossing of the magnetic equator.

2. *As to the Results of Observations on the Ship's external Polarity.*—The correctness of the last of the deductions above recorded obtained additional evidence in the general southerly polarity of the ship's stern and afterpart, and the gradual extension of that specific polarity forward with the advance of the ship into the magnetic equatorial regions.

3. *As to the Compass Comparisons.*—There were already indications of some reduction of the deviations of the adjusted compasses (as far as could be judged from observations on a few northerly and north-easterly points), and of some increase of the deviations of the standard compass; but everything went to show that the compass aloft retained its excellence of action and reliability for correct and safe guidance in the navigation of the ship.

We had in the evening, from 7·30 to 9 P.M., the finest exhibition of the phosphorescence of the water of the sea I ever remember to have witnessed. The wake of the ship, for perhaps a half of a quarter of a mile astern, was brilliantly illuminated as if by a vivid white light shining from beneath, and in a stream corresponding with the breadth of the portion of water displaced or disturbed by the passage of the ship. But the most striking part of the general phenomenon consisted in flashes of the colour and almost vividness of electric light from every white crest of the waves which the wind in its very moderate force threw over. These were beautifully flashing and luminous to the extent of a mile's distance or more from the ship. On the forecastle there was exhibited an additional feature—the dotting over in momentary specks of light, like that emitted by the firefly, the general surface of the water about the bows—the result obviously of the shower of drops of the phosphorescent water thrown off from the bows by the passage of the ship and the flashing of the waves against her. The phenomenon resulted, no doubt, from the presence of innumerable animalcules—phosphorescent medusæ or molluscs—with which the surface water must have been crowded. What the combination of circumstances, however, is for the best development of the phenomenon, besides that of darkness, has not, I believe, been determined. From its being a popular impression, however, that the phenomenon is more usual with some winds than others, and those warm or summer winds, it seems not improbable but that the atmospheric electricity may have some effective influence in the highest developments of it. On the rising of the moon the phosphorescence ceased to be visible.

In the evening, the passage of the "line" homeward was the occasion of general joyousness amongst the passengers. The oft described proceedings generally associated, until of late years, with this transit, were not allowed, nor was any sign of them, on the part of those "forward," exhibited. But performances and exhibitions of various kinds were got up amongst the gentlemen "aft," which were the occasion of much lively excitement; and, from the universal good feeling and propriety of everything, of no small real gratification and enjoyment to the general body of our associates. Dancing followed, and was continued for a considerable period.

This was not all. The night was mild and delicious, and the moonlight singularly lovely. Miss Catherine Hayes kindly sung on deck to the assembled passengers and crew "God save the Queen" and "Rule Britannia;" hundreds of hearty sympathising voices joining in the chorus. "Home, sweet Home," was kindly given us in conclusion. This was sung as a solo—and with a feeling, pathos, and charming simplicity, which seemed to reach every heart.

The second and third class passengers had their own recreations in the parts of the main deck and spar deck appropriated to them.

It is satisfactory to say that the cheerfulness of the occasion was marked by an uniform expression of friendly and social consideration, and unmarred, as was repeated to me by the officers of the ship, by a single known example of intemperance or an ungenerous or angry word.

Throughout the day we enjoyed a moderate pleasant trade wind of refreshing coolness, the temperature never rising above 77°,—under which we made an average progress of about 8½ knots on a N. ½ E. to N. N. E. compass course. For the nautical day we made 205 miles of distance, which brought us at noon to latitude 18′ N., and longitude 23° 49′ W. The night was brilliant. The stars of principal magnitude stood out as it were from the sky, and the moon shone resplendently. Up to this time the south-easterly *trade* had exhibited no sign of departure, and my own impression was that, because of the advanced position (lat. 19° to 20° S.) in which we first met with the trade wind, and because of the far northerly position of the sun, we might not improbably carry the existing pleasant south-easterly or easterly breeze for a day or two longer.

Tuesday, July 22 (S. E.)—A brilliant sunrise succeeded a most lovely night. We were greeted on reaching the deck with the announcement of a ship on the lee-bow. An attempt to communicate with her all but failed, as she passed to leeward of us at the distance of three or four miles. She appeared to be a Chilian barque. To our inquiry, "Peace or War?" we got no reply.

The atmospheric temperature was yet moderate—for the position cool—the thermometer scarcely rising above 78°, whilst the temperature of the sea was more than a degree higher. The trade wind continued with remarkable steadiness, giving us a progress up to noon of 192 miles of distance, on a course N. 16° E. true. Lat. 3° 20′ N., long. 23° 4′ W.

So little disturbed are the waters of this tropical region, that we have no seas except very moderate waves; and no action on the clinometer for several days extending to more than an angle of maximum heeling to port of 5½°, to starboard 0. We had a rather fine sunset—the sun setting clear and yielding a capital amplitude, but by no means equal in richness or variety of tints to those already described. The courses steered were N. E. by N., easterly, by compass, with the view of getting in with or going to the eastward of the Cape de Verde islands, under the expectation of a better N. E. trade.

Wednesday, July 23 (S. E., S., S. W.)—The wind veering more to the southward and in the afternoon coming aft—and tending south-westerly as of the changing of the trades— continued to avail us with the usual effect. At noon our observations gave 6° 17′ N. latitude, and 21° 31 W. longitude. The day was generally cloudy, but the weather fine and enjoyable, the temperature of the air still only 79°. The sea was smooth and the ship of course steady, the maximum heeling being only 5½° to port for three days, and only 2° for the night, and nothing to starboard; registered number, 4711.

Having now reached to the magnetical position of dip 30° N., I tried the effect in respect of changes in the polarity of the upright iron standard, stanchions, etc., about the gunwales and spar deck. Already the northerly polarity of the upper ends of such masses of iron was found to be changing, and in many cases had been reversed. The standards of the gunwale chains on the forecastle were found on the port side (ship's head N. E. true magnetic, and angle of heeling 1° to port) to have retained strong northern polarity at the top, the same in denomination as at Melbourne, but those of the starboard side were weaker. An upright iron anchor-stock on the forecastle was also of northern polarity at the upper extremity. But on the deck-house, abaft the foremast, a change of polarity to southern at the top was found already to have commenced; whilst on the bridge or gangway leading from the poop-deck to the deck-house, the change of polarity to southern was already fully effected. The large iron capstan on the forecastle had at the top southern polarity towards the ship's stern, or southward, and northern towards the head. Its magnetism corresponded mainly in direction and distribution with that of the earth's force. The top of the transverse iron plating furthest forward, 324 feet from the stern, still had strong northern polarity.

Observations of the ship's external magnetism at Station No. II. indicated, even with a N. E. compass course, an extension of the southern polarity.

Ship's Head (true Magnetic).	Compass over the side.	Side of Ship.
N. 51° E.	N. 79° E.	Starboard.
N. 53° E.	N. 47° E.	Port.

The night was fine, cloudy in part, with brilliant clear patches of stars.

Thursday, July 24 (S. W.)—A fine fresh and refreshing breeze the whole day—a boon and somewhat of a novelty in this ordinarily warm or oppressive position betwixt the two trade winds. The thermometer in our cabin was 81° at night and 79° in the morning; but on deck the temperature was 78° at 8 A.M., and at 2·30 P.M. 75½°, and had not yet reached 80°. The sea blue. No appearance of living creatures beyond the precincts of the *Royal Charter*. The continued abandonment of the tropical waters by birds, so abundant generally throughout the other regions of the globe, gives to them a singularly desolate character. Looking abroad for days and sometimes weeks together, nothing appears but the sky above and a sterile, desolate area of waters, unchanging in magnitude, form, and character, bounded always with the same sharp circle of the horizon. The fresh breeze and smooth water are, with pleasant society and many resources for occupation or recreation, truly enjoyable; but the scene abroad, except at sunset or sunrise, or under the brilliant garnishing of the stars above, is anything but picturesque or interesting for a continuance.

Steering a north-easterly compass course, or, corrected for deviation and variation, making N. E. by N. or N. N. E. true, we made continual advance eastward of the meridian of the Cape de Verde islands—it being Captain Boyce's intention, if the wind may allow, to go to the eastward (an uncommon route, and particularly so for a homeward voyage) of these islands. Our rate of going was generally seven to nine knots, which brought us, at noon, to latitude 8° 56′ N., and longitude 20° W. The ship sailed badly, however, from the foul state of her bottom, for our usual fresh breeze should have given a speed for a ship of this class of ten knots or occasionally more.

The decks, besides the enlivenment of numbers of passengers carrying on various exercises or recreations, exhibited now, daily, numerous cages with Australian and other birds. Of one species, the shell parrot, a beautiful and lively little creature, sometimes from 50 to 100 might be seen at once about the deck, besides others, in which a great traffic has been carried on by some of the ship's servants. With all the risks to which they are here exposed many have been sold as high as two guineas a pair—the price in England being generally very much higher. A great number perished in the early parts of our voyage, chiefly by fighting among themselves, numbers being put in one cage, and from other càuses, whilst not a few have escaped from their confinement, but only for a worse fate. Some, indeed, exhausted by vain endeavours to find out a resting place in the ship's rigging or spars, have been recovered.

An unusual entertainment was got up by a few of the saloon passengers, and carried out very successfully in the evening, to the no small entertainment of a large proportion of the passengers and seamen who could obtain footing on or about the poop deck. This consisted of the enactment of a court of law in the imaginary case of a breach of promise of marriage. The arrangements were extremely ingenious and effective. The judge's chair, covered with an oppossum rug, having a crimson lining outward; the places for the witnesses; the wigs of the barristers, made of canvass, thrummed with short white rope yarns, were admirable imitations;—bands, robes, etc., seemed to be complete. The young lady, the plaintiff in the case, was well got up in the person, in gay and fitting costume, of one of the gentleman passengers. The barristers, four in number, the judge, the jury, were all formally ordered, and the entire proceedings conducted in grave burlesque. The speeches of counsel, examination of witnesses, etc., was carried through with excellent tact and cleverness, and everything so well arranged that the attempt, as an amusing and entirely unobjectionable recreation, was completely successful.

Another series of dipping-needle observations was made in the afternoon, giving results quite accordant, in the changes observed, with the indications and deductions of those of former positions.

The sea was again luminous at night with the sparkling firefly appearance of the surface near the ship's side, on which drops of water by the splashing of the waves were thrown off; but the total phenomena were by means so brilliant as what we observed on the evening of the 21st.

Friday, July 25 (S. W. to W.)—Our pleasant breeze, giving indeed but a speed of seven knots during the night, began to fail us after daylight. Steam was got up in aid of it, and the engine started at 10 A.M. We immediately obtained a gain of 3 to $3\frac{1}{2}$ knots from the previous rate of $5\frac{1}{4}$. The course steered was N. N. E. and N. by E., giving us about N. by E. and N. true. About 2 P.M. the wind veered to the W. S. W., and freshened in force. The sky was covered generally with uniform cloudiness; the temperature, for the region and season, was yet moderate or rather low— the greatest heat yet experienced being 80°.

Our position at noon was latitude 11° 36' N., and longitude 19° 8' W.; Cape Roco, on the coast of Gambia, bearing E. N. E., distant 140 miles; Cape Verde, N. N. E., 207 miles. No living thing to be seen except some flying fish, one of which fell on the ship forward and was captured. It was a large one, being of the bulk of an undersized mackerel, to which in its general form of body it bears a considerable

resemblance. The absence of birds when now so near the
coast of a vast continent strikes one with surprise. The sea,
however, had become somewhat more animated—some shoals
of albacore having been seen.

Saturday, July 26 (W. S. W., W.)—We had a fine breeze
and smooth sea during most of the night, enabling us, with
the aid of the engine, to make nine knots or upwards; but in
the morning it began to decline, bringing our rate of going
down to about seven knots. The first and only flock of birds
yet seen (as far as I have heard) since entering the tropics,
and while passing through 39 degrees of latitude, was this
morning observed in pursuit of fish, breaking in for a moment upon the general aerial desolateness. They immediately
passed away, and left the element which they are wont " to
people" and cheer as desolate as before.

At noon, Cape Verde, according to our observations, was
only 60 miles distant, bearing E. ½ S.—our latitude being
14° 53' N., and longitude 18° 32' W. Several persons thought
they saw land; but the vision I doubt not was unreal.

The favourable westerly breeze now for some days enjoyed,
in the region considered as one of calms passing into the
N. E. trade winds, has been occasion of some surprise; but
so far as it is unusual in the experience of navigators it may
probably be due to our unwonted near approach to the coast
of Africa.

An admirable amplitude of the sun was obtained at its setting. The sun setting pretty clear, and descending in an arc so
near to the vertical, the amplitude becomes not only more precise in its direction, but changing in no very perceptible quantity for several minutes of time, it affords the opportunity of
taking a number of sights, and thus obtaining a very accurate
mean result. In this case I took no less than eight sights,
and rejecting from these three taken whilst the ship's head
was changing with some rapidity, five were left which were
closely referable to the same direction or course. They
varied only from N. 48° 50' W. to N. 48° 30' W.—the mean
being N. 48° 42' W. The ship's head on the mean of the set
was E. ½ N. The latitude at the time being 15° 38' N., and

the sun's declination, reduced for our 22¾ hours difference of time, being 19° 31′ N., we find the true amplitude to have been N. 69° 42′ W., and the difference (69° 42′—48° 42′) or 21° W., the error of the standard compass for variation and deviation. The deviation of the standard compass on an east course was àt Melbourne 2½° westerly, in which we have no reason to believe that any sensible alteration so near the point of change has taken place, and the variation of the compass, therefore, may be inferred to be 18° 30′ W.

The wind which had fallen away to a light breeze became variable in the afternoon and evening, casting the ship's head on a variety of courses between N. N. E. and E. About 10 P.M. all the square sails were taken in, and the ship steamed within four or five points of the wind, making generally a compass course of N. E., or true course of about N. E. by N.

Sunday, July 27 (N.)—As the wind freshened, and blew indeed a strong breeze in the night, we made but little way; and our course was scarcely better than N. E. or N. E. ½ N. true. This, notwithstanding the receding of the land to the northward of Cape Verde, gradually drew us towards the African shore, till at noon our position by observations was found to be 16° 38′ N. and 17° 9 W., showing an approach within about 40 miles of the coast of Senegal—a northern branch of the river Senegal emptying itself just opposite to our position into the North Atlantic. The weather being somewhat hazy and the coast low, no land was seen; but our position being considered as near enough to the shore, the ship was steamed round on the starboard tack, and at 2 P.M. topsails and courses set. The ship on this tack lay no better than W. N. W.—making, with variation, deviation, and leeway, about a west course true. But blowing strong as it did we could do no better; and as the engine could not under the circumstances help us, it was stopped soon after sunset and the screw hove up.

Though we had now advanced so far into the northern hemisphere the polar star had been but rarely seen, and that only by the Captain, I believe, when we had retired for the night. It was interesting to me, therefore, to look out for and in the early evening it was pleasant to recognise the guiding star of the northern hemisphere. The constellation of Ursa Major had lost nothing of its importance or interest in our eyes by recent familiarity with the stars of the south. Nothing there, indeed, can rival the glory of *our* Orion, nor to my eyes did any newly-seen southern constellation exceed or even equal in beauty the one with the humble designation of "Charles's Wain!"

Divine service, notwithstanding the very fresh breeze, was celebrated on deck under the screen of the awning. But the force of the wind rendered it difficult to address so as to be heard the numerous congregation. The subject of discourse was taken from second lesson—St. John xv. 5-6.

Monday, July 28 (N.)—The wind abating about midnight, steam was again got up, and the engine about 2 P.M. put in motion; and with all sail taken in except some of the staysails and spanker, and the yards braced sharp up, the ship was steamed on a N. W. and then N. N. W. compass course, or about W. N. W. and N. W. ½ N. true, a direction approaching the best of these courses being maintained throughout the latter part of the day and night. Our "day's work" up to noon, in consequence of the head wind and with it an unusually strong northerly sea, was necessarily a very poor one. Indeed, in the direction of our port we gained nothing, our observed latitude being only 17° 3′ N., with an increase of westerly longitude to 19° 10′.

It was some compensation that the weather was cool—singularly so. In the night the thermometer was at 75°, and at 8 A.M., 77°. The sky was "mousy," a dull continuous sheet of cloud, just broken into occasionally by the sun, overspreading the heavens; the breeze was moderate or light from the north. A course of N. N. W., or N. W. ½ N. true, was steadily maintained by the action of our little engine, with an average rate of speed of five to six knots. The northerly sea, however, which was occasionally considerable, indicated plainly enough a far stronger wind at some distance northward—of which a fall yesterday of one-tenth of an inch in

the column of mercury had admonished us; for a fall even
of that extent is considered large and may be admonitory in
these tropical regions. The usual elevation, however, was
found to be restored this morning. The effect of the sea
and fresh breeze of the previous day had moved the indices
of the clinometer to 8° towards both sides—starboard and
port,—with a numerical register of 4729.

Though no birds were seen to give animation to the dull,
cloud-canopied atmosphere, the sea was more full of life
and variety of life than we have yet seen within the tropics.
Sharks, cetaceous dolphins, flying fish, bonetos, were seen—
the flying fish in numerous shoals, and the dolphins, especially
towards evening, in extraordinary abundance, were playful and
unusually bold in their approach to the ship, sporting, leaping out of the water in numbers at a time within a few yards
of us, and many of them keeping us company for a considerable distance on our track. Their number must have exceeded a hundred. Their habit is to move very much abreast
of each other or in oblique rows in their particular shoals—
which may number six or eight to a dozen individuals. No
water was ejected from their blowholes, nor was their blowing very discernible. In hardly any case did they content
themselves with coming up to the surface and rolling longitudinally over like the porpoises, but, as if overstocked with
energy and life, they rose completely clear of the water,
assuming a bulky crescent-like shape and descending snout
foremost again into a brief concealment below. Their active
gambols made them irresistibly amusing to our passengers,
whose impulsive shouts and exclamations drew all from their
seats or amusements to watch the play. Those nearest to us
were scarcely hidden when they went under water, being so
close to the ship's side that they could be seen turning
towards the wonderful moving mass which sailed so steadily
in the water, and then suddenly perhaps turning off at a
sharp angle before resuming their usual peaceful course. The
Royal Charter at the time was steaming about six knots, a
rate with which they could easily keep up with, though leaping upward two or three feet above the surface, at intervals

of only a few seconds, as if engaged in a sort of hurdle race. I judged their ordinary length to be six or eight feet, and their diameter about a third part. The dorsal fin was rather large and but moderately curved backward. The snout was considerably elongated and had (rudely compared) the appearance of the neck of a long-necked bottle. The side fins appeared to be of the usual proportion to the size of the animal. The back was of a dark slate colour, being less dark on the sides, and lighter, but not white, below. In this particular alone it seemed to differ from that described in Sir Wm. Jardine's admirable manual on "Whales" of the *Delphinus Pernettii*. It differed, however, in its gambols; for our dolphin, whilst imitating that of Pernetti in its liveliness and leapings, did not, as he described, perform somersets in the air, but strictly adhered, varying only in the energy and vivacity of the action, to the performance of a sort of semi-circular projection of its body corresponding with the curvature of the back which was exactly adjusted in the line of motion. Such a freak or performance, however, as Pernetti describes, Captain Boyce mentions as having been seen by him; and that these were of the same kind and species as those described by Pernetti one could hardly have a doubt,—seeing that they occurred in the same region near the Cape Verde islands, were engaged, as to some seen in the morning, in the same pursuit—one being seen with a captured flying fish in its mouth,—and that they so greatly resembled in the main characteristics the dolphin referred to.

We watched these creatures for about half an hour with much interest and amusement, when the several little shoals which had paid their courtesies to the *Royal Charter* gradually drew off, no doubt for the more interesting object to them—the pursuit of the flying fish. The sailors call them, characteristically, " skip jacks."

Ordinarily, it is no easy task to denominate or determine the exact species of the variety of cetaceous creatures met with in the course of a long voyage. In our case, indeed, with the exception of a sperm-whale seen near to the ship, and these playful dolphins, the various other cetacea which

U

from time to time came within view, could not with any certainty be named. In the arctic regions, familiarity with the cetacea ordinarily met with there enables the whaler easily to distinguish them, and for the most part to name them with certainty. At least with regard to the mysticetus—the great rorqual, the razorback, the narwal, and the beluga, —they are all easily and with certainty denominated; and the first are known at any distance by their respective peculiarities and differences in " blowing." So, some of the commoner kinds of phocœna or porpoises, of the delphinus or dolphin, of the globocephelus or deductor dolphin, as well as the mysticetus or right whale, and the cachalot or sperm whale of the southern hemisphere or tropics, may frequently, with scarcely any uncertainty, be pronounced upon.

Sometimes the manner of congregating and swimming, in the more gregarious or sociable tribes, enables us to discriminate the species at very considerable distances. Thus, when on one occasion I was crossing the Atlantic to the United States, the attention of the passengers was called to a very striking assemblage of the whale tribe, pursuing a sportive course nearly parallel to that of the ship, though they were at a considerable distance. One series or society was arranged in a straight line extending for near a quarter of a mile as in Indian file, and they rose to the surface much about the same time, and pursued for some distance a somewhat systematic undulating course. This peculiar manner and aspect suggested the conviction that they were of the globocephelus species, the delphinus deductor, described in my account of the Arctic regions; of which species the forming in line, or following a leader, is a characteristic, and was suggested by my friend Professor Traill as a good demonstrative trait in the habits of the animal. The same appearance struck me as likely to give the impression of the sea serpent. It is not, however, here my intention to express any decided opinion against the various statements which have been made of sea serpents of prodigious magnitude having been seen.

Tuesday, July 29 (N., N. N. E., variable.)—Moderate or

light, sometimes fresh variable breezes from the N. to N. N. E. prevailed the whole of this day, which, with a considerable northerly swell not yet gone down, offered much resistance to our small power of steam. Still it was useful, giving a rate of 5½ to 6 knots. This brought us at noon almost exactly beneath a vertical sun. A deviation of only 16 minutes from the vertical being found by observation, the southern edge of the sun's disc must have touched the line of the zenith. All shadow, therefore, as far as under the dull sky a shadow could be cast, was confined to the horizontal dimensions of objects—our ladies, for the most part, as once before noticed, casting no shadow from their persons or figure —all being concealed and embraced within the circular out-line of the prevailing wide-brimmed hat! Lat. 18° 38′ N.; long. 20° 58′ W.

Strangely enough, the weather, to our feelings, was cold. The thermometer at 8 A.M. was at 73°, and never higher than 75°, falling again to 73° in the evening. With a smart breeze generally blowing fore and aft along the poop deck, the warmer and more sheltered parts of the deck were sought; the light thin coats which for some time we had been wearing were replaced by the warmer and ordinary cloth coat, and many of the ladies resumed for the occasion their discarded shawls.

An extraordinary condition of the sky attended this un-usual depression of temperature. Instead of the fine sky, patchy, with small masses of the cumulus kind of cloud, and generally clear and brilliant, with sunshine,—we had now had for about four days a canopy of dark slate-coloured clouds, overspreading, in tame and unpicturesque sameness, the whole of the sky. In the forenoons, indeed, occasional sights of the sun, just available for altitudes, were had, suffi-cient for the determination of the ship's position; but ordi-narily no cheering ray broke through the dark screen above.

We have also met with striking deviations from the usual quality and direction of the winds in the tropical regions. The S. E. trade was not encountered until we had entered about four degrees within the tropic. Again, the calms usually

prevailing between the two trade winds were not met
with at all, but in place of them pleasant and prosperous
breezes from the S. W. and W., which ran us up to latitude
15° N.   After that, instead of favourable winds, as we expected from being so far to the eastward, we have now had
only N. or directly adverse winds for a space of above four
days.   Possibly, and not improbably, indeed, our near approach to the African coast, and the influence of its highly
heated soil on the aerial currents above and around, may
have modified or nullified the ordinary action of the tropical sun, and subdued or prevented the regular development
of the N. E. trade wind.

Numerous shoals of flying fish were seen starting from
the surface of the sea, and advancing in parallelism at a little
elevation, like a flock of small sea birds just rising and commencing their flight; but I did not observe any other living
creature around our position.

A series of dipping-needle observations and of compass
comparisons was made in the course of the afternoon, under
the expectation that a change in the ship's magnetism greater
than that due to the change in the terrestrial dip might have
been caused by the ship having gone for some time (about
two days) on courses westward of north, and extending to or
beyond the direction in which the ship's head lay on the
stocks whilst building.   This circumstance, I imagined, with
a not inconsiderable pitching motion in the ship from the
strong northerly swell which prevailed, would solicit a more
rapid return to the original magnetic lines of oblique direction.   And so it turned out, when having written the above
I referred to and calculated the relative changes of the
terrestrial dip and the dips observed in the four principal
stations of the ship's deck, that a much larger proportional
change had taken place within the interval of the last two
series of experiments than formerly, and more particularly
than in the experiments of July 21st and 24th.   The following changes in the ship's stational dip were found to have
taken place within the limit of a change of about 11½ degrees
in the direction of the earth's magnetic force :—

Changes in the Terrestial Dip.	Changes in the Angle of the Dipping needle on Deck of the *Royal Charter.*			
	At Station I. near the Stern.	At Station II. 100 ft. from Stern.	At Station III. the Standard.	At Station IV. on Forecastle.
From 23½° N. to 35° N. } 11½°	5¾°	11°	7½°	17¼°
From 35° N. to 49° N. } 14°	9¾°	11¼°	19½°	22¼°
Reducing the last in the proportion of 14° to 11½°.				
Reduced ... 11½°	8°	9° 2′	16°	18° 3′

From hence, combining the whole of the differences in the two series, we find a greater change for the proportion of 11½° of terrestrial dip in the second case of difference than in the first, in the ratio of about 5 to 4. The greatest change, it will be observed, was at the standard station.

The generally good performance of the adjusted compasses, except in respect of a sudden and transient change under the heavy gales experienced in our first outset off the Bay of Biscay, during almost one-half of our outward voyage—that is for 6400 miles, on the particular courses steered, out of about 13,300 of distance accomplished—was a circumstance suggesting the inquiry,—How far this might be due to the good and effective operation of the system of adjustment by fixed magnets, or how far to the favourable position of the compasses, in respect to their general remoteness from particular magnetic attractions, or their fortunate place amid naturally compensating influences of the ship's antagonistic attractive forces? The maximum errors of the two adjusted compasses, as ascertained in Port Phillip, viz., about 19° and 15° seemed to indicate that the ship's disturbing influence was not particularly large. One mode of obtaining an insight into the existing disturbance of the ship's magnetism on the companion compass, or foremost compass, which always performed the best, presented itself in observing a favourable position for the occasional placing of a testing compass, only 11 feet distant

from the companion compass, and sufficiently removed from
the influence of its compensating magnets. This position was
on the top of the main companion. Observations in this
position had been now made for a period of three to four
weeks, on such differences of course as were accidentally presented, when, so far as those limited results went—embracing
the extreme points, north about, of N. 40° E. to N. 70° W.—
it was found that the probable maximum error of this unadjusted compass hardly exceeded a point. It was consequently
inferred, as I had been led to suppose, that these compasses
had the advantage of very favourable positions, and that their
corrective appliances, therefore, had not very much to correct.
Originally, or before the sailing of the ship, the ship's disturbing action was, I believe, much greater; but this disturbing force became generally, as we have seen, considerably
reduced in the progress of the voyage, whilst the antagonistic
action of the adjusting magnets, remaining pretty nearly permanent, occasioned errors, in the steering compass particularly,
probably greater than what the iron of the ship, so relatively
disposed, would of itself have produced. But this question,
it is hoped, may, on the arrival of the ship in port, be more
conclusively determined than the limited and incidental
changes of the direction of the ship's head now afford the
means of doing. Of the truth of this proposition I have a
strong conviction, though without any extensive means of
verifying it,—that whilst adjusted compasses may act generally well, and to the great convenience of the navigator,
within the hemisphere where an iron ship was built, that in
many cases their good performance is in no small degree due
to the naturally conpensating tendency in the different sides
and ends of an iron ship, so that the adjusting magnets have
comparatively but little to correct. Moreover, that where
the adjustments by permanent magnets extend to the correction of large deviations, they can never be depended upon,
and the less so as a general rule, as the deviations become
greater. Thus, in a case communicated to me by a frank
intelligent compass adjuster, Captain Andrew Small, of
Glasgow, in which he was required to adjust a compass far aft,

in a new iron steam yacht, having some 12 points of error, against which he made objection as to the possibility of a good result,—the adjustment, though acting tolerably for a short time, soon failed during a voyage from the Clyde round to Leith, by the north, so that the steering compass became absolutely useless, having indeed changed to an extent of about a quarter of the circle, or eight points of error. On readjustment, however, when the original magnetism of the vessel had been shaken down into a somewhat normal condition, it was found to act usefully and tolerably well.

Wednesday, July 30 (Northerly, N. N. Easterly)—Light airs or breezes, occasionally falling almost to a calm, prevailed throughout this day, but always in a direction so nearly against our proper course, that when steaming N. or N. by E. by compass (N. by W. or N. true), the staysails could rarely be employed to advantage; our progress, therefore, head to wind, with all the resistance of the large spars and heavy rigging of the *Royal Charter*, was restricted to an average of six knots; a fair rate, indeed, to be effected under much resistance by an engine of only 200 horse-power. This gave us 145 miles of distance for the nautical day; the latitude, at noon, being 20° 28′ N., and longitude 22° 17′ W.

The previously prevailing leaden sky, which yesterday had become considerably broken, completely changed for awhile into the natural-looking clear tropical sky, in the day, and continued, but with growing cloudiness, till sunset. The temperature, however, continued cool, being at 8 A.M. 73°, and never rising to 80° at the hottest period, and falling again to 75° or 76° at sunset. The surface of the water was also cool, being 73½° Whilst steaming head to wind with so chill a temperature in the forenoon and evening, cloth coats came into requisition; and I could well bear and enjoy this dress in returning on deck after our rather early dinner—3·30 P.M. It was pleasant once again to get the sun to the southward at mid-day—the position and aspect most natural to our feelings.

Having been very frequently disappointed, sometimes for days together, in catching an evening amplitude of the sun, by reason of its sudden and often unexpected obscuration behind a bank of cloud, I prepared myself this evening with a pocket sextant, which rendered me independent of assistance for obtaining the sun's altitude at a low elevation; and where the sun's descent was so moderately oblique, it required no particular precision in adjusting the intervals of the altitudes and azimuths to obtain a good and sufficiently perfect mean of a set of sights. The arrangement was found to be very convenient, and answered well. A set of three azimuths, with altitudes betwixt them, gave for the true altitude of 4° 50′ the azimuth of N. 46° 43′ W., the true azimuth coming out S. 108° 8′ W., or N. 71° 52′ W., shows the amount of deviation and variation together in the standard compass 25° 9′ W. on a course (by it) of N. 12½° E. Comparing the standard compass by the one aloft, the deviation, corrected for 1° E, the deviation of the standard at Melbourne, appeared to be 4° 15′ W., which, being deducted from the total errors, leaves 20° 54′ W. for variation, but omitting the last correction, which I think is in excess, we have 5° 15′ deviation and 19° 54′ variation. On referring to General Sabine's chart of magnetic declinations, for the year 1840, the variation was found to be laid down at 18½° W.; but in the Admiralty chart for 1850, being corrected for change, at 19¾° W. A further proportional change for six years might probably yield a close agreement.

In most of my other observations for the variation, the sights were fortunately obtained in directions of the ship's head where the standard was near its points of change, or had but a trifling and known deviation, but here, under a N. by E. course, where a deviation existed, when at Melbourne, of about 8° westerly, reduced apparently by the ship's magnetism to about 4°, the same confidence could not be placed in the results. The eliminating, out of the total errors of the compass, the quantity due to the ship's local attraction becomes necessarily difficult in proportion to the angular distance in the direction of the ship's head from either of the points of change. But the agreement now obtained would appear to give room for considerable confidence in the practicability of eliminating the local attraction on a compass favourably placed in an iron ship, and thus obtaining tolerably accurate information as to the true declination of the needle under terrestrial action alone.

Thursday, July 31 (N. to N. N. E.)—A cloudy night again with a light or moderate breeze from the northward, never coming above a point or two to the eastward and that but rarely. Dependent on the engine alone, we pursued a course, with all sails taken in, of N. by E. or N. $\frac{1}{2}$ E., making about N. $\frac{1}{4}$ W. true, and with a very steady progress of six knots. Soon after daylight we passed a steamer steering to the southward, and about 7 A.M. came in sight of a sailing ship, lying sufficiently in our track to be communicated with. This, by means of our steam and the light wind, was easily accomplished, the stranger ship heaving her main-topsail to the mast to facilitate the communication. We approached her twice. She proved to be the French ship *St. Vincent de Paul*, a nearly new vessel, 18 days from Bordeaux, bound for Valparaiso. Our greatest anxiety was for the answer to the question—" Peace or War ? " The reply—"Peace—Peace," called forth a general cheer from our crowd of passengers. The birth of a Prince to the Empress of the French, was another piece of information of cheering interest, not only on account of the general kind sympathies with the feelings and joys of the ruler of France, but of the influence of such an event, under Providence, for good to the nations, as an apparent support to the existing dynasty in that country. The grateful news were celebrated at dinner in an abundant supply of champagne for general use, contributed by one or two of the passengers, and after dinner by a salute of 21 guns, by order of Captain Boyce, followed by hearty cheers of those on deck, the resounding of which yet rings in my ears whilst writing this part of my journal. May the peace be graciously consecrated and sustained, under the Divine blessing, for the benefit of the world, the advancement of civil and religious liberty among the oppressed nations, the controlling of ambitious selfishness and wrong in the mighty of the earth, and the extension of the Redeemer's kingdom.

The communicator, on our part, with the friendly stranger

was a French gentleman among the passengers. Perceiving
a number of the sisters of charity on deck, he informed the
Captain that we had on board a prelate of the Roman-catholic
church. This information led to a solicitation for his blessing. As the Archbishop appeared at the stern of the *Royal
Charter*, then scarcely 50 yards distant from the *St. Vincent
de Paul*, sisters of charity, missionaries (for there appeared
to be two on board), captain and crew, all dropped respect-
fully and devoutly on their knees, whilst the prelate, crossing
his arms, made the wanted motion of salutation and blessing,
and finished by wishing them a "bon voyage." The scene
was impressive; so much so that a Protestant minister, though
far from envying the authority, frequently so injuriously
exercised by the clergy of the Church of Rome, and far from
desiring, but rather repudiating, a blind or humiliating submission of the people to their ecclesiastics, could scarcely fail
to feel the too prevalent reverse of all this amongst our Protestant churches, where, in many cases, not only is no respect
paid to the minister's official position, but not unfrequently a
less value attached to his judgment, on subjects, it may be,
professional and critical, than would generally be conceded
to a master in any other profession, such as that of medicine
or law.

Like the days for a week past, this was characterised by a
frequently, but not constantly, dull sky, gloomy indeed in
the forenoon, with the prevalently low temperature for the
region we were in of 73° to 75°, most unlike what is described of the tropical regions.

At noon we were in latitude 22° 31′ N. and longitude
22° 34′ W. Soon after 10 P.M. we crossed the tropic of
Cancer, and, according to geographical conventionalities,
passed from the torrid into the temperate zone, and what
appeared to be a novelty, having traversed the whole of the
northern tropic without experiencing any absolute calm, or
true and characteristic N. E. trade wind. The winds we
have had since emerging out of the south-eastern trade in the
seventh degree of N. latitude, having, without exception,
prevailed from the S. to W., N. W., and N., and in no in-

stance for more than an hour or two at a time having entered the N. E. quarter beyond N. by E. true, whilst generally, since it veered to the northward, it has preserved very nearly a meridian direction.

A splendid luminous meteor crossed over us about 7·20 P.M., which excited great interest and surprise in those who were fortunate enough to witness it. It was described as having a large and conspicuous nucleus and tail of a blueish brilliant white colour, and starting, when first observed, in the N. W. passing obliquely to the west, and disappearing behind a bank of cloud in the S. E. quarter. Another of equally imposing character and splendour was witnessed at about eight in the evening about a fortnight ago, which appeared in the N. E. quarter, and making a rapid transit eastward, to the S. E., seemed to burst and scatter abroad numerous igneous fragments, to appearance like the bursting of a prepared rocket. The colour of this too was described as of blueish white, and no doubt was a true aerolite.

Another entertainment was got up by a few of the saloon passengers, assisted by a clever musical performance by two of the stewards, under the shut-in awning, betwixt 8 and 10 o'clock:—Dramatic recitations, the story of William Tell, and portions from Shakspere's Julius Cæsar. The characters were capitally dressed in classic costume, and the parts were very creditably, some of them really well, performed.

In the same manner as in the preceding afternoon I took sights in azimuth and altitude of the sun, near to the setting. These gave (the mean of two azimuths) the sun's apparent azimuth N. 48½° W., and the true azimuth S. 108° 48' W., or N. 71° 12' W., showing when the direction of the ship's head by the standard compass was N. 21° E., an amount of errors in variation and deviation of 22° 52' W. On a correction for deviation, as indicated by a single set of comparisons with the compass aloft, giving 4° W. for the deviation of the standard compass, we obtain 18° 52' W. for the variation. The same position in General Sabine's chart has its 19° variation line, and in the Admiralty chart the variation is about 20¼°W.

Friday, August 1 (N., N. N. E.)—A cloudy night and

day, with a gentle or moderate breeze from the north to
N. N. E.   The ship steaming, sometimes assisted by the fore-
and-aft sails on a course of about N. N. W., occasionally N. by
W. ½ W. true, and with a pretty regular rate of six knots.
At noon, we were in latitude 24° 48′ N., longitude 23° 1′ W.
Two ships were seen, one at a great distance, the other steer-
ing about S. W. near enough to discern that she displayed
the French flag.

The temperature of the air continued unusually cool, being
72½° at 8 A.M., and about the same in the evening.   The
clinometer gave a register of maxima of heeling for two days
of 2⅓° to starboard and 1½° to port.

There being somewhat more wind in the evening, the topsails were tried; but, with a gain of only about a knot in speed,
we were deflected two points further from our course, that is
to N. N. W., so that our gain in speed was a loss.   Before
long the topsails were withdrawn, and steam and fore-and-aft
sails resorted to as before, under which we made about six
knots, on a course of about N. by W. true.   The evening
was again cloudy and rather dark.

Saturday, August 2 (N. N. E., N. E. by N.)—Light or
moderate breezes prevailing all night, we made a N. by W.
course nearly as before, which, at the usual rate of about six
knots, was maintained throughout the day.   Temperature 71½°
at 8 A.M., 72½° at 3 P.M., scarcely varying.

Nothing could be more monotonous than the weather, sea,
sky, and our proceedings in the way of progress, for nearly a
week.   A scant or head wind of the N. E. trade, a misnomer
or an apology for what is usually met with; no strong, commanding and advantageous breezes; dull, cloudy, often a mousy
sky; light or moderate breeze; steaming either with all sails
furled or all but staysails; sea quiet and tame; the air without a bird (except the single flock formerly noticed near the
African coast), in a progress of 50° of latitude; no rain or
lightning, of so ordinary occurrence hereabout, or within the
tropic; no clear brilliant sky by day, or heavens bespangled
gorgeously with stars at night; and, fortunately for our comfort, no hot weather.   Usually, I believe, the experience of

the navigator in passing upward from the equator to our present position is, in almost every particular, different. The latitude at noon was 27° 1′ N., longitude 23° 58′ W.

Observations for the dip, at the several principal stations, were again made. The dip and magnetic force were greatest near the stern (ship's head nearly north, true magnetic), and least on the forecastle. The series of dips commencing aft were 72½°, 71¾°, 60¾°, 60¼° N. The times of oscillation (in one direction) of the dipping-needle, gave the series of 2″, 2″·025, 2″·3, 2″·45. Whilst the terrestrial dip had advanced about 9¼° the dips at the several stations had moved on (commencing farthest aft) respectively, but somewhat irregularly, in the series 4¼°, 8¼°, 6¼°, 8¼°.

In continuation of the observations on the ship's magnetism externally, a new element was latterly taken in, viz., the observation of the time of oscillation (one way) of the compass-needle employed. This, though affording but a rough guidance to the magnetic force of the ship and earth acting on the compass, served to indicate remarkable differences, and considering the direction of the ship's head, and her long continuance on a northerly course, differences of much interest in their accordance with theoretic deductions. Thus for the stations at 23 feet, 100 feet, and 287 feet from the stern, the times of oscillation were, on the starboard side 3″·1, 4″, and 5″·5, and on the port side, 3″·8, 4″·5, and 5″. But the depths below the upper edge of the top plating were not the same, the first two being 3 ft. 9 in. down, and the last (on the forecastle) only 15 inches. Had the same level been taken, the probable mean series for the two sides would have been about 3″·2, 4″, and 5″·2, or the ratio of the magnetic forces inversely, as 10, 16, 27,—showing the intensity near the stern of the resultant of the ship's southern polarity near the gunwale and the earth's intensity to be greater than that forward in the proportion of 27 to 10.

Sunday, August 3 (N. N. E., or N. E. by N.) — No change in the wind or sky during the night. We continued, however, to make a steady progress of about six knots to the N. by W. or N. by W. ½ W. The morning was again

cloudy and dull, but it cleared away in the forenoon, and was
followed by a clear, fine sky, summer-like, of more tropical
aspect than hitherto. The temperature of the air, which at
8 A. M. was 72°, increased considerably as the day advanced.
No living creature to be seen in air or water. The latitude
at noon was 29° 14′, longitude 24° 52′W., our position being
about W. by N., 360 miles distant of Palma, the nearest of
the Canary islands. An excellent amplitude was obtained
at sunset, the bearing of the sun being N. 41° 45′ W., or corrected for the deviation of the standard compass (5° 15′ W.)
with the ship's head N. 12¼° E., N. 47° W. The true amplitude being N. 69° 35′ W., the variation appeared to be
22° 35′ W. In the Admiralty chart it is laid down at about
22° 20′ W.

About 10 P.M. a light appeared nearly on the line of our
course, obviously that of a ship. The wind having gone
down to nearly a calm, we steered towards the light. The
sky was dull, and the ship first appeared when within half a
mile of us looming obscurely at that distance. Gradually it
assumed a more definite shape, and excited great interest as
we slowly approached it and brought out in succession first
one feature and then another, until it stood well revealed as
a stout ship crowded with sails, and steering or pointing
toward the southward. We passed her majestically at the
distance only of about half our ship's length, first hailing, and
then asking questions or giving answers, until again we were
too distant for the voice. It was the *Ashland*, 21 days from
Liverpool, had spoken the *Oliver Lang* six or seven days
ago, which left Melbourne 24 days before us. The news
of peace were confirmed with some further facts of interest.
We parted first with cheers and then stimulated by the *Ashland's* example, with a display of blue lights and rockets,
which, being of good quality, had a fine effect in the dark
still sky above.

Divine service was performed in the saloon in the forenoon.
My address was from the first lesson, 2 Kings, v. 13—40,
the story of Naaman. The Sacrament of the Lord's Supper
was afterwards celebrated in the ladies' boudoir, at which

about ten saloon passengers and three or four, including a coloured man, of the third class attended.

Monday, August 4 (N. variable, calm.)—We had light or gentle breezes all night, turning to calm, and subsequently in the day a return of the northerly breeze, as usual of late, right ahead of us. Three or four times an attempt was made to improve our progress by setting the topsails (lower), but in all cases the disadvantage in direction was far more than the gain in speed. They seldom remained on trial for more than an hour or two. The steam-power, small as it was, was our grand resource, giving us generally a six-knot speed, and ordinarily yielding us a progress of 140 miles or more in distance, and a gain of $2°$ or $2\frac{1}{4}°$ in north latitude. For the day up to noon it brought us up to the position of $31°$ $35'$ N. and $24°$ $53'$ W.; the island of Madeira being 390 miles E. by N. of us; St. Michaels, N. $\frac{1}{2}$ W., 400 miles.

At 10 A.M., having been quite calm and the sea of a glassy surface for some time, with singularly small disturbance by undulations, a remarkable and beautiful colouring came over the sea in the S. W. quarter, or rather extending round from S. E. to W., of which I do not remember before to have seen any counterpart, as there was no coloured cloud above, either to transmit its own tint, or optically to produce a complimentary or consequential colour.

It may be mentioned that along with glassy smoothness of the sea, in the early morning, there was a bright and almost cloudless sky. About 9 A.M., the sky to the northward became cloudy, and a low stratum of dark grey cloud rapidly came over us, covering the entire sky, except a low arch near the horizon to the southward or south-westward. A breeze from the north brought this stratum of cloud and, being a little brisk in force, rippled the entire surface of the sea. Betwixt us and a sharp and regularly curved line about halfway between us and the horizon, the rippled surface had a sort of slate grey or leaden tint, but beyond the curved line as far as the bounding horizon, the curved segment of the horizontal arc was of the richest cerulean blue, marked, indeed, with some parallel streaks of a more glossy character. The appearance was singular and strikingly beautiful.

A brig showing Spanish colours passed us in the forenoon steering to the south-westward.

A turtle was seen by the Captain, at the surface of the sea, which, had we had no propelling power, in the calm, might probably have been captured. Their appearance in or near this region is a familiar occurrence, and sometimes a considerable number cautiously approached by a boat have been taken by the hand and secured.

Some deals and a large balk of timber, apparently long in the water, were passed.

Another excellent amplitude was obtained at the evening sunset. The apparent bearing of the sun by the standard compass was N. 43° 20′ W., with the ship's head at N. 33⅔° E., heeling to starboard 2°. The correction for deviation was estimated by the aid of the compass aloft (and with reference to an observed reduction of the deviation since leaving Melbourne of about one-third part on north or north-easterly courses) to be 3° W. The true amplitude for the latitude 32° 11′ N. was found to be N. 69° 22′ W., and the variation thus freed from deviation appeared to be 23° 2′ W., that in the Admiralty chart for the same position being about 23° 20′ W. Where the exact deviation of the standard compass can only thus be estimated approximately, the near agreement of the various variations deduced from recent observations with the variation charts of General Sabine and that published by the Admiralty is not a little striking. It would require, however, the speedy swinging of the ship again to be quite sure of the exactness of our results.

A rather fresh breeze ahead prevailed at night, retarding us somewhat in our usual progress. The sky at 8—11 P.M., was quite clear and full of stars; but they seemed to have retired into their usual place on or in the firmament above, instead of being let down, as one might imagine in the tropical and more transparent skies.

Tuesday, August 5 (N. E. by N. to N. E.)—The breeze from the northward inclining N. N. E. and N. E., continued still through the most of the day. Steaming head to wind, we made a somewhat less progress than usual making little more than two degrees of latitude in the 24 hours up to noon, course about N.N.E. The wind fell in the evening to nearly calm, so that our speed was accelerated.

We enjoyed, however, a refreshing and beautiful day, the temperature genial (74° to 75°), the sun generally shining, and the clouds consisting of irregular masses, sometimes large but not dense or heavy. The sea was still beautifully smooth.

A phenomenon of interest to most of our passengers was witnessed in the evening at 5·45 P.M., in the case of a distant water-spout in the S. W. quarter. The evening was fine and serene, and, with a generally bright sun, patches and masses of cloud, as just described, were distributed in considerable but not obscuring quantity. In the south-western quarter the sky was clear about the horizon, but at an elevation of 2° 20′ commenced, with its lower edge straight and in parallelism with the horizon, a broad and uniformly dense, but not dark or threatening, stratum of cloud, of a dull grey colour. It might be a point of the compass or more in horizontal spread, and, measured by the sextant, 2° 40′ in breadth vertically. Within this darker cloud, and extending a little obliquely through it from top to bottom, was seen a narrow white line or bend, thickest at the top and pointing, though apparently terminating at the lower edge of the clouds to the sea at the horizon. This was the water-spout, the character of which was made sufficiently distinctive by the raising of a white hillock, as it were, of water beneath it. The first appearance of it was straight, but in a short time it became more oblique and waving in form—in both cases resembling a mineral vein running nearly vertically through a bed of rock. Finding that the water elevated on the surface of the sea was almost exactly on the horizon, as viewed at an elevation of about 24 feet from the poop deck, the distance of the water-spout must have been about 5½ miles from us. By means of this distance, and the angular breadths obtained of the belt of cloud and the clear space below, we find the width of the cloud penetrated by the water-spout and the space below to have been about 2700 feet—the

x

visible trunk of the water spout, within the cloud, being about 1500 feet! As the angular measurements were taken by a pocket sextant, the only error likely to be involved in this estimate should arise, I apprehend, from misjudgment of the distance of the elevated (hillock of) water. But as this was observed more than once with a good opera-glass, the distance could hardly be materially less than the estimate; though if the result be correct, it conveys a surprising impression of the extent at which this singular influence may act on the surface of the sea! Possibly the water raised might have been beyond the horizon, and so the extent of influence greater.

Good sights were again obtained of the sun at setting. The ship's head being N. 55° E. by the standard compass, an allowance of $\frac{3}{4}°$ westerly for the deviation taken from the Melbourne tables and curves reduced, as on courses between N. and E. the change appears to have been in the proportion of one-third. This gave the variation 22° 31' W., or, by a final sight of the sun, 23° 6' W., the mean being 22° 49' W. The variation by the Admiralty chart appears to be about 24° 20' W.[*]

These results, connected with those before recorded, serve to show that the eliminating of the local attraction, even in iron ships, so as to obtain useful and satisfactory determinations for the variation, is by no means so difficult as might have been apprehended, and as in my personal anticipations I thought it would be. The modes adopted for the eliminating of the deviation of the standard compass have been three:— First, endeavouring to get azimuths or amplitudes on or near one of the points of change in the compass deviations. Secondly, referring the standard compass when out of the most favourable position of the ship's head to the compass aloft, and taking that with such small corrections as the swinging of the ship at Melbourne suggested as a true reference; and this

---

[*] From the observations made on the return to Liverpool (Introduction, p. xxxviii), the deviation at N 55° E seems to have changed from West to East. Applying the correction in this way the variation approaches still more nearly to that of the Admiralty chart.—ED.

mode, had the compass aloft been properly adapted for correct observations by a bold graduation of the card into degrees, and for being seen by a telescope or an opera-glass from the deck, would doubtless have been very satisfactory. Thirdly, by taking the Melbourne deviations of the standard compass as a ratio, and applying the proportional change, which from a great number of compass comparisons appears to have taken place, very good results appear to be yielded. The mean change appears to be a reduction of the deviations obtained at Melbourne (a second reduction since the first swinging of the ship at Liverpool) to an extent of about one-third. In this adjustment the deviation is best obtained by means of the curve of deviations. By the use of this the original observations themselves are corrected, and the deviation is obtained by measuring off the quantity corresponding to the exact azimuth on which the ship's head may happen to be. For this, and indeed either of the other processes of eliminating the compass deviations, the Admiralty azimuth compass possesses peculiar advantages in the moveable horizontal rim carrying the sight vanes, by the graduation of which the exact deviation of the ship's head is derivable in her precise position at the time of taking the sun's bearing.

The evening was brilliantly fine, and the moon for some hours very resplendent as a crescent, the dark portion of the body being very conspicuous. The phenomenon, I presume, must be familiar to most observers of the moon, of the greater illumination of the dark body of the satellite at the extreme edge, this illumination being rapidly shaded off to a less distinct exhibition of the main part of the disc, and of the somewhat surprising excess in size of the bright crescent over the dark part, giving the crescent an appearance as if it were a portion of a globe of larger diameter than that of the dark part embraced by it and exhibited by the light thrown upon it by the earth, when doing service as a magnificent full moon to its own satellite.

Wednesday, August 6 (Calm, northerly airs.)—It was nearly calm, or calm but slightly disturbed by occasional " cats' paws " on the water, or, as once or twice happened,

by the general minute ruffling of the surface by a gentle
breeze, during the whole of this 24 hours.　The engine continued to do us respectable and valuable service in a general
speed of 6 to 6½ knots.　Of the value of this appliance
we became additionally sensible this morning on speaking a
French brig, the *Indiana*, bound for Marseilles, almost alongside of which we ranged in passing, which had been 40 days
in her passage from Senegal to her present position (by reason
of light contrary winds and calms), a distance which our
auxiliary steam had enabled us to accomplish (having been
very near to the Senegal river on Sunday, the 27th of July)
in an interval of less than ten days.　And what was additionally inspiriting to us, tardy as our passage measured by our
too sanguine expectations or hopes had been, was the remem-
brance of the fact that when the *Indiana* was starting northward from Senegal (June 27th, we presume), the *Royal
Charter* was just passing Cape Horn, in latitude 56° 43′ S.;
and hence while this sadly delayed vessel had been making a
distance of about 19° of latitude towards the north, we had
made a transit from the South Atlantic into the North to an
extent of 92° of latitude, or a gain upon the brig of above
4000 miles.　In other words, whilst she had been making
about 2000 miles of latitude we had made above 6000.　After
we had spoken her we were soon far ahead, and in the course
of about two hours had left her out of sight.

This case has been worked out in detail to show the immense advantage of the system of auxiliary steam-power for
ships of a large class, and though it may appear to be an
extreme case, the great advantage of possessing this peculiar
appliance admits of perpetually recurring proofs.　Thus, in
the instance of the fine and admirably navigated ship the
Kent, Captain Coleman, a passenger now with us here, informed me that he had made a passage in her from Australia
to England last year, the results of which had a striking bearing on the point now insisted on; for leaving Melbourne on
the 1st of March, the *Kent* made a splendid passage round the
Horn and northward to the equator, accomplishing the distance in the singularly short space of 40 days, but on reaching

our present position nearly, her progress was arrested by light winds and calms extending to 17 days, during which time she accomplished almost nothing ; and so occasioning a passage which, with the *Royal Charter's* small steam power, might probably have been completed in 65 to 70 days, to be extended to 81.

The position obtained at noon was found to be latitude 35° 55′ N., and longitude 23° 36′ W.; our distance from the nearest land, St. Mario, of the Azores group, was about 90 miles, and from Liverpool about 1400.

The day was brilliantly fine, rather hot for our position, but on deck, under the awning, pleasant and enjoyable. Having advanced to the extent of another of the intervals which I had decided on as stages for dip observations, the usual series was gone through, and with results very much corresponding with those of the last observations on the 2nd. Thus the series of dips, commencing with the station nearest the stern, was $75\frac{1}{3}°$, $76\frac{1}{2}°$, $68\frac{1}{4}°$, $68\frac{1}{4}°$, the terrestrial dip being about $64\frac{1}{4}°$; and whilst the terrestrial dip had advanced $64\frac{1}{4}°$, the dip on board had advanced at the several stations to the varying extent of $2\frac{2}{6}°$, $4\frac{3}{4}°$, $7\frac{1}{2}°$, $8°$. The ship's head, it should be noted, was N. 40° E. when the observations were made, and the ship upright.

A number of small cetacea were seen in the afternoon. They appeared a good deal like those seen on the 28th July, but I think paler in colour, and few examples were observed of their leaping out of the water; I did not see one, but the case, I was told, did occur. They were not sufficiently near for the determination of the species. No other living thing was observed.

Four vessels came within sight during the day.

Thursday, August 7 (Calms, W., S. W.)—It was calm all night, but at 7 A.M. a gentle breeze sprung up from the westward, veering first aft and then back again on the port side. It tempted the spreading of our canvass, which added a knot or two to our rate of steaming, and enabled us to reduce the force of the steam and quantity of fuel, an object now to us of considerable importance, from its having been found that our coals were getting very low in quantity. This was the more matter of regret on account of the decided benefits which now for many days we had been continually reaping from our little engine. It was thought on the first investigation that we had but four days' coal left, but on shifting some of the residue from the hold into the proper bunkers, the quantity happily appeared to be somewhat but not very much extended.

The breeze, varying much in its direction, but always feeble, continued the whole of the day. A small but abrupt sea from the north-westward showed that a much stronger wind prevailed at no considerable distance. The weather being cooler on deck (thermometer 71° to 75°), with clear sunshine generally above, with cloudiness near the horizon, was refreshing and enjoyable. Most of the time with our passengers was spent as usual on deck. So still had the weather for some weeks been, and so smooth the sea, that the clinometer had rarely indicated a measure of heeling beyond 5°. Our latitude, at noon, was 38° 8′ N., longitude 22° W.

The course generally steered was N. E. by the wheel compass, or about N. N. E. ½ E. true.

The light favourable breeze of 24 hours' continuance so improved in strength after daylight that the steam could be satisfactorily discontinued, and the screw, after a perpetual action, excepting for a few hours during 14 days, was raised. During that interval the wind had never been fair—or if for an hour or two approaching to fair, never adapted for doing any effective service alone. For some days, indeed, none of our canvass, except occasionally some of the fore-and-aft sails, with feeble advantage, could be used. The steam, therefore, during a progress through 25° of latitude might be considered an absolute gain of 1200 miles of distance in a region where under the circumstances we could have hardly accomplished by sailing a single degree of direct distance.

Some small petrels (the stormy petrel) this morning appeared near us; the first birds I had seen for many days. The sea, however, had not been so barren of visible life as the atmosphere. Cetaceous dolphins again appeared in considerable numbers not far from us. Of those which I happened to see none were sufficiently near for the discrimination of the species. Their action was in certain respects similar to that of the delphini seen near the Cape Verde islands—some of them leaping clear out of the water. There appeared, however, according to my impressions, to be some differences which interfered with the supposition of their being of the same species; for instance, only some of them appeared to quit the water, and these, in leaping, appeared to describe an arc of a larger circle, whilst their colour, as far as I could determine, appeared to be still more pale and whiter underneath.

The breeze, which had relieved our engine-men from their long spell of exhausting duties, soon freshened. At noon our observations gave the latitude 40° 42', longitude 20° 16' W.

In the evening, a concert was most kindly and gracefully given by Miss Catherine Hayes, assisted by several of the passengers, for the benefit of a poor widow on board with six small orphan children, who by the death of her husband had been left all but destitute, and whose small funds had been nearly exhausted by the payment of passage money for herself and little ones: the proceeds were 38*l*. 12*s*. 6*d*.

Saturday, August 9 (W. S. W., S. W., S. to S. W.)—We had a fine breeze at W. S. W. all night, which continued, with some variations in strength and direction, during the day. Our rate of going was 9 to 9½ knots, sometimes more. In the evening the breeze freshened, and shifting from right aft to the starboard quarter, the ship for a time resumed her outward habit of running at a speed of 13 knots or upwards. There was something very exhilarating, and, to a sailor's feelings, I may say charming, in this quiet performance of what in other classes of ships would, if possible, be deemed wonderful, and the more wonderful seeing that with the wind on the quarter and smooth water, this splendid feat of going is accomplished without apparent effort, and without any apparent strain on ship or spars. The fast clipper, in such performances reminds me of the high-bred and high-conditioned horse, which in bounding with the swiftness of the wind over the downs or firm sward expends no apparent effort.

The course now steered was necessarily more and more easterly, and the progress for the nautical day brought us to latitude 43° 58′ N. and longitude 18° 18′ W. at noon. Towards midnight we had the (to us) rare phenomenon of rain which came in heavy showers. In this particular, indeed, our experience in passing from the antarctic side of the southern tropic, through the whole of the tropical regions, and for about 20° beyond, comprising the region within the usual north-east trades at the rainy season, and through the section technically, from the prevalent deluges, called "the rains," with scarcely a shower, is, if not all but unprecedented, extremely rare.

The almost daily comparison of the compasses has yielded within the last day or two further curious and instructive results, only, I think, to be explained on the principle theoretically deduced, and published in 1852 in the " Magnetical Investigations," of the tendency of the magnetism in iron ships produced by the position of the ship whilst building, to assume after considerable usage what I had assumed to be a normal condition, viz., a distribution as to the retentive magnetism, having a vertical magnetical axis, with an equatorial plane running horizontally fore and aft, and, except as transiently modified by simple terrestrial induction when acting obliquely, having the neutral lines externally alike in position on the two sides.

The organic changes in the magnetism of the *Royal Charter*, as indicated by the alterations in the deviations of the several fixed compasses, all, I think, point to this approximation of the ship to the normal condition. Thus, the obliquity of the ship's magnetism, always and necessarily in the extreme when on the stocks and first put afloat, which gave the results, in compass deviations already elucidated, became obviously moderated by the voyage to Australia, as shown by the great reduction of the deviations of the unadjusted compasses; whilst the reduced deviation has been found, up at least to the present time, to have become gradually more and more reduced. But a new feature, as intimated, has recently been developed in the relative or comparative deviations of the two adjusted compasses,—the companion or foremost compass, which always, until very recently, was the most accurate, was now found (whilst pursuing an E. N. E. compass course) to have the largest deviations. Hence I cannot but look with anticipations of much interest to the reswinging of the ship on (please God) our arrival at Liverpool. It is, indeed, difficult to predict what might be the fruits in scientific and practical information of obtaining another series of deviations at Liverpool, in connection with the interesting results yielded by the observations at Melbourne, so as conclusively to ascertain the extent and nature of the organic magnetic changes caused by sea-going vibration, straining, and blows from the waves, during five months' voyaging from 70° of north magnetic dip to 75° south dip, and back again.

The clinometer at 8 A.M. noted maxima of heeling of 11° to starboard and $2\frac{1}{2}$° to port, with a registry of 4722; the prevalent heeling at this time was 4° to port.

The " donkey engine " broke down in an early stage of its reinstalment in the duty of pumping the ship, and threw a heavy burden, in consequence of the great leakage forward, into the hands of the seamen, who for several hours, at two or three spells, were engaged in this operation.

Heavy rain commenced towards midnight, which caused a sudden shift of the wind, at first in the direction of the starboard quarter, as already spoken of, but in its continuance the fresh breeze subsided to its ordinary strength. There was now a regular sea from the south-westward.

Our progress under a fresh breeze of wind continued satisfactory, though not rapid, the foulness of the ship's bottom retarding her very seriously—probably from 1 to $1\frac{1}{2}$ knots— the growth of weeds, besides animal attachments, it is expected, will be like a bit of rich ground rampant with coarse weeds. The courses steered were from N.N.E. to N.E by E. true, and, with an average rate of about eight knots, gave our noon position, latitude 46° 44′ N. and longitude 16° 3′ W. The day altogether was fine and the temperature cool (66° at 8 A.M.), but some haziness was almost continually prevalent

near the horizon. Approaching our shores and channels southward, ships were met with in increasing numbers, chiefly outward bound, but one was passed pursuing a course somewhat
parallel with ours. Two sharks appeared very near to the ship.

[Here Dr. Scoresby's journal breaks off. The log of the
Royal Charter gives the following particulars of the conclusion of the voyage :—

Sunday, August 10.—Distance run, 190 miles; latitude,
46° 43′; longitude, 16° 3′.

Monday, August 11.—Distance run, 193 miles; latitude,
49° 20′; longitude, 13° 28′.

Tuesday, August 12.—Distance run, 168 miles; latitude,
51°. At 10 o'clock sighted Old Head of Kinsale, N. 9 E.
10 miles.

Wednesday, August 13 (or rather Tuesday, August 12, at
home.)—At 12 a tug steamer came alongside ; mails put on
board, and several passengers.

Thursday, August 14 (Wednesday, 13).—Had to delay
our progress for the rise of the tide. At 3 cleared up and
furled sails, using only the engine. Passed the Rock Fort
at — P.M., firing some guns, and at — brought up in the
river Mersey.

The deviations of the compasses were observed when the
Royal Charter was lying in the Mersey, and afterwards in
entering the dock. I have been favoured by Mr. W. W.
Rundell, who assisted Dr. Scoresby in making them, with
the following account of the process :—

" The deviations about which you inquire were obtained
in the Mersey, by means of the magnetic bearings of the
Vauxhall chimney, which had then been recently painted,
on the Dock walls. About 22 observations were made, as
opportunity allowed, the same day the *Royal Charter* arrived,
and it was agreed by Dr. Scoresby and myself that we
would sleep on board, so as to continue them at daylight
in the morning, as the ship was to be docked very early.
By the Doctor's ready thought in hailing a steam tug to
pull the ship's stern round as far as possible against the

tide, additional points were observed; and by good fortune we were able to observe the remainder as the ship was turned into the London Basin, on her way to the Wellington Dock. As you may suppose, it was a subject of much congratulation that, by a little energy and perseverance, the reswinging the ship on her return to Liverpool has been effected; and Dr. Scoresby repeatedly remarked that had it not been for the bearings printed on the walls his experiments would have remained incomplete.

" Dr. Scoresby had charge of the Admiralty compass, and gave the signal for each observation, while I undertook to give the ship's head correct magnetic for the same moment, and to read the steering compass. One of the subordinate officers read the companion compass, and a seaman was sent aloft to read the mast compass, but through the sluggishness of the last compass the readings were of no service."

The results of these observations, and the conclusions derived from them, will be found in the " Introduction."]

LONDON :
PRINTED AT 12, IVY LANE, PATERNOSTER ROW.

Printed in the United States
By Bookmasters